Circuit Design for Audio, AM/FM, and TV

TEXAS INSTRUMENTS ELECTRONICS SERIES

The Engineering Staff of
Texas Instruments Incorporated ■ CIRCUIT DESIGN FOR AUDIO, AM/FM, AND TV

The Engineering Staff of
Texas Instruments Incorporated ■ SOLID-STATE COMMUNICATIONS

The Engineering Staff of
Texas Instruments Incorporated ■ TRANSISTOR CIRCUIT DESIGN

Runyan ■ SILICON SEMICONDUCTOR TECHNOLOGY

Sevin ■ FIELD-EFFECT TRANSISTORS

Circuit Design for Audio, AM/FM, and TV

Prepared by the Engineering Staff of
Texas Instruments Incorporated

Edited by
Wm. A. Stover
Technical Publications

Contributors

K. G. Cherry
R. C. Grimes
W. C. Harrison
R. S. Henley
S. H. Kuehler
D. N. Leonard
D. E. Pippenger
M. P. Schreiner
R. E. Smith
S. M. Weaver
D. W. Whitten
D. L. Wilcox

McGRAW-HILL BOOK COMPANY
New York San Francisco Toronto London Sydney

CIRCUIT DESIGN FOR AUDIO, AM/FM, AND TV

 Library of Congress Catalog Card Number 67-12332.

63740

234567890-HDMB-987

Preface

This book was originally published by the Texas Instruments Microlibrary as two separate paperback books, "Audio and AM/FM Circuit Design Handbook" and "Television Circuit Design Handbook." The result of several years of research and development by Texas Instruments communications applications engineers; the book provides a practical guide for practicing circuit designers and a valuable supplemental text for the advanced engineering student.

Time- and cost-saving procedures have been stressed throughout. Design examples have been chosen to suggest the broad application of the procedures. Inquiries regarding material in this book may be addressed to Communications Applications Branch, Semiconductor-Components Division, Texas Instruments Incorporated, Post Office Box 5012, Dallas, Texas 75222.

Texas Instruments Incorporated
Semiconductor-Components Division

Contents

PART I

Audio Design

1

Audio Design Considerations

As an introduction to the following chapters on practical audio amplifier design, this chapter discusses the merits of Class A versus Class B amplifiers, characteristics of various coupling schemes, and criteria for semiconductor device selection. Following this discussion, a list of symbols used in this and subsequent chapters is given. Finally, a bibliography of reference works used in the preparation of this text is presented.

Subjects which do not directly affect design parameters have been omitted in the discussions which follow. Many excellent texts are available to those interested in pursuing subjects such as semiconductor physics and rigorous stability criteria.

AMPLIFIER CIRCUIT SELECTION

Class A Versus Class B. Class A amplifiers generally provide lower distortion but lower power output than Class B. Class A amplifiers feature the highest power gain per stage but provide poor power efficiency. This low efficiency seriously limits Class A application where power consumption is a limiting factor; for example, in portable radios. Class B design provides higher power efficiency than Class A, but requires more components per stage. Class B is suitable for high power outputs, and in applications where power consumption is a limiting factor. The foregoing comments provide a guide for choosing between Class A and Class B amplifiers. After selection of the class, the designer must choose a coupling scheme. Selection of the coupling circuit depends on the performance and economy required. The following description gives an outline of coupling characteristics. These characteristics are not hard and fast rules. Indeed, a careful design can blend the virtues of several schemes into one, but usually at higher cost.

Class A Amplifier Coupling Characteristics

Transformer-coupled Input, Transformer-coupled Output design provides fair frequency response and distortion. Bias stability is very good and easily obtained and the design is relatively inexpensive below three watts.

RC-coupled Input, Transformer-coupled Output design provides fair frequency response and distortion. Bias stability is nearly as easily obtained as for the transformer input design. The design is less expensive because the transformer is replaced with a capacitor.

Direct-coupled Input, Direct-coupled Output design provides excellent frequency response and low distortion. Bias stability is not as easily obtained as in the first two designs. This design is the least expensive for outputs below 0.1 watt.

Direct-coupled Input, Transformer-coupled Output design provides good frequency response and fair distortion. Bias stability is not as good as in the transformer and RC input designs. This design is the least expensive for outputs above 0.1 watt.

Class B Amplifier Coupling Characteristics

Transformer-coupled Input, Transformer-coupled Output design provides fair frequency response and distortion. Bias stability is good and easily obtained and power efficiency is high. Relative cost is high below one watt.

Transformer-coupled Input, RC-coupled Output design provides good frequency response and fair distortion. Bias stability is good and easily obtained and power efficiency is high. Relative cost is high below 1 watt, but low above 20 watts.

TRANSISTOR SELECTION

The performance and reliability of an amplifier depend upon careful selection of the transistor type. Parameter considerations which are most important in making this selection are discussed here.

Power Considerations

$P_{(Rate)}$—The maximum power rating of the device, normally referenced to 25°C. The power rating is always related to the device temperature (case or ambient) at which this power can be delivered. Data sheets generally specify a derating factor (K_1) from which $P_{(Rate)}$ at temperatures above 25°C may be calculated.

$P_{(Diss)}$—The power dissipated by the device while delivering the specified power output to the load. The device temperature, while dissipating this power, must be maintained below the device temperature rating T_1 calculated at $P_{(Diss)}$.

Safe Operating Curve. At a given temperature, the permissible V_{CE}-I_c combinations are determined by the hyperbola $P_{(Rate)} = V_{CE}I_c$.

At high voltages, secondary breakdown phenomena require that the power be further derated below the hyperbolic curve. The new derating line is called the safe operating curve. For circuits with heavy inductive loading, the a-c load line becomes an ellipse with an appreciable minor axis. The ellipse may exceed the safe operating curve and result in device failure unless precautions are taken. The manufacturer includes a safe operating curve on the data sheet of devices which require this consideration.

Current Considerations

I_{CM}—The maximum collector current of the device while operating at maximum output power. I_{CM} must be smaller than the maximum collector current rating of the device. The value of I_{CM} may correspond to impractical circuit values of h_{FE} and $V_{CE(sat)}$. The designer should select a transistor which not only provides sufficient I_{CM} but also maintains practical gain and saturation characteristics at that current. In Class B design, the peak current output greatly exceeds the average current. I_{CM} is then determined by the peak current requirement of the circuit.

I_{CBO}—The collector-base reverse current with the emitter open. I_{CBO} is generally specified near the maximum collector-base voltage rating of the device. This reverse leakage current consists of two components: saturation current and surface current. Saturation current increases with an increase in base resistivity and voltage, and it is the dominant leakage mode at high temperatures. Surface current is voltage dependent and prevails at low temperatures. I_{CBO} is important in determining the circuit quiescent point stability. The stability criteria should be evaluated using the high-temperature I_{CBO} specification. This assures stability under worst-case leakage conditions.

Voltage Considerations

$V_{(BR)EBO}$—Emitter-base reverse breakdown voltage with the collector open. This parameter is of importance in Class B amplifiers since the emitter-base junctions are alternately reverse biased each half-cycle. $V_{(BR)EBO}$ should always be specified greater than the peak voltage input V_{ip} delivered to the base.

$V_{(BR)CEO}$—Collector-emitter reverse breakdown voltage with the base open. This specification gives the lowest collector-emitter voltage rating of the device.

$V_{(BR)CES}$—Collector-emitter reverse breakdown voltage, with the emitter shorted to the base. This specification can be as high as $V_{(BR)CBO}$, but is normally slightly lower.

$V_{(BR)CER}$—Collector-emitter reverse breakdown voltage with a resistance R terminated between the base and emitter. This parameter is of importance in circuits having emitter-base resistance terminations. For large values of R, $V_{(BR)CER}$ approaches $V_{(BR)CEO}$. For low resistances, $V_{(BR)CER}$ appraoches $V_{(BR)CES}$.

$V_{(BR)CEV}$—Collector-emitter reverse breakdown voltage with the base returned to a negative bias back to the emitter terminal. $V_{(BR)CEV}$ approaches $V_{(BR)CBO}$ on most devices with a few volts reverse bias.

Other Considerations

h_{FE}—The static value of the common-emitter short-circuit current gain. h_{FE} may usually be assumed to be equal to h_{fe} for audio frequencies. This approximation does not hold true at low current levels where reverse saturation current becomes appreciable. At higher collector currents, the device emitter efficiency falls off due to conductivity modulation. This drop in emitter efficiency reduces the h_{FE} of the device at higher currents.

As junction temperature increases, h_{FE} increases. However, the h_{FE} normally starts to fall off at a lower collector current level with increasing temperature. Because several effects are present, calculation of collector-to-base current ratios should be performed using the $h_{FE(min)}$ value. This assures sufficient gain under all operating conditions.

y_{FE}—The plot of V_{BE} versus I_c is called the y_{FE} curve. For collector currents above one milliampere the y_{FE} curve is fairly linear. V_{BE} varies with individual devices and semiconductor materials. The V_{BE} voltage falls off with increasing temperature. For silicon, the reduction is about 2.0 millivolts per °C; for germanium, about 1.3 millivolts per °C. The y_{FE} curve is very useful for selecting the operating Q point. The base voltage drive requirements and peak collector current linearity can be evaluated also.

In Class A operation, the Q point is chosen so that the peak collector current excursion is within the linear region of the curve. In Class B operation, the y_{FE} curve is used to determine V_{BEQ}. By selecting proper V_{BEQ}, crossover distortion is reduced and the quiescent power dissipation is ar a minimum.

$V_{CE(sat)}$—The collector-to-emitter saturation voltage with both emitter-base and collector-base junctions forward biased. This parameter is generally measured with several times the minimum amount of base current required by the device h_{FE} for the required collector current. This base current overdrive ensures the device saturation. The ideal value of $V_{CE(sat)}$ would be zero, allowing maximum voltage swings on the a-c load line and maximum power output. This ideal voltage is never achieved in practice and actual voltages are usually from 0.2 to 2.0 volts. Certain high-voltage devices are found with $V_{CE(sat)}$ as high as 15 volts measured with usable base drives. This can be a considerable problem in direct-coupled Class B circuits where one of the driver pairs must saturate to a very low value in order to drive the output stage to the maximum collector swing.

SYMBOLS

A_v	Voltage Gain
C_1	Bypass Capacitor
C_c	Coupling Capacitor
C_o	Output Capacitor
η	Transformer Efficiency
f	Frequency
h_{fe}	Device a-c Short Circuit Current Gain, Common Emitter
h_{FE}	Device d-c Short Circuit Current Gain, Common Emitter
$h_{FE(min)}$	Minimum h_{FE} Specified on Data Sheet
h_{ie}	Device Input Impedance, Common Emitter
Hz	Hertz (cycles per second)
I_{BB}	Circuit d-c Base Bias Current
I_{BM}	Maximum Base Current, Referenced to d-c Zero

I_{bp}	Peak Base Current, Referenced to a-c Zero
I_{BQ}	Base Current, Quiescent Point
I_c	Collector Current
I_{CBO}	Reverse Collector-to-base Current, Emitter Open
ΔI_{CBO}	Change In Collector-to-base Current Due to Temperature Variations
I_{CM}	Maximum Collector Current, Referenced to d-c Zero
I_{cp}	Peak Collector Current, Referenced to a-c Zero
I_{CQ}	Collector Current, Quiescent Point
K_1	Thermal Derating Factor
K_2	Base-emitter Temperature-dependent Term
K_4	Empirical Material-dependent Constant
K_T	Empirical Temperature-dependent Constant
$P_{(Diss)}$	Device Power Dissipation
$P_{i(rms)}$	RMS Power Input
$P_{o(rms)}$	RMS Power Output
$P_{(Rate)}$	Device Maximum Power Rating
P.G.	Power Gain
$P.G._{(T)}$	Total Power Gain
Q Point	Quiescent Point
R_{cc}'	Collector-to-collector Reflected Load Impedance
R_D	Decoupling Resistance
R_E	External Emitter Resistance
r_e	Device Internal Emitter Diffusion Resistance
R_i	Circuit Input Resistance
R_L	Load Impedance
R_L'	Reflected Load Impedance
R_o	Equivalent to R_1, R_2 in Parallel
S	Siemans $(1/\Omega)$
T_1	Maximum Device Operating Temperature At Full Load
T_2	Temperature Related to Maximum Device Power Rating
T_3	Maximum Temperature I_{CBO} Specified on Data Sheet
T_4	Minimum Device Operating Temperature
T_A	Ambient Temperature
V_B	Base-to-ground Voltage, Referenced to d-c Zero
V_{BE}	Base-to-emitter Voltage
ΔV_{BE}	Change in Base-to-emitter Voltage Due to Temperature Variations
V_{BEM}	Maximum Base-to-emitter Voltage
V_{BEQ}	Base-to-emitter Voltage, Quiescent Point
$V_{(BR)CBO}$	Reverse Breakdown Voltage, Collector-to-base
$V_{(BR)CEO}$	Reverse Breakdown Voltage, Collector-to-emitter, Base Open
$V_{(BR)CER}$	Reverse Breakdown Voltage, Collector-to-emitter, Base Resistor termination to emitter
$V_{(BR)CES}$	Reverse Breakdown Voltage, Collector-to-emitter, Base shorted to emitter
$V_{(BR)CEV}$	Reverse Breakdown Voltage, Collector-to-emitter, Base reverse-biased to emitter
$V_{(BR)EBO}$	Reverse Breakdown Voltage, Emitter-to-base, Collector Open
V_{cc}	Circuit Supply Voltage
V_{CE}	Collector-to-emitter Voltage

$V_{CE(sat)}$ Collector-to-emitter Saturation Voltage
V_{CEM} Maximum Collector-to-emitter Voltage, Referenced to d-c Zero
V_{cep} Peak Collector-to-emitter Voltage, Referenced to a-c Zero
V_{CEQ} Collector-to-emitter Voltage, Quiescent Point
V_{ip} Peak Input Voltage, Referenced to a-c Zero
V_{R1} Voltage Across Resistance R_1
V_{RE} Voltage Across the Emitter Resistance
y_{FE} D-C Transconductance
y_{fe} A-C Transconductance
Z_b Base Secondary Impedance, Input Transformer
Z_{bb} Base-to-base Secondary Impedance, Input Transformer
Z_c Collector Primary Impedance, Input Transformer

The notation (D) in parentheses means the symbol has the above meaning but is referenced to the driver circuit. All symbols without the notation (D) are referenced to the output stage.

REFERENCES

1. Wheatley, C. F.: Reliability Considerations In The Application of Power Transistors To Consumer Products, *IRE Transactions,* Vol. BTR-8, July 1962.
2. Engineering Staff of Texas Instruments Incorporated: "Transistor Circuit Design," McGraw-Hill Book Co., Inc., New York, 1963.
3. Shea, Richard F.: "Transistor Applications," John Wiley & Sons, Inc., New York.
4. Kuehler, Hank: Small Signal Audio Design, Applications Report, Texas Instruments Incorporated.
5. Reference Data for Radio Engineers, International Telephone and Telegraph Corporation, 4th Edition, American Book—Stratford Press, Inc.
6. Corcoran, G. F., and H. W. Price: "Electronics," John Wiley & Sons, Inc., New York.

2

Class A Output and Driver Design Procedures

This chapter presents design procedures for four common combinations of amplifier coupling between the driver device and the power output stage and between the power output stage and the load. In providing complete procedures for the most common combinations, practical, simple solutions to design problems are given. This practical approach requires many approximations and assumptions based on design experience. All assumptions are noted in the text.

To make the design procedure complete for each coupling combination, some of the design equations and the comments prefacing these equations are repeated. Boldface numbers are used to designate equations appearing for the first time. Where equations are repeated, lightface numbers are used. Derivations of important equations are given in Chapter 6.

TRANSFORMER-COUPLED INPUT, TRANSFORMER-COUPLED OUTPUT

A circuit diagram is shown in Fig. 1 and dynamic characteristics are given in Fig. 2 for this coupling combination.

1. Maximum dissipation occurs in Class A amplifiers at the zero-signal-input quiescent point. Normally, Class A stages are biased for a 25% overload current to account for a shift in the circuit quiescent point caused by temperature variations. Power output $P_{o(rms)}$ is therefore related to the device power dissipation $P_{(Diss)}$ as shown in Eq. (1):

$$P_{(Diss)} = 3.12\ P_{o(rms)} \qquad \textbf{(1)}^*$$

2. This power dissipation capability of the device must be satisfied at the operating temperature conditions. The maximum permissible device temperature that can be tolerated without exceeding the device ratings is calculated

*Equation derived in Chapter 6.

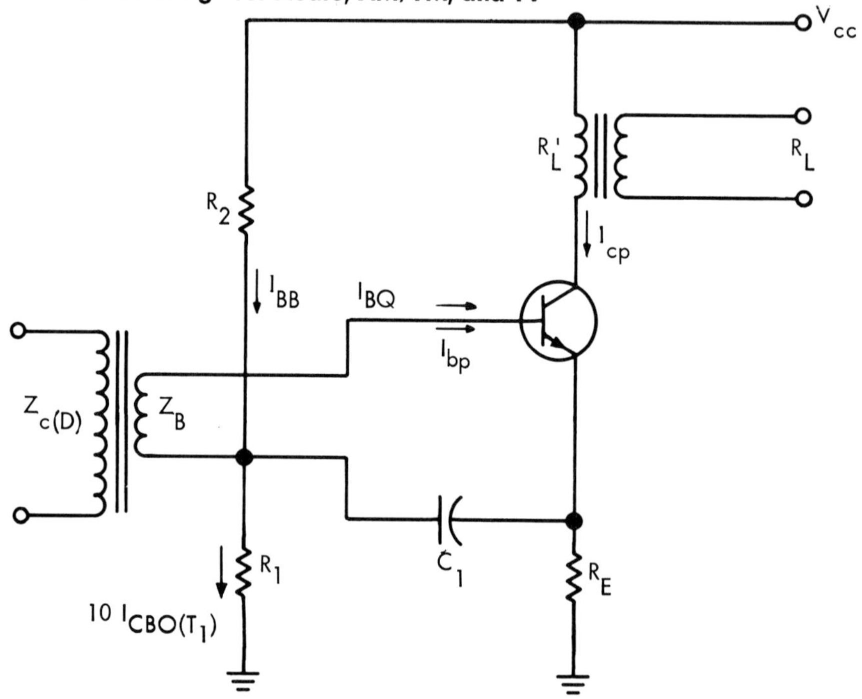

Fig. 1. Class A Transformer-coupled Input, Transformer-coupled Output Schematic

in Eq. (2). The transistor derating slope, the maximum device power rating, and corresponding temperature T_2 are all taken from the device data sheet.

$$T_1 = \frac{P_{(Rate)} - P_{(Diss)} + K_1 T_2}{K_1} \qquad \textbf{(2)*}$$

where

K_1 = Transistor Derating Slope — mW/°C
T_1 = Maximum Device Operating Temperature at $P_{(Diss)}$ — °C
T_2 = Temperature Related to Maximum Device $P_{(Rate)}$ — °C
$P_{(Rate)}$ = Maximum Device Power Rating — mW.

If the device is in a stud-mounted package, the heat sink should be evaluated to determine if it is able to dissipate the required $P_{(Diss)}$ while maintaining the device temperature at or below T_1. This requirement must be calculated at the highest desired ambient temperature.

3. Now consider the device voltage requirements. In power stages, higher load impedances result in higher power gain per stage. Therefore, the device voltage capabilities should be determined from the data sheet to allow the selection of optimum circuit voltages. Class A circuits with transformer loads require:

---

*Equation derived in Chapter 6.

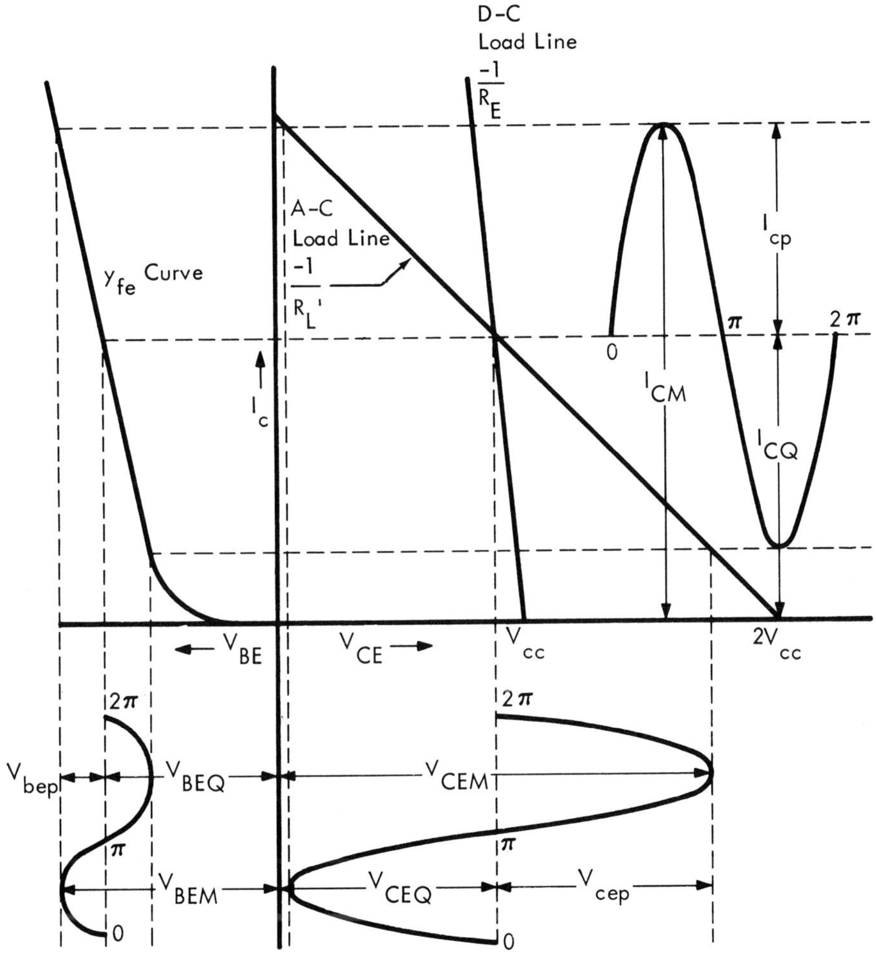

Fig. 2. Class A Transformer-coupled Input, Transformer-coupled Output Dynamic Characteristics

$$V_{cc} < \frac{V_{(BR)CER}}{2} \tag{3}$$

While this relationship represents the absolute maximum value of V_{cc} possible without exceeding the device maximum ratings, it is generally good design procedure to select V_{cc} equal to about 90% of the V_{cc} derived in Eq. (3). Additional care should be exercised in selecting the maximum circuit voltages for power devices that exhibit a reduction in the power rating of the device at high voltage levels. This derating is normally referred to as the Safe Operating Curve (Fig. 3). The a-c load line must be checked to make sure that a value of collector voltage has been chosen that will allow the load line to remain below the safe operating derating curve.

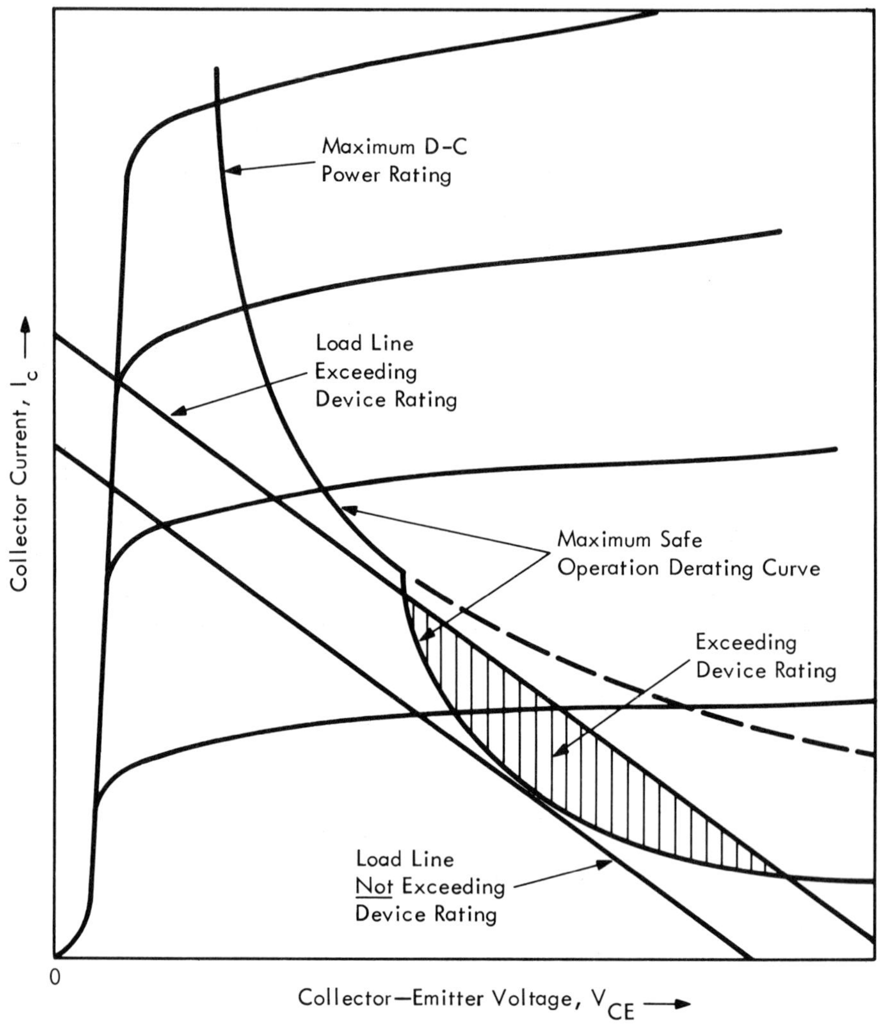

Fig. 3. Safe Operating Curve

4. One of the first considerations given worst-case circuit design is the effect of temperature variations on the critical device parameters. One of the most important parameters to evaluate for temperature stability is the change in the base-emitter voltage over the operating temperature range. This ΔV_{BE} is calculated in Eq. (4):

$$\Delta V_{BE} = K_2 (T_1 - T_4) \tag{4}$$

where

K_2 = Base-Emitter Temperature-Dependent Term
Silicon – 2.0×10^{-3} V/°C
Germanium – 1.3×10^{-3} V/°C
T_4 = Minimum Device Operating Temperature.

5. With V_{cc} established, the reflected load impedance R_L' can be calculated. The load voltage V_{cep} is approximated by $(V_{cc} - 8\Delta V_{BE} - V_{CE(sat)})$. The ΔV_{BE} term comes from assuming a maximum permissible change in collector quiescent current of 10% over the operating temperature range. This condition sets a value for the voltage drop across the emitter resistance R_E in terms of ΔV_{BE}.

$$R_L' = \frac{(V_{cc} - 8\Delta V_{BE} - V_{CE(sat)})^2\ \eta}{2P_{o(rms)}} \tag{5)*$$

6. The peak a-c collector current can now be calculated in terms of the power output and the V_{cep} equivalent.

$$I_{cp} = \frac{2P_{o(rms)}}{(V_{cc} - 8\Delta V_{BE} - V_{CE(sat)})\ \eta} \tag{6}$$

7. The quiescent collector current should be designed to be 25% greater than the peak a-c collector current to ensure operation in the linear portion of the y_{FE} curve. Linear operation on the device y_{FE} curve is necessary to prevent output distortion.

$$I_{CQ} = 1.25\ I_{cp} \tag{7}$$

8. The maximum d-c collector current requirement is the sum of the quiescent point current and the a-c peak collector current during maximum signal operation. This maximum collector current I_{CM} must be less than the maximum collector current rating of the device and within the range of collector currents having a useable h_{FE} value.

$$I_{CM} = 2.25\ I_{cp} \tag{8}$$

9. The maximum circuit requirements of power, voltage, and current have now been described, and the transistor type can be confirmed for the particular application in question.

10. The collector-base leakage current I_{CBO} is important in evaluating the circuit stability criteria. This leakage is a function of temperature and can be calculated at any temperature by assuming that the current doubles for an increase in temperature of 8°C for germanium and 12°C for silicon where

$$I_{CBO(T_1)} = I_{CBO(T_3)} 2^{(T_1 - T_3)/K_T} \tag{9}$$

where

*Equation derived in Chapter 6.

K_T = Empirical temperature-dependent constants
Silicon – 12
Germanium – 8
T_3 = Maximum temperature I_{CBO} specified on data sheet
$I_{CBO(T_3)}$ = Maximum I_{CBO} specified on data sheet at temperature T_3

11. The quiescent point base current I_{BQ} is calculated from the relationship of the quiescent collector current I_{CQ} and the device minimum h_{FE} specification taken from the data sheet.

$$I_{BQ} = \frac{I_{CQ}}{h_{FE(min)}} \tag{10}$$

12. The base bias that will provide the desired quiescent currents can now be calculated. To ensure minimum temperature effects on the Q-Point stability caused by I_{CBO} increasing the V_{BEQ} quiescent point, the base bias current I_{BB} is made considerably larger than the maximum $I_{CBO(T_1)}$ calculated in Eq. (9).

$$I_{BB} = I_{BQ} + 10\ I_{CBO(T_1)} \tag{11}$$

13. The Q-point base-emitter voltage V_{BEQ} may be obtained from the y_{FE} specification or curve on the data sheet. The value of V_{BEQ} is read off the curve at the quiescent collector current I_{CQ}.

$$V_{BEQ} = \text{Data Sheet Value at } I_{CQ} \tag{12}$$

14. Series resistances are used in the emitter circuit of Class A amplifiers to provide d-c stability in the output stage. There are two main sources that cause shifts in the d-c Q point with temperature. The base-emitter voltage ΔV_{BE} decreases with temperature and the collector-base current I_{CBO} increases with temperature. I_{BB} has been designed 10 times $I_{CBO(T_1)}$ so that I_{CBO} through R_1 cannot change V_{BEQ} more than 10% due to ΔI_{CBO}. If also $I_{CQ}R_E$ is calculated to be 10 times the ΔV_{BE} contribution, the collector current I_{CQ} should not change more than 10% due to increasing temperature effects on the device parameters. Equation (13) shows the minimum value of emitter resistance necessary to maintain the ΔI_{CQ} within 10% of the nominal value calculated:

$$R_E \geq \frac{100\Delta V_{BE} + V_{BEQ}}{10\ I_{CQ}} \tag{13}^*$$

If more regulation is required in the collector current stability point, the resistance R_E may be increased with a resultant increase in power loss from the d-c power source. Calculated values of R_E may become unreasonably large in circuits requiring strict control of the Q-point stability. In such instances, a compromise may be necessary between the permissible power loss in R_E and the collector-current Q-point stability. Feedback arrangements other than R_E have not been considered here, but this is a possible solution to improve the compromise should better Q-point stability be required.

15. The peak a-c base current I_{bp} is calculated from the peak collector current and device h_{FE} relationship in Eq. (14):

*Equation derived in Chapter 6.

$$I_{bp} = \frac{I_{cp}}{h_{FE(min)}} \tag{14}$$

16. The bias resistor R_1 is calculated to provide the required Q-point base-emitter voltage and corresponding collector current as described by the device y_{FE} curve.

$$R_1 = \frac{I_{CQ}R_E + V_{BEQ}}{10\ I_{CBO(T_1)}} \tag{15}$$

17. The bias resistance R_2 can now be determined from the base bias current and R_1.

$$R_2 = \frac{V_{cc} - 10\ I_{CBO(T_1)}R_1}{I_{BB}} \tag{16}$$

18. To prevent degeneration of the a-c circuit gain caused by the emitter resistance R_E, a capacitor C_1 is used to bypass R_E and R_1 back to the common input transformer terminal. The necessary value of capacitance is determined by the amount of gain reduction that can be tolerated over the desired frequency range limits. The equation shown assumes an output falloff of three dB at the lowest desired frequency.

$$C_1 \geq \frac{I_{cp}}{2\pi f K_4} \tag{17}^*$$

where

K_4 = Empirical constant dependent on material
Silicon -35×10^{-3}
Germanium $- 25 \times 10^{-3}$
f = Lowest desired frequency.

19. The power gain of the stage can now be calculated in terms of the rms power output and the input base-emitter parameters.

$$P.G. = \frac{2P_{o(rms)}}{V_{bep}\ I_{bp}} \tag{18}$$

20. The secondary impedance of the input transformer to the power stage is required for transformer design and later for calculation of the primary driver impedance. Since both series elements R_E and R_1 in the input circuit have been bypassed, the input transformer secondary impedance Z_b may be calculated from the peak base currents and voltages:

$$Z_b = \frac{V_{bep}}{I_{bp}} \tag{19}$$

All of the necessary components and device parameters for the output stage have now been determined. Normally, a power output stage requires a driver input, and the collector circuit constants of a driver stage will now be evaluated.

---

*Equation derived in Chapter 6.

21. Maximum power gain in the driver stage is accomplished by maximum collector load impedance. The driver power output requirement is usually low and this allows a high-impedance primary input transformer. Since with a high-impedance load, the collector voltage swing requirements are normally easily obtained, a practical high-impedance value for the driver transformer primary $Z_{c(D)}$ may be assumed. Practical values are as high as 20 kilohms. A compromise may be necessary in applications where the driver supply voltage $V_{cc(D)}$ is extremely limited and a sacrifice in maximum power gain is necessary. A value of $Z_{c(D)}$ may be approximated in Eq. (20) in terms of the $V_{cc(D)}$ available.

$$Z_{c(D)} = \frac{(V_{cc(D)} - 3V_{CE(sat)(D)})^2\ \eta}{V_{bep}\ I_{bp}} \tag{20}$$

22. The peak collector current of the driver is dependent upon the power input requirements of the output stage and the transformer primary impedance $Z_{c(D)}$. This peak current requirement may be calculated in Eq. (21).

$$I_{cp(D)} = I_{bp}\sqrt{\frac{Z_b}{\eta Z_{c(D)}}} \tag{21}$$

23. The driver collector-current quiescent point is determined by the peak collector current calculated above. The quiescent current must be greater than the peak collector current to prevent operation of the device in the nonlinear portion of the y_{FE} curve. Distortion of the driver signal will result if the device is allowed to operate to cut-off, so a minimum value of collector current is chosen that will always be above the knee of the y_{FE} curve. The collector quiescent current may be calculated in Eq. (22):

$$I_{CQ(D)} = I_{cp(D)} + 1.0 \times 10^{-3}\text{A} \tag{22}$$

24. The driver supply voltage requirements can be considerably smaller than the power output stage requirements due to the smaller power output of the driver. The minimum driver supply voltage may be calculated in Eq. (23) in terms of the base-emitter parameters of the output stage and the collector saturation voltage of the driver device. This driver supply voltage may then be obtained by dropping the output stage supply voltage through a decoupling resistor or by providing a separate voltage source.

$$V_{cc(D)} = \frac{V_{bep}\ I_{bp}}{\eta I_{cp(D)}} + 3V_{CE(sat)(D)} \tag{23}^*$$

25. The critical driver collector requirements have now been calculated and the remainder of the driver parameters and components may be determined by using the appropriate equations derived for the same coupling scheme in a power output stage. The power output requirement of the driver device may be calculated in Eq. (24) in terms of the collector requirements.

$$P_{o(rms)(D)} = \frac{I_{cp(D)}^2\ Z_{c(D)}\ \eta}{2} \tag{24}$$

*Equation derived in Chapter 6.

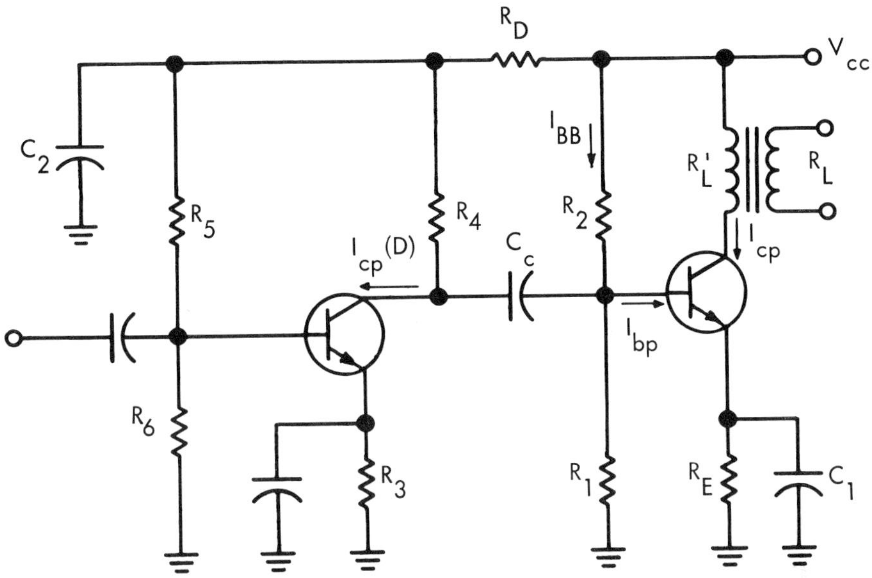

Fig. 4. Class A RC-coupled Input, Transformer-coupled Output Schematic

RC-COUPLED INPUT, TRANSFORMER-COUPLED OUTPUT

A circuit diagram is shown in Fig. 4 and dynamic characteristics are given in Fig. 5 for this coupling combination.

1. Maximum dissipation occurs in Class A amplifiers at the zero-signal-input quiescent point. Normally, Class A stages are biased for a 25% overload current to account for a shift in the circuit quiescent point caused by temperature variations. Power output $P_{o(rms)}$ is therefore related to the device power dissipation $P_{(Diss)}$ as shown in Eq. (1):

$$P_{(Diss)} = 3.12\ P_{o(rms)} \tag{1*}$$

2. This power dissipation capability of the device must be satisfied at the operating temperature conditions. The maximum permissible device temperature that can be tolerated without exceeding the device ratings is calculated in Eq. (2). The transistor derating slope, the maximum device power rating, and corresponding temperature T_2 are all taken from the device data sheet.

$$T_1 = \frac{P_{(Rate)} - P_{(Diss)} + K_1 T_2}{K_1} \tag{2*}$$

where

K_1 = Transistor Derating Slope — mW/°C
T_1 = Maximum Device Operating Temperature at $P_{(Diss)}$ — °C
T_2 = Temperature Related to Maximum Device $P_{(Rate)}$ — °C
$P_{(Rate)}$ = Maximum Device Power Rating — mW.

*Equation derived in Chapter 6.

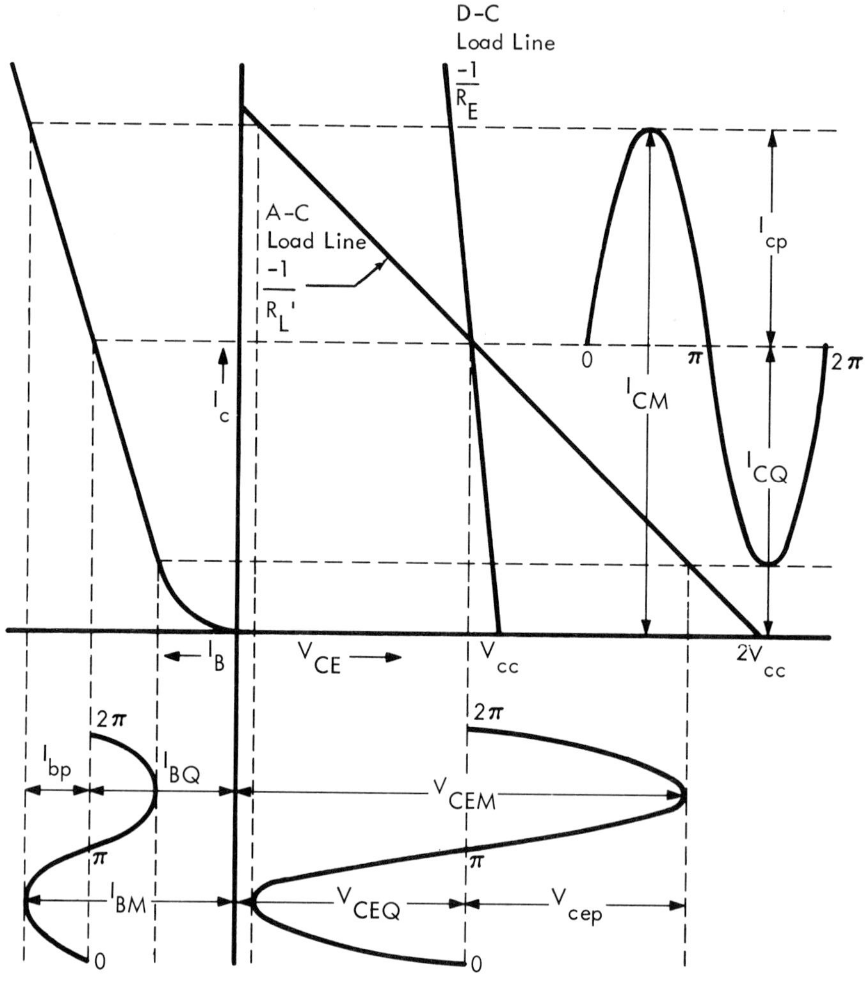

Fig. 5. Class A RC-coupled Input, Transformer-coupled Output and Direct-coupled Input, Transformer-coupled Output Dynamic Characteristics

If the device is in a stud-mounted package, the heat sink should be evaluated to determine if it is able to dissipate the required $P_{(Diss)}$ while maintaining the device temperature at or below T_1. This requirement must be calculated at the highest desired ambient temperature.

3. Now consider the device voltage requirements. In power stages, higher load impedances result in higher power gain per stage. Therefore, the device voltage capabilities should be determined from the data sheet to allow the selection of optimum circuit voltages. Class A circuits with transformer loads require:

$$V_{cc} < \frac{V_{(BR)CER}}{2} \tag{3}$$

While this relationship represents the absolute maximum value of V_{cc} possible without exceeding the device maximum ratings, it is generally good design procedure to select V_{cc} equal to about 90% of the V_{cc} derived in Eq. (3). Additional care should be exercised in selecting the maximum circuit voltages for power devices that exhibit a reduction in the power rating of the device at high voltage levels. This derating is normally referred to as the Safe Operating Curve (Fig. 3). The a-c load line must be checked to make sure that a value of collector voltage has been chosen that will allow the load line to remain below the safe operating derating curve.

4. One of the first considerations given worst-case circuit design is the effect of temperature variations on the critical device parameters. One of the most important parameters to evaluate for temperature stability is the change in the base-emitter voltage over the operating temperature range. This ΔV_{BE} is calculated in Eq. (4):

$$\Delta V_{BE} = K_2 (T_1 - T_4) \tag{4}$$

where

K_2 = Base-emitter temperature-dependent term
Silicon $- 2.0 \times 10^{-3}$ V/°C
Germanium $- 1.3 \times 10^{-3}$ V/°C
T_4 = Minimum device operating temperature.

5. With V_{cc} established, the reflected load impedance R_L' can be calculated. The load voltage V_{cep} is approximated by $(V_{cc} - 8\Delta V_{BE} - V_{CE(sat)})$. The ΔV_{BE} term comes from assuming a maximum permissible change in collector quiescent current of 10% over the operating temperature range. This condition sets a value for the voltage drop across the emitter resistance R_E in terms of ΔV_{BE}.

$$R_L' = \frac{(V_{cc} - 8\Delta V_{BE} - V_{CE(sat)})^2\,\eta}{2P_{o(rms)}} \tag{5}^*$$

6. The peak a-c collector current can now be calculated in terms of the power output and the V_{cep} equivalent.

$$I_{cp} = \frac{2P_{o(rms)}}{(V_{cc} - 8\Delta V_{BE} - V_{CE(sat)})\,\eta} \tag{6}$$

7. The quiescent collector current should be designed to be 25% greater than the peak a-c collector current to ensure operation in the linear portion of the y_{FE} curve. Linear operation on the device y_{FE} curve is necessary to prevent output distortion.

$$I_{CQ} = 1.25\ I_{cp} \tag{7}$$

8. The maximum d-c collector current requirement is the sum of the quiescent point current and the a-c peak collector current during maximum signal operation. This maximum collector current I_{CM} must be less than the maximum collector current rating of the device and within the range of collector currents having a useable h_{FE} value.

---

*Equation derived in Chapter 6.

$$I_{CM} = 2.25\ I_{cp} \tag{8}$$

9. The maximum circuit requirements of power, voltage, and current have now been described, and the transistor type can be confirmed for the particular application in question.

10. The collector-base leakage current I_{CBO} is important in evaluating the circuit stability criteria. This leakage is a function of temperature and can be calculated at any temperature by assuming that the current doubles for an increase temperature of 8°C for germanium and 12°C for silicon.

$$I_{CBO(T_1)} = I_{CBO(T_3)} 2^{(T_1 - T_3)/K_T} \tag{9}$$

where

K_T = Empirical temperature-dependent constants
Silicon — 12
Germanium — 8
T_3 = Maximum temperature I_{CBO} specified on data sheet
$I_{CBO(T_3)}$ = Maximum I_{CBO} specified on data sheet at temperature T_3

11. The quiescent point base current I_{BQ} is calculated from the relationship of the quiescent collector current I_{CQ} and the device minimum h_{FE} specification taken from the data sheet.

$$I_{BQ} = \frac{I_{CQ}}{h_{FE(min)}} \tag{10}$$

12. The base bias that will provide the desired quiescent currents can now be calculated. To ensure minimum temperature effects on the Q-point stability caused by I_{CBO} increasing the V_{BEQ} quiescent point, the base bias current I_{BB} is made considerably larger than the maximum $I_{CBO(T_1)}$ calculated from Eq. (9).

$$I_{BB} = I_{BQ} + 10\ I_{CBO(T_1)} \tag{11}$$

13. The Q-point base-emitter voltage V_{BEQ} may be obtained from the y_{FE} specification or curve on the data sheet. The value of V_{BEQ} is read off of the curve at the quiescent collector current I_{CQ}.

$$V_{BEQ} = \text{Data Sheet Value at } I_{CQ} \tag{12}$$

14. Series resistances are used in the emitter circuit of Class A amplifiers to provide d-c stability in the output stage. There are two main sources that cause shifts in the d-c Q-point with temperature. The base-emitter voltage ΔV_{BE} decreases with temperature and the collector-base current I_{CBO} increases with temperature. I_{BB} has been designed 10 times $I_{CBO(T_1)}$ so that ΔI_{CBO} through R_1 cannot change V_{BEQ} more than 10% due to ΔI_{CBO}. If also $I_{CQ}R_E$ is calculated to be 10 times the ΔV_{BE} contribution, the collector current I_{CQ} should not change more than 10% due to increasing temperature effects on the device parameters. Equation (13) shows the minimum value of emitter resistance necessary to maintain the ΔI_{CQ} within 10% of the nominal value calculated:

$$R_E \geq \frac{100\ \Delta V_{BE} + V_{BEQ}}{10\ I_{CQ}} \tag{13}^*$$

*Equation derived in Chapter 6.

If more regulation is required in the collector current stability point, the resistance R_E may be increased with a resultant increase in power loss from the d-c power source. Calculated values of R_E may become unreasonably large in circuits requiring strict control of the Q-point stability. In such instances, a compromise may be necessary between the permissible power loss in R_E and the collector-current Q-point stability. Feedback arrangements other than R_E have not been considered here, but this is a possible solution to improve the compromise should better Q-point stability be required.

15. The peak a-c base current I_{bp} is calculated from the peak collector current and device h_{FE} relationship in Eq. 14:

$$I_{bp} = \frac{I_{cp}}{h_{FE(min)}} \tag{14}$$

16. The bias resistor R_1 is calculated to provide the required Q-point base-emitter voltage and corresponding collector current as described by the device y_{FE} curve.

$$R_1 = \frac{I_{CQ}R_E + V_{BEQ}}{10\ I_{CBO(T_1)}} \tag{15}$$

17. The bias resistance R_2 can now be determined from the base bias current and R_1.

$$R_2 = \frac{V_{cc} - 10\ I_{CBO(T_1)}\ R_1}{I_{BB}} \tag{16}$$

18. The input impedance to the output stage R_i is the parallel configuration of the device common-emitter input impedance and the equivalent resistance R_o.

$$R_i = \frac{h_{FE}\ R_o}{h_{FE} + y_{fe}\ R_o} \qquad \textbf{(25)}^*$$

where

$$R_o = \frac{R_1\ R_2}{R_1 + R_2}$$

19. Degeneration of the a-c gain of the stage is eliminated by bypassing the emitter resistance R_E. The value of capacitance necessary is determined by the amount of gain falloff that can be tolerated over the desired frequency range. Equation (17) assumes an output falloff of three dB at the lowest frequency.

$$C_1 \geq \frac{I_{cp}}{2\pi f K_4} \qquad (17)^*$$

where

K_4 = Empirical constant dependent on material
Silicon — 35×10^{-3}
Germanium — 25×10^{-3}
f = Lowest desired frequency.

*Equation derived in Chapter 6.

20. The power gain for the stage can then be calculated as a function of the device h_{fe} and the ratios of the collector reflected load impedance and the input impedance.

$$P.G. = \frac{h_{fe}^2 R_L' \eta}{R_i} \tag{26}$$

21. The decoupling resistance R_D in conjunction with capacitor C_2 (usually approximately 10 μF) prevents signal feedback through the internal resistance of the power source. The decoupling resistor is assumed to be at least equal to the internal resistance of the source delivering 20 mA at half the terminal voltage.

$$R_D = \frac{V_{cc} \times 10^3}{40} \tag{27}$$

22. The driver emitter resistance R_3 can now be approximated in terms of the decoupling resistance R_D. The value shown is an empirical value, and it is not normally very critical due to the low peak collector currents and small collector voltage swing present in the driver stage.

$$R_3 = 3R_D \tag{28}$$

23. The driver collector load is resistive and therefore the maximum collector-emitter voltage is equal to the supply voltage V_{cc}. The output of the driver stage is capacitive coupled to the input of the output stage. This requires V_{bep} to equal $V_{cep(D)}$. It is common practice to allow for a 50% voltage overload to ensure an undistorted driver signal. The maximum collector-emitter requirements for the driver device are given in Eq. (30) when the assumption of Eq. (29) is made:

$$V_{cep(D)} = V_{bep} = V_{BEM} - V_{BEQ} \tag{29}$$

$$V_{CEM(D)} = 3(V_{cep(D)} + V_{CE(sat)(D)}) \qquad \textbf{(30)*}$$

24. The collector-emitter quiescent voltage of the driver is calculated in terms of the driver saturation voltage and the peak collector-emitter voltage of the driver.

$$V_{CEQ(D)} = 2V_{cep(D)} + 3V_{CE(sat)(D)} \tag{31}$$

25. The driver stage is designed to be capable of delivering 180% of base current I_{bp} to the output stage in addition to the parallel resistance losses caused by R_o. This prevents driver overload under maximum output conditions:

$$I_{cp(D)} = 1.8 \left(I_{bp} + \frac{V_{BEQ}}{R_o} \right) \tag{32}$$

26. The driver quiescent collector current is calculated from $I_{cp(D)}$ using the same factor derived for ensuring linear operation in the output stage collector current.

*Equation derived in Chapter 6.

$$I_{CQ(D)} = 1.25\ I_{cp(D)} \tag{33}$$

27. The driver load resistance R_4 is calculated at the quiescent collector current $I_{CQ(D)}$.

$$R_4 = \frac{V_{cc} - V_{CEQ(D)}}{I_{CQ(D)}} - (R_D + R_3) \tag{34}$$

28. The output of the driver stage is coupled to the power output stage through the coupling capacitor C_c and the driver load resistance R_4. The minimum coupling capacitance required can be calculated in terms of the lowest desired frequency response of the circuit.

$$C_c \geq \frac{1}{2\pi f R_4} \tag{35}$$

29. The remainder of the driver circuit parameters may be calculated from the appropriate bias and quiescent equations derived for the power output stages.

DIRECT-COUPLED INPUT, DIRECT-COUPLED OUTPUT

A circuit diagram is shown in Fig. 6 and dynamic characteristics are given in Fig. 7 for this coupling configuration.

1. Class A direct-coupled amplifiers are becoming more popular where cost and frequency response are the most important considerations. While there are some inherent advantages there are also some disadvantages concerning stability criteria in direct-coupled amplifiers. Because the driver and output stages are coupled directly, any variations in one device parameter will be reflected to the other device with a resultant change in its operating point. One of the most severe causes of parameter drift is due to the device temperature during maximum power operation.

2. Maximum power dissipation occurs in Class A amplifiers at the zero-signal-input quiescent point. Normally, Class A stages are biased for a 25% current overload to account for a shift in the quiescent point as described above. Power output is therefore related to the device power dissipation as shown in Eq. (36).

$$P_{(Diss)} = 2.5\ P_{o(rms)} \tag{36}*$$

3. This maximum power dissipation capability of the device must be satisfied at the full-load temperature conditions. The maximum permissible device temperature that can be tolerated without exceeding the device rating is calculated in Eq. (2). The transistor derating slope, the maximum device power rating, and corresponding temperature T_2 are all taken from the transistor data sheet. Care must be observed to keep the rating units the same when the derating slope and the device power ratings are taken from the data sheet.

---

*Equation derived in Chapter 6.

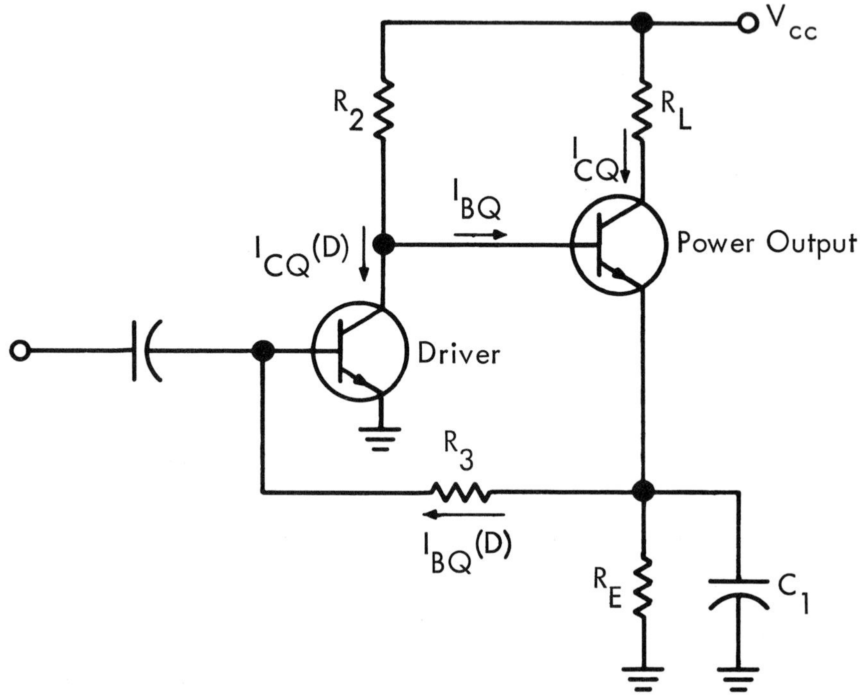

Fig. 6. Class A Direct-coupled Input, Direct-coupled Output Schematic

$$T_1 = \frac{P_{(Rate)} - P_{(Diss)} + K_1T_2}{K_1} \qquad (2)^*$$

where

K_1 = Transistor derating slope — mW/°C
T_1 = Maximum device operating temperature at $P_{(Diss)}$ — °C
T_2 = Temperature related to maximum device $P_{(Rate)}$ — °C
$P_{(Rate)}$ = Maximum device power rating — mW.

If the device is in a stud-mounted package, the heat sink should be evaluated to determine if it is able to dissipate the required $P_{(Diss)}$ while maintaining the device temperature at or below T_1. This requirement must be calculated at the highest desired ambient temperature.

4. The maximum collector current in a direct-coupled stage must be less than the maximum collector current rating of the device and within the range of collector currents having a useable h_{FE} value. The maximum current I_{CM} is equal to the sum of the peak load current and the quiescent operating current

*Equation derived in Chapter 6.

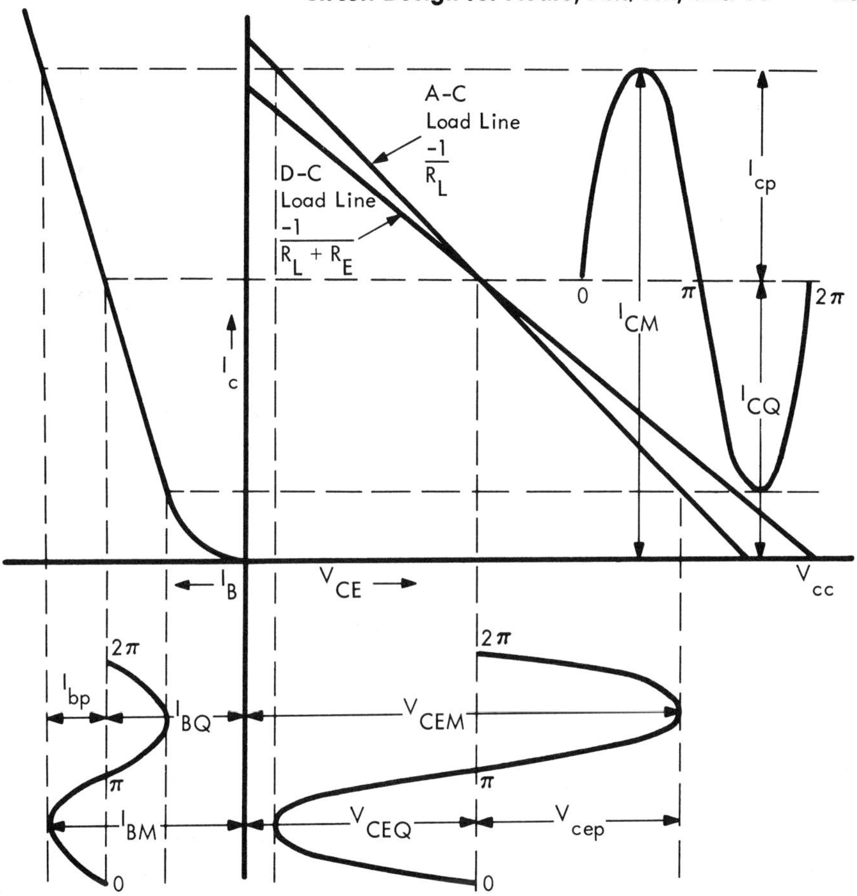

Fig. 7. Class A Direct-coupled Input, Direct-coupled Output Dynamic Characteristics

in the collector of the device. The quiescent collector current is increased 25% over the peak load current requirement to ensure device operation in the linear portion of the h_{FE} curve over the desired temperature range. The peak collector current is calculated in Eq. (37) and the maximum device collector current is calculated in Eq. (8):

$$I_{cp} = \sqrt{\frac{2P_{o(rms)}}{R_L}} \tag{37}$$

$$I_{CM} = 2.25\ I_{cp} \tag{8}$$

5. Now consider the device voltage requirements. In Class A power stages with resistive loads, the maximum collector-emitter voltage requirement of the device is:

$$V_{(BR)CER} > V_{cc} \tag{38}$$

Low-impedance speaker loads can be considered almost resistive as far as the device collector circuit is concerned and therefore the above relationship can be used for almost any direct-coupled speaker loads. This value represents the minimum device voltage rating that may be used in this application. The minimum collector voltage requirement is given in terms of the power output and load resistance in Eq. (39). This equation contains a maximum approximation for the voltage drop across the emitter resistance R_E. Therefore the calculated value represents a maximum approximation for estimating the collector voltage requirements of the circuit.

$$V_{(BR)CER} \geqq 2\sqrt{2P_{o(rms)}R_L} + 4V + 2V_{CE(sat)} \quad \textbf{(39)}$$

6. The quiescent collector current is set at 25% greater than the peak collector load current to ensure operation in the linear portion of the h_{FE} curve. This requirement allows maximum power output with minimum output distortion. The output-stage collector quiescent current is calculated in Eq. (7):

$$I_{CQ} = 1.25\ I_{cp} \quad (7)$$

7. The quiescent point base current in the output stage is calculated from the relationship of the quiescent collector current I_{CQ} and the device minimum h_{FE} specification taken from the data sheet.

$$I_{BQ} = \frac{I_{CQ}}{h_{FE(min)}} \quad (10)$$

8. The output stage quiescent base current is supplied from the collector circuit of the driver device. In Class A operation, the maximum base current is twice the base quiescent current. Therefore the minimum collector current requirement of the driver device is determined by the output-stage base quiescent current. In practice, the driver collector current must not be allowed to cut off completely or distortion from the driver stage will distort the output. This is prevented by increasing the driver collector current by an empirical factor equal to 1.25 times the minimum collector current requirement. For $I_{CM(D)}$:

$$I_{CM(D)} = 2.5\ I_{BQ} \quad \textbf{(40)}$$

9. The collector-base leakage current I_{CBO} is important in evaluating the circuit stability criteria. This leakage is a function of temperature and can be calculated at any temperature by assuming that the current doubles for an increase in temperature of 8°C for germanium and 12°C for silicon. An I_{CBO} specification at temperature T_3 is taken from the data sheet for calculating the I_{CBO} corresponding to the maximum device temperature at maximum dissipation. In calculating I_{CBO} for driver stages, the power dissipation in the driver stage is very small, and the device temperature can be assumed to be approaching the ambient temperature. Therefore a maximum ambient temperature value T_A is used in Eq. (41) to calculate the maximum I_{CBO} of the driver. The value of collector-base leakage calculated in the driver is very important in direct-coupled circuits because the value of feedback resistance R_3 is dependent upon this parameter for d-c stability.

$$I_{CBO(T_A)(D)} = I_{CBO(T_3)} 2^{(T_A - T_3)/K_T} \quad \textbf{(41)}$$

where

K_T = Empirical temperature-dependent constants
Silicon — 12
Germanium — 8
T_3 = Maximum temperature I_{CBO} specified on data sheet
$I_{CBO(T_3)}$ = Maximum I_{CBO} specified on data sheet at temperature T_3

10. The driver collector quiescent current must now be calculated to determine the driver base-emitter quiescent voltage. The quiescent collector current must be greater than the peak a-c collector current $I_{cp(D)}$ to prevent collector cut-off operation and distortion of the driver output. The quiescent collector current will be equal to the maximum collector current $I_{CM(D)}$ from Eq. (40) minus the base quiescent current of the output stage. The calculation may be made as shown in Eq. (42):

$$I_{CQ(D)} = 1.5\ I_{BQ} \quad \textbf{(42)}$$

11. The driver base-emitter quiescent voltage $V_{BEQ(D)}$ may be determined from the quiescent collector current $I_{CQ(D)}$ and the driver y_{FE} curve.

12. The driver base bias resistor R_3 provides a feedback path and stabilizing current from the output stage emitter resistance. It is desirable to make R_3 as large as possible, consistent with bias stability criteria to minimize the power loss of R_3 shunting the input. Bias stability would become a problem, however, if the I_{CBO} of the driver stage caused an appreciable change in the voltage across R_3 thereby resulting in a change in the driver quiescent current. The voltage across the emitter resistance is proportional to the output collector current, and this feedback provides an output collector-current stability factor. The voltage drop across the resistance R_3 is V_{RE} minus $V_{BEQ(D)}$. The criteria for stability is set in terms of the output quiescent current I_{CQ}. The assumption is made that the change in V_{BEQ} is held within 5% of the maximum value at the maximum device temperature. The maximum value of resistance R_3 is calculated in Eq. (43):

$$R_3 \leqq \frac{V_{BEQ(D)}}{20\ I_{CBO(T_A)(D)}} \quad \textbf{(43)}$$

13. The base quiescent current of the driver is determined from the collector quiescent current calculated in Eq. (42).

$$I_{BQ(D)} = \frac{I_{CQ(D)}}{h_{FE(min)(D)}} \quad \textbf{(44)}$$

14. The emitter resistance R_E may now be calculated in terms of the driver quiescent conditions and the output collector current. The emitter resistance is employed mainly for collector quiescent current feedback to the driver device. The output quiescent current will stabilize when the d-c closed-loop voltage gain approaches unity. This condition is satisfied when the driver d-c y_{FE} times the driver collector load impedance is equal to unity. The emitter resistance R_E is:

$$R_E = \frac{V_{BEQ(D)} + I_{BQ(D)}R_3}{I_{CQ}} \tag{45}$$

15. The collector supply voltage is the sum of the peak load voltage and the quiescent drop across the emitter resistance R_E. The collector voltage must not be allowed to operate in the saturated condition, so an empirical factor in terms of the device $V_{CE(sat)}$ is added to the minimum calculated value. The total supply voltage requirement is shown in Eq. (46):

$$V_{cc} = 2\sqrt{2P_{o(rms)}R_L} + I_{CQ}R_E + 2V_{CE(sat)} \tag{46*}$$

16. The driver collector is coupled directly to the base of the output device. This forces the base voltage of the output device to operate at the collector voltage of the driver. The total base voltage V_B of the output stage must be determined to ensure that the driver is not forced to operate in the saturation region. The driver collector resistance R_2 is also dependent upon the total base voltage of the output stage. The quiescent base-emitter voltage V_{BEQ} is obtained from the y_{FE} curve at the collector current I_{CQ}, and the total base voltage V_B is calculated in Eq. (47):

$$V_B = I_{CQ}R_E + V_{BEQ} \tag{47}$$

17. The driver collector resistance is determined by the difference between the supply voltage and the total base quiescent voltage with the total quiescent load current flowing. The maximum current in the collector of the driver must be equal to the sum of the collector quiescent current and the base quiescent current of the output stage. The resistance R_2 is calculated in Eq. (48):

$$R_2 = \frac{V_{cc} - V_B}{I_{CM(D)}} \tag{48}$$

18. The bypass capacitor C_1 must be calculated to prevent a-c gain degeneration at the minimum desired frequency. The minimum value of capacitance required is determined by assuming the output is down three dB at the minimum frequency. The relationship for the capacitance calculation is given in Eq. (17).

$$C_1 \geqq \frac{I_{cp}}{2\pi f K_4} \tag{17*}$$

where

K_4 = Empirical constant dependent on material
Silicon — 35×10^{-3}
Germanium — 25×10^{-3}
f = Minimum desired frequency.

19. The driver input resistance is required for the total power gain (Eq. (50)). The input resistance is proportional to the driver current gain and the feedback resistance R_3. The a-c y_{fe} of the driver is required for this calculation because of the relationship between the device input resistance, typical h_{FE}, and the y_{FE} characteristics. The y_{fe} of the driver is almost a constant over

*Equation derived in Chapter 6.

collector-current ranges, from about one mA up to the collector-current rating of the device. This enables an accurate value to be obtained from the y_{FE} curve even though the collector current excursions ΔI_c are very small. The input driver resistance R_i is calculated in Eq. (49):

$$R_{i(D)} = \frac{h_{FE(D)} R_3}{h_{FE(D)} + y_{fe} R_3} \qquad \textbf{(49)}$$

20. The amplifier design may now be completed by calculating the total power gain of the amplifier in terms of the input impedance of the driver and the output impedance of the power output stage. Equation (50) shows the power gain equation for calculating total power performance of the amplifier.

$$P.G._{(T)} = \frac{(h_{FE(D)} h_{FE})^2 R_L}{R_{i(D)}} \qquad \textbf{(50)*}$$

DIRECT-COUPLED INPUT, TRANSFORMER-COUPLED OUTPUT

A circuit diagram is shown in Fig. 8 and dynamic characteristics are given in Fig. 5 for this coupling combination.

1. Class A direct-coupled amplifiers are becoming more popular where cost and frequency response are the most important considerations. While there are some inherent advantages, there are also some disadvantages concerning stability criteria in direct-coupled amplifiers. Because the driver and output stages are coupled directly, any variations in one device parameter will be reflected to the other device with a resultant change in its operating point. One of the most severe causes of parameter drift is due to the device temperature during maximum power operation. Maximum power dissipation occurs in Class A amplifiers at the zero-signal-input quiescent point. Normally, Class A stages are biased for a 25% current overload to account for a shift in the quiescent point as described above. Power output is therefore related to the device power dissipation as shown in Eq. (1).

$$P_{(Diss)} = 3.12\ P_{o(rms)} \qquad (1)*$$

2. This maximum power dissipation capability of the device must be satisfied at the full-load temperature conditions. The maximum permissible device temperature that can be tolerated without exceeding the device ratings is calculated in Eq. (2). The transistor derating slope, the maximum device power rating, and corresponding temperature T_2 are all taken from the transistor data sheet. Care must be observed to keep the rating units the same when the derating slope and the device power ratings are taken from the data sheet.

$$T_1 = \frac{P_{(Rate)} - P_{(Diss)} + K_1 T_2}{K_1} \qquad (2)*$$

where

*Equation derived in Chapter 6.

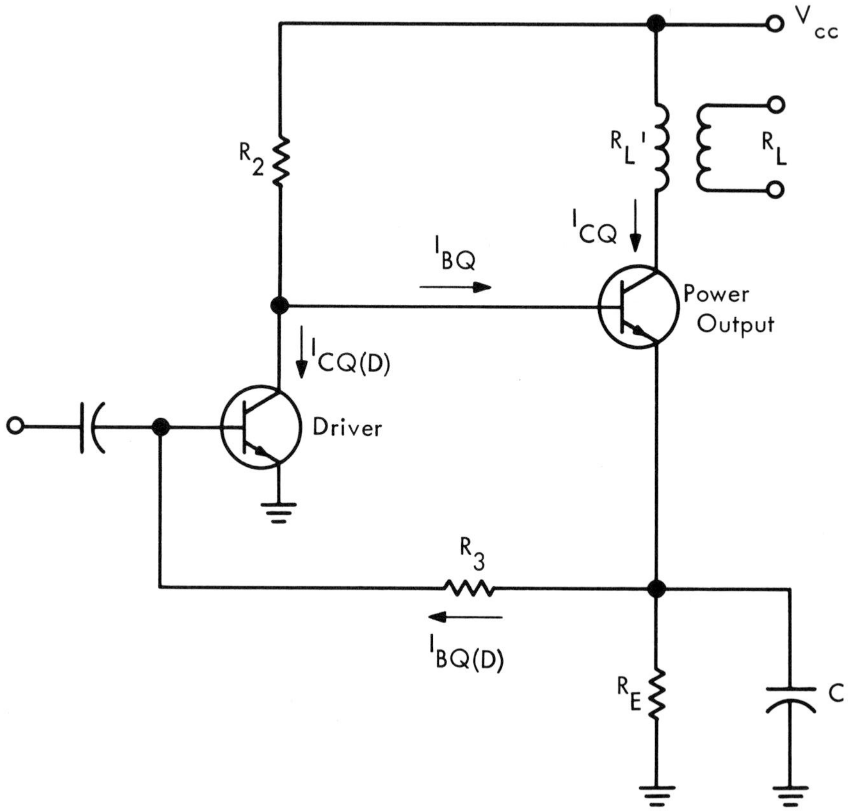

Fig. 8. Class A Direct-coupled Input, Transformer-coupled Output Schematic

K_1 = Transistor derating slope — mW/°C
T_1 = Maximum device operating temperature at $P_{(Diss)}$ — °C
T_2 = Temperature related to maximum device $P_{(Rate)}$ — °C
$P_{(Rate)}$ = Maximum device power rating — mW.

If the device is in a stud-mounted package, the heat sink should be evaluated to determine if it is able to dissipate the required $P_{(Diss)}$ while maintaining the device temperature at or below T_1. This requirement must be calculated at the highest desired ambient temperature.

3. Now consider the device voltage requirements. In power stages, high load impedances result in higher power gain per stage. Therefore the device voltage capabilities should be determined from the data sheet to allow the selection of optimum circuit voltages. Class A circuits with transformer loads require:

$$V_{cc} < \frac{V_{(BR)CER}}{2} \tag{3}$$

While this relationship represents the absolute maximum value of V_{cc} possible without exceeding the device maximum ratings, it is generally good design procedure to select V_{cc} equal to about 90% of the V_{cc} derived in Eq. (3). Additional care should be exercised in selecting the maximum circuit voltages for power devices that exhibit a reduction in the power rating of the device at high voltage levels. This derating is normally referred to as the Safe Operating Curve (Fig. 3). The a-c load line must be checked to make sure a value of collector voltage has been chosen that will allow the load line to remain below the safe operating derating curve.

4. One of the first considerations given worst case circuit design is the effect of temperature variations on the critical device parameters. One of the most important parameters to evaluate for temperature stability is the change in the base-emitter voltage over the operating temperature range. This ΔV_{BE} is calculated in Eq. (4):

$$\Delta V_{BE} = K_2 (T_1 - T_4) \tag{4}$$

K_2 = Base-emitter temperature-dependent term
Silicon — 2.0×10^{-3} V/°C
Germanium — 1.3×10^{-3} V/°C
T_4 = Minimum device operating temperature.

5. With the maximum V_{cc} established, the reflected collector load impedance R_L' may be calculated. The load voltage V_{cep} is approximated by $V_{cc} - 16\ \Delta V_{BE} - V_{CE(sat)}$. The ΔV_{BE} term comes from assuming a maximum permissible change of 5% in collector quiescent current over the operating temperature range. This condition sets a value for the voltage drop across the emitter resistance R_E in terms of ΔV_{BE}. The output reflected load impedance R_L' is calculated in Eq. (51) with the approximations noted here.

$$R_L' = \frac{(V_{cc} - 16\Delta V_{BE} - V_{CE(sat)})^2\ \eta}{2P_{o(rms)}} \tag{51}$$

6. The maximum collector current in a direct-coupled stage must be less than the maximum collector-current rating of the device and within the range of collector currents having a useable h_{FE} value. The maximum current I_{CM} is equal to the sum of the peak load current and the quiescent operating current in the collector of the device. The quiescent collector current is increased 25% over the peak load current requirement to ensure device operation in the linear portion of the y_{FE} curve over the desired temperature range. The peak collector current is calculated in Eq. (52) and the maximum device collector current is calculated in Eq. (8).

$$I_{cp} = \sqrt{\frac{2P_{o(rms)}}{R_L'\ \eta}} \tag{52}$$

$$I_{CM} = 2.25\ I_{cp} \tag{8}$$

7. The quiescent collector current is set 25% greater than the peak collector load to ensure operation in the linear portion of the y_{FE} curve. This requirement allows maximum power output with minimum output distortion. The output-stage collector quiescent current is calculated in Eq. (7):

$$I_{CQ} = 1.25\ I_{cp} \tag{7}$$

8. The quiescent point base current in the output stage is calculated from the relationship of the quiescent collector current I_{CQ} and the device minimum h_{FE} specification taken from the data sheet.

$$I_{BQ} = \frac{I_{CQ}}{h_{FE(min)}} \tag{10}$$

9. The output stage quiescent base current is supplied from the collector circuit of the driver device. In Class A operation, the maximum base-current is twice the base quiescent current. Therefore, the minimum collector current requirement of the driver device is determined by the output stage base quiescent current. In practice, the driver collector current must not be allowed to cut off completely or distortion from the driver stage will distort the output. This is prevented by increasing the driver collector current by an empirical factor equal to 1.25 times the minimum collector current requirement. For $I_{CM(D)}$:

$$I_{CM(D)} = 2.5\ I_{BQ} \tag{40}$$

10. The collector-base leakage current I_{CBO} is important in evaluating the circuit stability criteria. This leakage is a function of temperature and can be calculated at any temperature by assuming that the current doubles for an increase in temperature of 8°C for germanium and 12°C for silicon. An I_{CBO} corresponding to the maximum device temperature at maximum dissipation. In calculating I_{CBO} for driver stages, the power dissipation in the driver stage is very small, and the device temperature can be assumed to be approaching the ambient temperature. Therefore a maximum ambient temperature value T_A is used in Eq. (41) to calculate the maximum I_{CBO} of the driver. The value of the collector-base leakage calculated in the driver is very important in direct-coupled circuits because the value of feedback resistance R_3 is dependent upon this parameter for d-c stability.

$$I_{CBO(T_A)(D)} = I_{CBO(T_3)} 2^{(T_A - T_3)/K_T} \tag{41}$$

where

K_T = Empirical temperature-dependent constants
Silicon — 12
Germanium — 8
T_3 = Maximum temperature I_{CBO} specified on data sheet
$I_{CBO(T_3)}$ = Maximum I_{CBO} specified on data sheet at temperature T_3

11. The driver-collector quiescent current must now be calculated to determine the driver base-emitter quiescent voltage. The quiescent collector current must be greater than the peak a-c collector current $I_{cp(D)}$ to prevent collector cut-off operation and distortion of the driver output. The quiescent

collector current will be equal to the maximum collector current $I_{CM(D)}$ from Eq. (40) minus the base quiescent current of the output stage. The calculation may be made as shown in Eq. (42):

$$I_{CQ(D)} = 1.5\ I_{BQ} \tag{42}$$

12. The driver base quiescent voltage may be determined from the quiescent collector current $I_{CQ(D)}$ and the driver y_{FE} curve.

13. The driver base bias resistor R_3 provides a feedback path and stabilizing current from the output stage emitter resistance. It is desirable to make R_3 as large as possible, consistent with bias stability criteria to minimize the power loss of R_3 shunting the input. Bias stability would become a problem, however, if the I_{CBO} of the driver stage caused an appreciable change in the voltage across R_3 resulting in a change in the driver quiescent current. The voltage across the emitter resistance is proportional to the output collector current, and this feedback provides an output collector-current stability factor. The voltage drop across the resistance R_3 is V_{RE} minus $V_{BEQ(D)}$. The criteria for stability is in terms of the output quiescent current I_{CQ}. The assumption is made that the change in V_{BEQ} is held to within 5% of its nominal value while I_{CBO} of the driver increases to its maximum device temperature value calculated in Eq. (41). The maximum value of resistance R_3 is calculated in Eq. (43):

$$R_3 \leqq \frac{V_{BEQ(D)}}{20\ I_{CBO(T_A)(D)}} \tag{43}$$

14. The base quiescent current of the driver is determined from the collector quiescent current calculated in Eq. (42).

$$I_{BQ(D)} = \frac{I_{CQ(D)}}{h_{FE(min)(D)}} \tag{44}$$

15. The emitter resistance R_E may now be calculated in terms of the driver quiescent conditions and the output collector current. The emitter resistance is employed mainly for collector quiescent current feedback to the driver device. The output quiescent current will stabilize when the d-c closed-loop voltage gain approaches unity. This condition is satisfied when the driver d-c y_{FE} times the driver collector load impedance equals unity. The emitter resistance R_E is:

$$R_E = \frac{V_{BEQ(D)} + I_{BQ(D)}R_3}{I_{CQ}} \tag{45}$$

16. The driver collector is coupled directly to the base of the output device. This forces the base voltage of the output device to operate at the collector voltage of the driver. The total base voltage V_B of the output stage must be determined to ensure that the driver is not forced to operate in the saturation region. The driver load resistance R_2 is also dependent upon the quiescent base voltage of the output stage. The quiescent base-emitter voltage V_{BEQ} is obtained from the data sheet at the collector current I_{CQ}, and the total base voltage V_B is calculated in Eq. (47):

$$V_B = I_{CQ}R_E + V_{BEQ} \tag{47}$$

17. The driver collector resistance is determined by the difference between the supply voltage and the total base quiescent voltage with the total quiescent load current flowing. The total load current in the collector of the driver must be equal to the sum of the collector quiescent current and the base quiescent current of the output stage. The resistance R_2 is calculated in Eq. (48):

$$R_2 = \frac{V_{cc} - V_B}{I_{CM(D)}} \tag{48}$$

18. The bypass capacitor C_1 must be calculated to prevent a-c gain degeneration at the minimum desired frequency. The minimum value of capacitance required is determined by assuming the output is down three dB at the minimum frequency. The relationship for the capacitance calculation is given in Eq. (17).

$$C_1 \geqq \frac{I_{cp}}{2\pi f K_4} \qquad (17)^*$$

where

K_4 = Empirical constant dependent on material
Silicon — 35×10^{-3}
Germanium — 25×10^{-3}
f = Minimum desired frequency.

19. The driver input resistance is required for the total power gain (Eq. (53)). The input resistance is proportional to the driver current gain and the feedback resistance R_3. The a-c y_{fe} of the driver is required for this calculation because of the relationship between the device input resistance, typical h_{FE}, and y_{FE} characteristics. The y_{fe} of the driver is almost a constant over collector-current ranges, from about one mA up to the collector-current rating of the device. This enables an accurate value to be obtained from the y_{FE} curve even though the collector current excursions ΔI_c are very small. The input driver resistance R_i is calculated in Eq. (49):

$$R_{i(D)} = \frac{h_{FE(D)} R_3}{h_{FE(D)} + y_{fe} R_3} \tag{49}$$

20. The amplifier design may now be completed by calculating the total power gain of the amplifier in terms of the input impedance of the driver and the output impedance of the power output stage. Equation (53) shows the power gain equation for calculating total power performance of the amplifier.

$$P.G._{(T)} = \frac{(h_{FE(D)} h_{FE})^2 R_L' \,\eta}{R_{i(D)}} \tag{53}$$

*Equation derived in Chapter 6.

3

Class B Output and Driver Design Procedures

Procedures for two of the more common coupling combinations for Class B design are presented here. Both designs use a transformer-coupled input, one uses a transformer-coupled output and one an RC-coupled output. A practical approach incorporating a number of approximations and assumptions based on design experience is used to arrive at the simplest solutions to design problems.

Because of similarities in design, some of the procedural steps and equations are repeated for both coupling combinations. This repetition permits the complete presentation of each procedure without reference to the other. Derivations of key equations are given in Chapter 6.

TRANSFORMER-COUPLED INPUT, TRANSFORMER-COUPLED OUTPUT

A circuit diagram is shown in Fig. 1 and dynamic characteristics are given in Fig. 2.

1. Class B amplifier circuits dissipate minimum power at the zero-signal quiescent point. Maximum device dissipation for resistive loads normally occurs at 40% of the maximum power output. The power output may be calculated in terms of this maximum device dissipation as shown in Eq. (1).

$$P_{(Diss)}/\text{Device} = 0.312\ P_{o(rms)} \qquad \textbf{(1)}^*$$

2. This required dissipation capability of the device must be satisfied at the operating temperature conditions. The maximum permissible device temperature that can be tolerated without exceeding the device ratings is calculated in Eq. (2). The transistor derating slope, the maximum device power rating and the corresponding temperature T_2 are all taken from the device data sheet.

*Equation derived in Chapter 6.

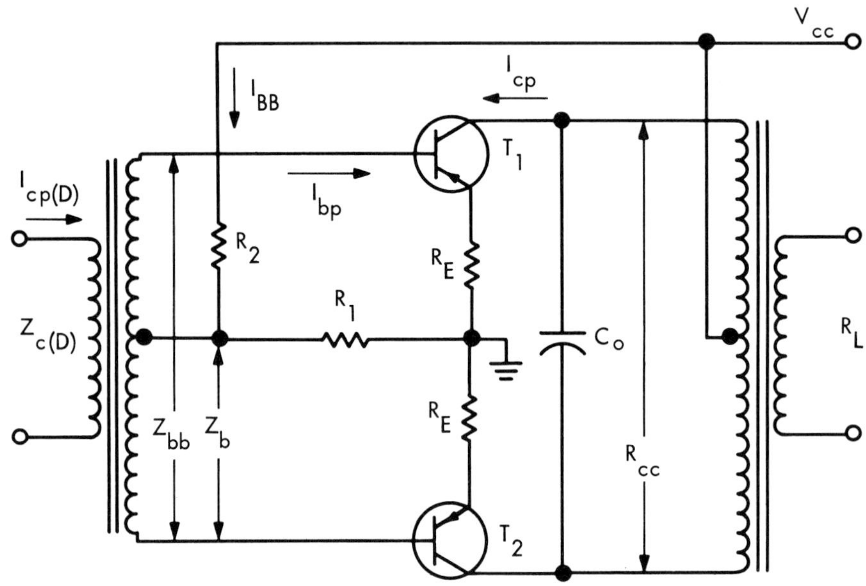

Fig. 1. Class B Transformer-coupled Input, Transformer-coupled Output Schematic

$$T_1 = \frac{P_{(Rate)} - P_{(Diss)} + K_1 T_2}{K_1} \tag{2}$$ *

where

K_1 = Transistor derating slope — mW/°C
T_1 = Maximum device operating temperature at $P_{(Diss)}$ — °C
T_2 = Temperature related to maximum device $P_{(Rate)}$ — °C
$P_{(Rate)}$ = Maximum device power rating — mW.

If the device is in a stud-mounted package, the heat sink should be evaluated to determine if it is able to dissipate the required $P_{(Diss)}$ while maintaining the device temperature at or below T_1. This requirement must be calculated at the highest desired ambient temperature.

3. Maximum device collector voltage requirements in Class B transformer-coupled output stages are more severe than Class B capacitor-coupled outputs. To obtain maximum power gain, high load-impedance values should be chosen; therefore, maximum circuit voltages within the maximum ratings of the device are desirable. During one half-cycle of operation, the device base-emitter is forward biased through an effective base-emitter resistance. The other half-cycle of operation reverse biases the base-emitter junction and the collector-emitter voltage reaches its maximum value. These device requirements are:

---

*Equation derived in Chapter 6.

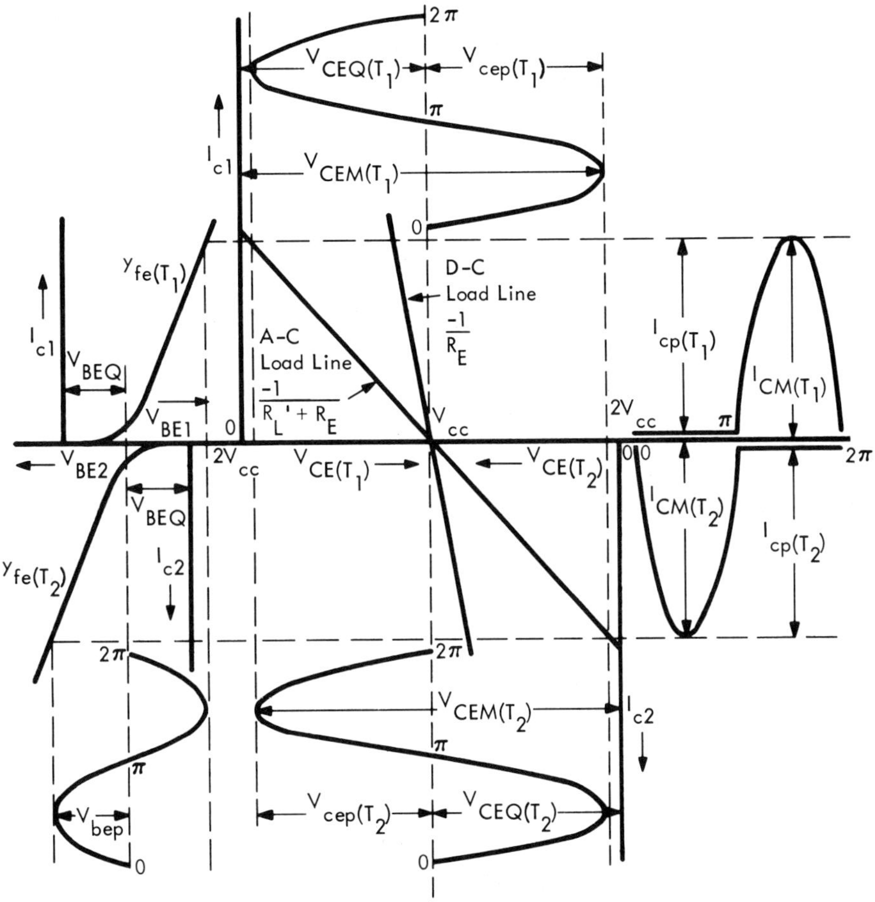

Fig. 2. Class B Transformer-coupled Input, Transformer-coupled Output Dynamic Characteristics

$$V_{(BR)CER} > V_{cc} \tag{3}$$

$$V_{(BR)CEV} > 2\ V_{cc} \tag{4}$$

While these relationships represent the absolute maximum value of V_{cc} possible without exceeding the device maximum ratings, it is generally good design procedure to select V_{cc} equal to about 90% of the V_{cc} calculated in Eq. (3) and (4).

4. One of the first considerations given worst-case circuit design is the effect of temperature variations on the critical device parameters. One of the most important parameters to evaluate for temperature stability is the change in the base-emitter voltage over the operating temperature range. This ΔV_{BE} is calculated in Eq. (5).

$$\Delta V_{BE} = K_2\ (T_1 - T_4) \tag{5}$$

where

K_2 = Base-emitter temperature-dependent term
Silicon — 2.0×10^{-3} V/°C
Germanium — 1.3×10^{-3} V/°C
T_4 = Minimum device operating temperature.

5. With V_{cc} established, the reflected load impedance R_L' can be calculated. The load voltage V_{cep} is approximated by $(V_{cc} - 25\Delta V_{BE} - V_{CE(sat)})$. The ΔV_{BE} term comes from assuming a maximum change in quiescent collector current over the operating temperature range. In Class B circuits, the quiescent current is designed to be as low as possible since any quiescent current represents a power loss. A small value of quiescent current is necessary however to eliminate cross-over distortion. Since the quiescent current level is so small, the I_{CQ} value can vary considerably without appreciably affecting the performance of the stage. The approximation of the load voltage is derived from the assumption that I_{CQ} may vary as much as five times the nominal value, and that the nominal I_{CQ} will not exceed 1% of the collector peak load current. This assumption sets a maximum value for the voltage drop across the emitter resistance R_E in terms of ΔV_{BE}.

$$R_L' = \frac{(V_{cc} - 25\Delta V_{BE} - V_{CE(sat)})^2 \, \eta}{2 \, P_{o(rms)}} \qquad \textbf{(6)}$$

6. The peak collector current can now be calculated in terms of the power output and the V_{cep} equivalent.

$$I_{cp} = \frac{2 \, P_{o(rms)}}{(V_{cc} - 25\Delta V_{BE} - V_{CE(sat)}) \, \eta} \qquad \textbf{(7)}$$

7. The quiescent collector current can normally be determined as a function of the peak collector load current. This quiescent current should be chosen from the y_{FE} curve to put the Q point just below the linear portion of the y_{FE} curve. This is a compromise, but generally the reduction in power loss outweighs the slight increase in distortion. Because the increasing power loss is the big disadvantage in operating at higher quiescent collector currents, the I_{CQ} value is not allowed to exceed 1% of the peak collector current.

$$I_{CQ} = 0.01 \, I_{cp} \qquad \textbf{(8)}$$

8. The maximum d-c collector current requirement is the sum of the quiescent current and the a-c peak collector current during full-load operation. This maximum collector current I_{CM} must be less than the maximum collector current rating of the device and within the range of collector currents having a useable h_{FE} value.

$$I_{CM} = 1.01 \, I_{cp} \qquad \textbf{(9)}$$

9. The maximum circuit requirements of power, voltage, and current have now been described and the transistor type can be confirmed for the particular application.

10. The quiescent-point base current I_{BQ} is calculated from the relationship of the quiescent collector current I_{CQ} and the device minimum h_{FE} specification taken from the data sheet.

$$I_{BQ} = \frac{I_{CQ}}{h_{FE(min)}} \quad (10)$$

11. The collector-base leakage current I_{CBO} is important in evaluating the circuit stability criteria. This leakage is a function of temperature and can be calculated at any temperature by assuming that the current doubles for an increase in temperature of 12°C for silicon and 8°C for germanium.

$$I_{CBO(T_1)} = I_{CBO(T_3)} 2^{(T_1 - T_3)/K_T} \quad (11)$$

where

K_T = Empirical temperature-dependent constants
Silicon — 12
Germanium — 8
T_3 = Maximum temperature I_{CBO} specified on the data sheet
$I_{CBO(T_3)}$ = Maximum I_{CBO} specified on the data sheet at temperature T_3

12. The base bias can now be calculated to provide the desired quiescent currents. To ensure minimum temperature effects on the Q-point stability caused by I_{CBO} increasing the V_{BEQ} quiescent point, the base bias current is made considerably larger than the maximum $I_{CBO(T_1)}$ calculated from Eq. (11).

$$I_{BB} = I_{BQ} + 10\ I_{CBO(T_1)} \quad (12)$$

13. The Q-point base-emitter voltage V_{BEQ} may be obtained from the y_{FE} specification or curve on the data sheet. The value of V_{BEQ} is read off the curve at the quiescent collector current I_{CQ}. The change in I_{CBO} due to temperature variations over the operating temperature range of the circuit is defined as ΔI_{CBO}. This value is calculated in Eq. (13).

$$\Delta I_{CBO} = I_{CBO(T_1)} - I_{CBO(25°C)} \quad (13)$$

14. The base bias resistor R_1 must be calculated to ensure minimum shift of the Q point over the operating temperature range. Equation (14) defines a value of R_1 that will prevent the quiescent collector current from increasing to more than five times the nominal value of I_{CQ}.

$$R_1 \leq \frac{5V_{BEQ} + \Delta V_{BE}}{5I_{BB}} \quad (14)^*$$

15. Series resistance is used in the emitter circuit of Class B amplifiers to provide both a-c and d-c stability. There are two main sources that cause shifts in the feedback provided by R_E. One of these sources is the decrease in V_{BE} with temperature and the other is the increase in I_{CBO} with temperature which changes the bias point through R_1. R_E is calculated with the assumption that the quiescent current will not be allowed to change more than five times the nominal value of I_{CQ}. Degeneration is present in this circuit which provides some a-c stability but the value of R_E calculated should be considered a minimum value.

---

*Equation derived in Chapter 6.

$$R_E \geq \frac{\Delta V_{BE} + \Delta I_{CBO}R_1}{5I_{CQ}} \tag{15}$$

If more regulation is required in the collector-current stability point, the resistance R_E may be increased with a resultant increase in power loss and therefore a reduction in the power output since R_E is not bypassed. Calculated values of R_E may become unnecessarily large in circuits requiring strict control of the Q-point stability. In such instances, a compromise may become necessary between the permissible power loss in R_E and the collector-current Q-point stability.

16. The output transformer impedance primary is defined as the reflected load impedance from collector to collector. This value may be determined from the expression for reflected collector load impedance given in Eq. (6). The collector-to-collector reflected load impedance is calculated in Eq. (16):

$$R_{cc}' = 4R_L' \tag{16}$$

17. The other base bias resistor R_2 in the divider network is calculated in Eq. (17).

$$R_2 = \frac{V_{cc}}{I_{BB}} - R_1 \tag{17}$$

18. The base input parameters can now be determined from the previous requirements of peak collector currents and quiescent bias voltages. The peak base current required to drive the transistor to I_{cp} in Eq. (7) is calculated in Eq. (18).

$$I_{bp} = \frac{I_{cp}}{h_{FE(min)}} \tag{18}$$

19. The peak voltage drop in the base circuit occuring from the peak base current through R_1 is calculated in Eq. (19).

$$V_{R1} = I_{bp}R_1 \tag{19}$$

20. The peak voltage drop in series with the input base signal as a result of the peak collector current through R_E is calculated in Eq. (20).

$$V_{RE} = I_{cp}R_E \tag{20}$$

21. From the data sheet y_{FE} characteristics, the maximum base-emitter voltage required to drive the collector current to I_{CM} is determined.

22. Now the peak signal voltage required at the secondary of the center-tapped input transformer can be calculated as shown in Eq. (21).

$$V_{ip} = V_{R1} + V_{RE} + V_{BEM} - V_{BEQ} \tag{21}$$

23. As noted in paragraph 7, the selection of the quiescent bias point is a compromise because of the rapid increase in power losses caused by increasing quiescent collector current. This compromise results in crossover distortion in the output. With the addition of a capacitor C_o across the output transformer, any remaining distortion can be eliminated. The maximum value of C_o is calculated in Eq. (22).

$$C_o \leq \frac{1}{2\pi f R_{cc}} \tag{22}$$

24. The input transformer to the power stage must be determined in terms of the peak voltage and current required for maximum power output. The total secondary impedance Z_{bb} may then be calculated from Eq. (23) and (24).

$$Z_b = \frac{V_{ip}}{I_{bp}} \tag{23}$$

$$Z_{bb} = 4Z_b \tag{24}$$

25. The total power gain of the output stage should now be considered before considering the driver requirements.

$$\text{P.G.} = \frac{V_{cc}h_{FE}\eta}{V_{ip}} \tag{25}^*$$

All of the necessary components and device parameters for the output stage have been determined. Normally, a power output stage requires a driver input and the collector-circuit constants of a driver stage can now be evaluated.

26. Maximum power gain in the driver stage is accomplished by maximum collector-load impedance. The driver power output requirement is usually low and this allows a high-impedance primary input transformer. Because the collector peak voltage requirements are normally easily obtained with a high-impedance load, a practical high-impedance value for the driver transformer primary $Z_{c(D)}$ may be assumed. Practical values are as high as 20 kilohms. In paragraph 29, $V_{cc(D)}$ is calculated as a result of the $Z_{c(D)}$ assumed here. Occasionally, this assumption results in an impractical value of $V_{cc(D)}$ and a compromise which sacrifices maximum power gain is necessary. For this, a value of $Z_{c(D)}$ may be calculated in Eq. (26) in terms of the $V_{cc(D)}$ available.

$$Z_{c(D)} = \frac{(V_{cc(D)} - 3V_{CE(sat)(D)})^2\eta}{V_{ip}I_{bp}} \tag{26}$$

27. The peak collector current of the driver is dependent upon the power input requirements of the output stage and the transformer primary impedance $Z_{c(D)}$. This peak current requirement may be calculated in Eq. (27).

$$I_{cp(D)} = I_{bp}\sqrt{\frac{Z_b}{\eta Z_{c(D)}}} \tag{27}$$

28. The driver collector-current quiescent point is determined by the peak collector current calculated in Eq. (27). The quiescent current must be greater than the peak collector current to prevent operation of the device in the non-linear portion of the y_{FE} curve. Distortion of the driver signal will result if the device is allowed to operate to cutoff; therefore, a minimum value of collector current is chosen that will always be above the knee of the y_{FE} curve. The collector quiescent current is calculated in Eq. (28).

*Equation derived in Chapter 6.

$$I_{CQ(D)} = I_{cp(D)} + 1.0 \times 10^{-3} A \qquad \textbf{(28)}$$

29. The driver supply voltage requirement can be much smaller than the power output-stage requirement because of the smaller power output of the driver. The minimum driver supply voltage is calculated in Eq. (29) in terms of the base-emitter parameters of the output stage and the collector saturation voltage of the driver device. This driver supply voltage may then be obtained by dropping the output-stage supply voltage through a decoupling resistor or by providing a separate voltage source.

$$V_{cc(D)} = \frac{V_{ip} I_{bp}}{\eta I_{cp(D)}} + 3V_{CE(sat)(D)} \qquad \textbf{(29)}$$

30. The critical driver collector requirements have now been calculated and the remainder of the driver parameters and components are determined by using the appropriate equations derived for the power output stage. The power output requirement of the driver device is calculated in Eq. (30) in terms of the collector requirements.

$$P_{o(rms)(D)} = \frac{I_{cp(D)}^2 Z_{c(D)} \eta}{2} \qquad \textbf{(30)}$$

TRANSFORMER-COUPLED INPUT, RC-COUPLED OUTPUT

A circuit diagram is shown in Fig. 3 and dynamic characteristics are given in Fig. 4.

1. Class B amplifier circuits dissipate minimum power at the zero-signal quiescent point. Maximum device dissipation for resistive loads normally occurs at 40% of the maximum power output. The power output may be calculated in terms of this maximum device dissipation as shown in Eq. (31).

$$P_{(Diss)}/\text{Device} = 0.25\, P_{o(rms)} \qquad \textbf{(31)*}$$

2. This required dissipation capability of the device must be satisfied at the operating temperature conditions. The maximum permissible device temperature that can be tolerated without exceeding the device ratings is calculated in Eq. (2). The transistor derating slope, the maximum device power rating, and the corresponding temperature T_2 are all taken from the device data sheet.

$$T_1 = \frac{P_{(Rate)} - P_{(Diss)} + K_1 T_2}{K_1} \qquad (2)*$$

where

K_1 = Transistor derating slope — W/°C
T_1 = Maximum device operating temperature at $P_{(Diss)}$ — °C
T_2 = Temperature related to maximum device $P_{(Rate)}$ — °C
$P_{(Rate)}$ = Maximum device power rating — mW.

*Equation derived in Chapter 6.

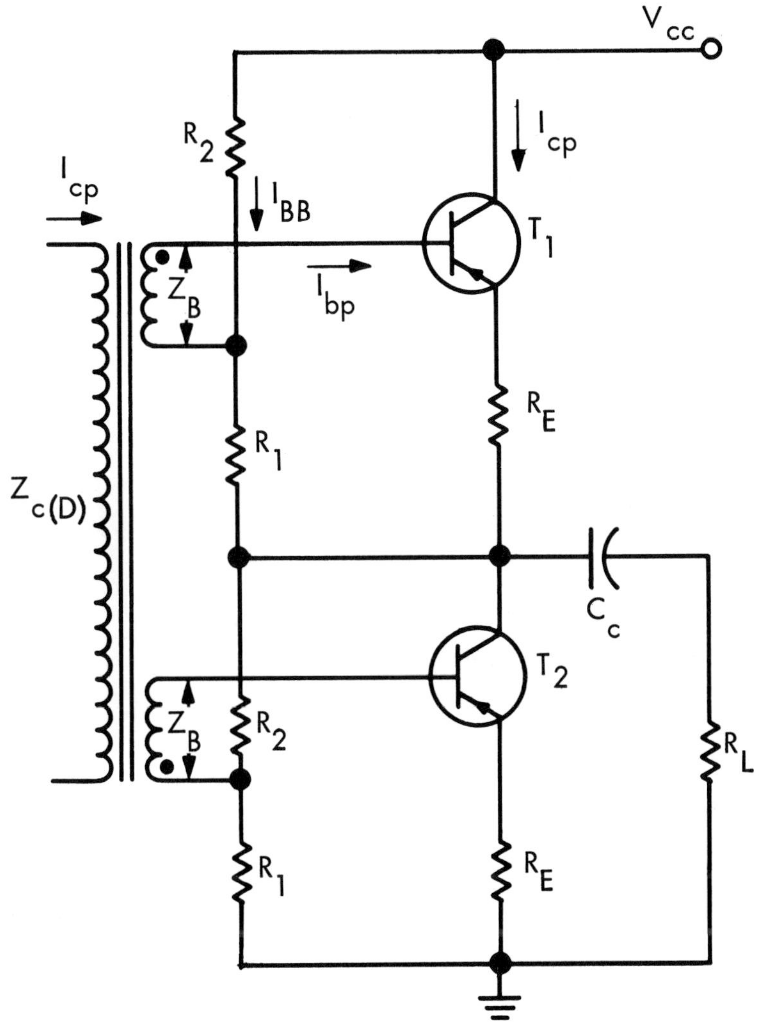

Fig. 3. Class B Transformer-coupled Input, RC-coupled Output Schematic

If a device is in a stud-mounted package, the heat sink should be evaluated to determine if it is able to dissipate the required $P_{(Diss)}$ while maintaining the device temperature at or below T_1. This requirement must be calculated at the highest desired ambient temperature.

3. One of the first considerations given worst-case circuit design is the effect of temperature variations on the critical device parameters. One of the most important parameters to evaluate for temperature stability is the change in the base-emitter voltage over the operating temperature range. This ΔV_{BE} is calculated in Eq. (5).

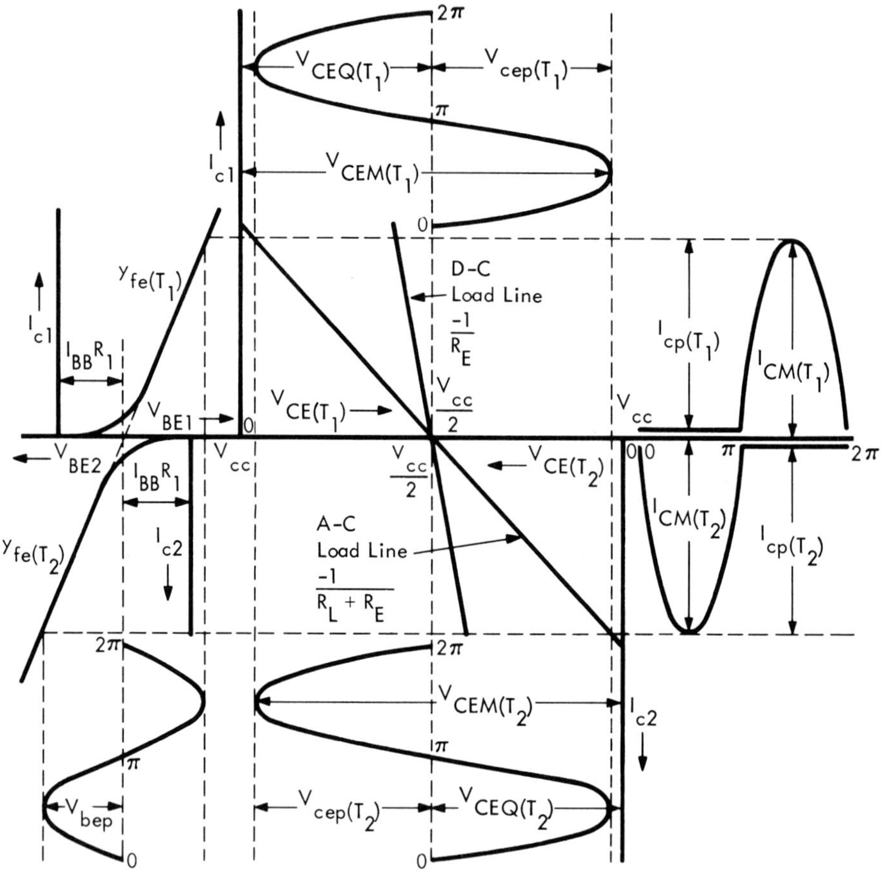

Fig. 4. Class B Transformer-coupled Input, RC-coupled Output Dynamic Characteristics

$$\Delta V_{BE} = K_2 (T_1 - T_4) \quad (5)$$

where

K_2 = Base-emitter temperature-dependent term
Silicon — 2.0×10^{-3} V/°C
Germanium — 1.3×10^{-3} V/°C
T_4 = Minimum device operating temperature.

4. The circuit voltage requirement may now be determined in terms of the power output and load resistance R_L, because the collector reflected load impedance in capacitor-coupled output stages is equal to the true load impedance. The V_{cc} equation (32) contains a ΔV_{BE} expression which comes from an approximation that defines the voltage drop across the emitter resistance R_E when the collector quiescent current is not allowed to exceed five times its nominal value under worst-case temperature conditions. This criterion is

reasonable because the quiescent current in Class B circuits is designed to be a very low value and the small absolute current change does not represent an appreciable increase in either power loss or distortion.

$$V_{cc} = \sqrt{8\ P_{o(rms)}R_L} + 50\Delta V_{BE} + 2V_{CE(sat)(max)} \qquad \textbf{(32)*}$$

5. The maximum device collector voltage must now be considered. In Class B capacitor-coupled circuits, the collector voltage requirements are much less severe than in Class B transformer-coupled stages. This allows lower-voltage devices to be used in capacitor-coupled circuits for the same power gain. In capacitor-coupled stages, one device base-emitter junction is reverse-biased while the other output device is conducting. This allows the device to block the maximum circuit voltage during its reverse-biased interval. The maximum circuit voltage across either device is equal to V_{cc}; therefore, the following equations describe the device requirements in capacitor-coupled stages as a function of the V_{cc} calculated in Eq. (32).

$$V_{(BR)CEV} > V_{cc} \qquad \textbf{(33)}$$

$$V_{(BR)CER} > \frac{V_{cc}}{2} \qquad \textbf{(34)}$$

These relationships represent the absolute maximum value of V_{cc} possible without exceeding the device maximum ratings. In practice however, it is generally good design procedure to select the collector rating to be about 110% of the value calculated in Eq. (33) and (34). Additional care should be exercised in selecting the maximum circuit voltages for power devices that exhibit a reduction in the power rating of the device at high voltage levels. This derating is normally referred to as the Safe Operating Curve. (See Fig. 3 in Chapter 2.) The a-c load line must be checked to make sure a value of collector voltage has been chosen that will allow the load line to remain below the safe operating derating curve.

6. The peak collector current may be calculated from Eq. (35) in terms of the power output and load resistance R_L.

$$I_{cp} = \sqrt{\frac{2P_{o(rms)}}{R_L}} \qquad \textbf{(35)}$$

7. The quiescent collector current can normally be determined as a function of the peak collector load current. This quiescent current should be chosen from the y_{FE} curve to put the Q point just below the linear portion of the y_{FE} curve. This is a compromise, but generally the reduction in power loss outweighs the slight increase in distortion. Because the increasing power loss is the big disadvantage in operating at higher quiescent collector currents, the I_{CQ} value is not allowed to exceed 1% of the peak collector current.

$$I_{CQ} = 0.01\ I_{cp} \qquad (8)$$

*Equation derived in Chapter 6.

8. The maximum d-c collector current requirement is the sum of the quiescent current and the a-c peak collector current during full-load operation. This maximum collector current I_{CM} must be less than the maximum collector current rating of the device and within the range of collector currents having a useable h_{FE} value.

$$I_{CM} = 1.01\ I_{cp} \tag{9}$$

9. The maximum circuit requirements of power, voltage, and current have now been described and the transistor type can be confirmed for the particular application.

10. The quiescent-point base current I_{BQ} is calculated from the relationship of the quiescent collector current I_{CQ} and the device minimum h_{FE} specification taken from the data sheet.

$$I_{BQ} = \frac{I_{CQ}}{h_{FE(min)}} \tag{10}$$

11. The collector-base leakage current I_{CBO} is important in evaluating the circuit stability criteria. This leakage is a function of temperature and can be calculated at any temperature by assuming that the current doubles for an increase in temperature of 12°C for silicon and 8°C for germanium.

$$I_{CBO(T_1)} = I_{CBO(T_3)} 2^{(T_1 - T_3)/K_T} \tag{11}$$

where

K_T = Empirical temperature-dependent constants
Silicon — 12
Germanium — 8
T_3 = Maximum temperature I_{CBO} specified on the data sheet
$I_{CBO(T_3)}$ = Maximum I_{CBO} specified on data sheet at temperature T_3

12. The base bias can now be calculated to provide the desired quiescent currents. To ensure minimum temperature effects on the Q-point stability caused by I_{CBO} increasing the V_{BEQ} quiescent point, the base bias current is made considerably larger than the maximum $I_{CBO(T_1)}$ calculated from Eq. (11).

$$I_{BB} = I_{BQ} + 10\ I_{CBO(T_1)} \tag{12}$$

13. The Q-point base-emitter voltage V_{BEQ} may be obtained from the y_{FE} specification or curve on the data sheet. The value of V_{BEQ} is read off of the curve at the quiescent collector current I_{CQ}.

14. The change in I_{CBO} due to temperature variations over the operating temperature range of the circuit is defined as ΔI_{CBO}. The value is calculated in Eq. (13).

$$\Delta I_{CBO} = I_{CBO(T_1)} - I_{CBO(25°C)} \tag{13}$$

15. The base bias resistor R_1 must be calculated to ensure minimum shift of the Q point over the operating temperature range. Equation (14) defines a value of R_1 that will prevent the quiescent collector current from increasing to more than five times the nominal value of I_{CQ}.

$$R_1 \leq \frac{5V_{BEQ} + \Delta V_{BE}}{5I_{BB}} \tag{14)*$$

16. The other base bias resistor R_2 in series with R_1 to form a divider network is calculated in Eq. (36).

$$R_2 = \frac{V_{cc}}{2I_{BB}} - R_1 \quad \textbf{(36)}$$

17. Series resistance is used in the emitter circuit of Class B amplifiers to provide both a-c and d-c stability. There are two main sources that cause shifts in the feedback provided by R_E. One of these sources is the decrease in V_{BE} with temperature and the other is the increase in I_{CBO} with temperature which changes the bias point through R_1. R_E is calculated with the assumption that the quiescent current will not be allowed to change more than five times the nominal value of I_{CQ}. Degeneration is present in this circuit which provides some a-c stability but the value of R_E calculated should be considered a minimum value.

$$R_E \geq \frac{\Delta V_{BE} + \Delta I_{CBO} R_1}{5I_{CQ}} \quad (15)$$

If more regulation is required in the collector-current stability point, the resistance R_E may be increased with a resultant increase in power loss and therefore a reduction in the power output since R_E is not bypassed. Calculated values of R_E may become unnecessarily large in circuits requiring strict control of the Q-point stability. In such instances, a compromise may be necessary between the permissible power loss in R_E and the collector-current Q-point stability.

18. The peak a-c base current I_{bp} is calculated from the peak collector current and device h_{FE} relationship in Eq. (18).

$$I_{bp} = \frac{I_{cp}}{h_{FE(min)}} \quad (18)$$

19. The peak voltage drop in the base circuit occurring from the peak base current through R_1 is calculated in Eq. (19).

$$V_{R1} = I_{bp} R_1 \quad (19)$$

20. The peak voltage drop in series with the input base signal as a result of the peak collector current through R_E is calculated in Eq. (20).

$$V_{RE} = I_{cp} R_E \quad (20)$$

21. From the data sheet y_{FE} characteristics, the maximum base-emitter voltage required to drive the collector current to I_{CM} is determined.

22. Now the peak signal voltage required at each of the secondary windings on the input transformer is determined by writing voltage loop equations around the circuit. The peak input voltage is calculated in Eq. (21).

---

*Equation derived in Chapter 6.

$$V_{ip} = V_{R1} + V_{RE} + V_{BEM} - V_{BEQ} \tag{21}$$

23. The load is coupled to the output stage through capacitance C_c. The frequency response of the amplifier is affected by this capacitance value at the minimum desired frequency limit. The minimum value of capacitance therefore, is calculated in Eq. (37) which assumes the output has decreased three dB at the minimum frequency limit.

$$C_c \geq \frac{1}{2\pi f R_L} \tag{37}$$

24. Each of the two secondary windings is determined by the peak base input circuit. The secondary impedance is calculated in Eq. (23).

$$Z_b = \frac{V_{ip}}{I_{bp}} \tag{23}$$

25. The power gain of the output stage may now be calculated to complete the power output stage requirements. Equation (38) shows the circuit power gain in terms of the device h_{FE} and previously calculated circuit values.

$$P.G. = \frac{V_{cc} h_{FE}}{2V_{ip}} \tag{38}^*$$

All of the necessary components and device parameters for the output stage have been determined. Normally, a power output stage requires a driver input and the collector circuit constants of a driver stage can now be evaluated.

26. Maximum power gain in the driver stage is accomplished by maximum collector-load impedance. The driver power output requirement is usually low and this allows a high-impedance primary input transformer. Because the collector voltage requirements are normally easily obtained with a high-impedance load, a practical high-impedance value for the driver transformer primary $Z_{c(D)}$ may be assumed. Practical values are as high as 20 kilohms. In paragraph 29, $V_{cc(D)}$ is calculated as a result of the $Z_{c(D)}$ assumed here. Occasionally, this assumption results in an impractical value of $V_{cc(D)}$ and a compromise which sacrifices maximum power gain is necessary. For this, a value of $Z_{c(D)}$ may be calculated in Eq. (26) in terms of the $V_{cc(D)}$ available.

$$Z_{c(D)} = \frac{(V_{cc(D)} - 3V_{CE(sat)(D)})^2 \eta}{V_{ip} I_{bp}} \tag{26}$$

27. The peak collector current of the driver is dependent upon the power input requirements of the output stage and the transformer primary impedance $Z_{c(D)}$. The peak current requirement may be calculated in Eq. (27).

$$I_{cp(D)} = I_{bp} \sqrt{\frac{Z_b}{\eta Z_{c(D)}}} \tag{27}$$

28. The driver collector-current quiescent point is determined by the peak collector current calculated in Eq. (27). The quiescent current must be greater

*Equation derived in Chapter 6.

than the peak collector current to prevent operation of the device in the non-linear portion of the y_{FE} curve. Distortion of the driver signal will result if the device is allowed to operate to cutoff; therefore, a minimum value of collector current is chosen that will always be above the knee of the y_{FE} curve. The collector quiescent current is calculated in Eq. (28).

$$I_{CQ(D)} = I_{cp(D)} + 1.0 \times 10^{-3} A \tag{28}$$

29. The driver supply voltage requirements can be much smaller than the power output-stage requirements because of the smaller power output of the driver. The minimum driver supply voltage is calculated in Eq. (29) in terms of the base-emitter parameters of the output stage and the collector saturation voltage of the driver device. This driver supply voltage may then be obtained by dropping the output stage supply voltage through a decoupling resistor or by providing a separate voltage source.

$$V_{cc(D)} = \frac{V_{ip} I_{bp}}{\eta I_{cp(D)}} + 3V_{CE(sat)(D)} \tag{29}$$

30. The critical driver collector requirements have now been calculated and the remainder of the driver parameters and components are determined by using the appropriate equations derived for the power output stage. The power output requirement of the driver device is calculated in Eq. (30) in terms of the collector requirements.

$$P_{o(rms)(D)} = \frac{I_{cp(D)}^2 Z_{c(D)} \eta}{2} \tag{30}$$

4

Class A Design Examples

Three design examples are presented in this chapter. Each example illustrates a corresponding step-by-step Class A design procedure outlined in Chapter 2. Both paragraph and equation numbers match those used in Chapter 2. The circuits chosen for the examples represent the simplest and most straightforward approach to practical circuit design. More elaborate circuits often provide advantages over the simple examples shown here, but the basic design principles are generally the same. Circuit modifications which provide additional advantages are suggested throughout the text. The transistors used and the power output levels of the amplifiers were chosen to show the capabilities of the design procedures for several common requirements.

At the end of the chapter, additional Class A circuits are shown complete with component values and performance data.

THREE-WATT TRANSFORMER-COUPLED INPUT, TRANSFORMER-COUPLED OUTPUT

1. The required rms power output in this example is three watts. The circuit diagram is shown in Chapter 2, Fig. 1. The power dissipation requirement is determined as a starting point in the selection of the output transistor.

$$\begin{aligned} P_{(Diss)} &= 3.12\ P_{o(rms)} \\ &= 3.12\ (3\ W) = 9.4\ W \end{aligned} \tag{1}$$

2. The output transistor must dissipate 9.4 watts under operating conditions. To meet this requirement, a device must be selected which is rated at considerably more power at the standard rating conditions (usually 25°C case temperature). The SP-2158 silicon power transistor is evaluated in this example. The maximum case temperature at which the transistor can be operated and remain within its power rating is calculated in Eq. (2). This calculation is needed because it is not practical to maintain a 25°C case temperature while dissipating

several watts. The temperature T_1 computed in Eq. (2) is also the upper limit of the transistor case temperature to be maintained by the selected heat sink.

$$T_1 = \frac{P_{(Rate)} - P_{(Diss)} + K_1T_2}{K_1} \qquad (2)$$

$$= \frac{15\text{ W} - 9.4\text{ W} + 0.1\text{ W/°C }(25°\text{C})}{0.1\text{ W/°C}}$$

$$= 81°\text{C}$$

3. The optimum or highest supply voltage V_{cc} at which the device can safely remain within its rating must now be calculated. From the SP-2158 data sheet, the maximum voltage for the transistor $V_{(BR)CER}$ is 250 volts. V_{cc} should be set at no more than one-half $V_{(BR)CER}$ because a transformer-coupled Class A stage can stress the transistor collector-emitter junction to two V_{cc} under maximum signal conditions. To allow for tolerances in power supply voltage, V_{cc} is selected at about 90% of this limit.

$$V_{cc} < \frac{V_{(BR)CER}}{2} \qquad (3)$$

$$< \frac{250\text{ V}}{2} < 125\text{ V}$$

To allow for V_{cc} tolerance, 110 volts is selected for the output stage power supply.

4. To ensure sufficient bias stability over the operating temperature range, the change in base-emitter voltage ΔV_{BE} is now considered.

$$\Delta V_{BE} = K_2\,(T_1 - T_4)$$

$$= 2.0 \times 10^{-3}\text{ V/°C }(81°\text{C} - 25°\text{C})$$

$$= 0.112\text{ V}$$

5. If no precautions are taken, ΔV_{BE} will change the quiescent operating point as the temperature changes. However, ΔV_{BE} is compensated for with d-c degeneration provided by R_E. Since this degenerative feedback voltage represents a loss to the supply voltage, it must be considered in the R_L' equation. The proper load impedance (R_L') of the collector for maximum undistorted power output is determined by the following equation, taking $V_{CE(sat)}$ from the data sheet and assuming a transformer efficiency η of 0.80.

$$R_L' = \frac{(V_{cc} - 8\Delta V_{BE} - V_{CE(sat)})^2\,\eta}{2P_{o(rms)}} \qquad (5)$$

$$= \frac{[110\text{ V} - 8\,(0.112\text{ V}) - 1\text{ V}]^2\,0.80}{2\,(3\text{ W})}$$

$$= 1550\ \Omega$$

6. In addition to the proper load impedance, the peak collector current I_{cp} is determined:

$$I_{cp} = \frac{2P_{o(rms)}}{(V_{cc} - 8\Delta V_{BE} - V_{CE(sat)})\ \eta} \tag{6}$$

$$= \frac{2\ (3\ \mathrm{W})}{[110\ \mathrm{V} - 8\ (0.112\ \mathrm{V}) - 1\ \mathrm{V}]\ 0.80}$$

$$= 70\ \mathrm{mA}$$

7. In a practical Class A stage, the quiescent collector current I_{CQ} must be somewhat larger than the I_{cp} to ensure undistorted output due to the non-linearity of the y_{FE} curve at very low currents and to ensure an allowance for the shift in Q point.

$$I_{CQ} = 1.25\ I_{cp} \tag{7}$$

$$= 1.25\ (70\ \mathrm{mA}) = 88\ \mathrm{mA}$$

8. The maximum collector current I_{CM} is I_{cp} plus I_{CQ}. The transistor considered for the output stage must be rated for at least this maximum collector current to ensure reliable and linear operation.

$$I_{CM} = 2.25\ I_{cp} \tag{8}$$

$$= 2.25\ (70\ \mathrm{mA}) = 157\ \mathrm{mA}$$

9. Since the SP-2158 is rated at 500 mA, the indicated I_{CM} of 157 mA is well within rating.

10. The collector cutoff current I_{CBO} is approximately exponentially related to temperature. I_{CBO} is an important consideration for bias stability. The $I_{CBO(T_3)}$ is taken from the data sheet.

$$I_{CBO(T_1)} = I_{CBO(T_3)} 2^{(T_1 - T_3)/K_T} \tag{9}$$

$$= (0.05\ \mathrm{mA})\ 2^{(81°C - 25°C)/12°C}$$

$$= 1.3\ \mathrm{mA}$$

11. I_{CBO}, at the maximum case temperature T_1, is 1.3 milliamperes. This quantity plays an important part in the bias network design which includes R_1, R_2 and R_E. The first consideration in biasing the output stage is the required base quiescent current I_{BQ}.

$$I_{BQ} = \frac{I_{CQ}}{h_{FE(min)}} \tag{10}$$

$$= \frac{88\ \mathrm{mA}}{40} = 2.2\ \mathrm{mA}$$

12. The total current I_{BB} through the biasing resistors can now be determined.

$$I_{BB} = I_{BQ} + 10\ I_{CBO(T_1)} \tag{11}$$

$$= 2.2\ \mathrm{mA} + 10\ (1.3\ \mathrm{mA})$$

$$= 15.2\ \mathrm{mA}$$

13. From the y_{FE} curve (Fig. 1), the quiescent base emitter voltage V_{BEQ} is determined at the quiescent collector current of 88 mA:

$$V_{BEQ} = 0.69 \text{ V} \tag{12}$$

14. The primary function of R_E in this circuit is to provide d-c degeneration for bias stability. The amount of stability required is dependent on V_{BEQ} and ΔV_{BE}, both of which have been determined. The R_E resulting from the following equation is designed to control I_{CQ} to $\pm 10\%$ for changes in each I_{CBO} and V_{BEQ}. Larger values of R_E may be used for additional I_{CQ} stability with the only penalty being higher V_{cc} requirements with associated circuit power and voltage rating increases.

$$R_E \geq \frac{100\Delta V_{BE} + V_{BEQ}}{10 I_{CQ}} \tag{13}$$

$$\geq \frac{100\,(0.112\text{V}) + 0.69\text{V}}{10\,(88 \times 10^{-3}\text{A})}$$

$$\geq 13.5\Omega$$

Thirteen and one-half ohms should provide adequate bias stability up to a case temperature of 81° C for the output transistor.

15. The peak base current I_{bp} is required to compute the driver transformer secondary impedance and finally the driver power requirements.

$$I_{bp} = \frac{I_{cp}}{h_{FE(min)}}$$

$$= \frac{70 \text{ mA}}{40} = 1.75 \text{ mA} \tag{14}$$

16. R_1 in the bias network is required to develop the voltage drop for proper biasing.

$$R_1 = \frac{I_{CQ}R_E + V_{BEQ}}{10 I_{CBO\,(T_1)}} \tag{15}$$

$$= \frac{(88 \times 10^{-3} \text{ A})\ 13.5\ \Omega + 0.69 \text{ V}}{10\,(1.3 \times 10^{-3} \text{ A})}$$

$$= 145\ \Omega$$

17. R_2 supplies the proper current from the V_{cc} supply to develop I_{BB}. In conjunction with R_1, the correct base quiescent voltage is generated for biasing the output stage.

$$R_2 = \frac{V_{cc} - 10 I_{CBO(T_1)}\, R_1}{I_{BB}} \tag{16}$$

$$= \frac{110\text{V} - 10\,(1.3 \times 10^{-3} \text{ A})\ 145\ \Omega}{15.2 \times 10^{-3} \text{ A}}$$

$$= 7.1 \text{ k}\Omega$$

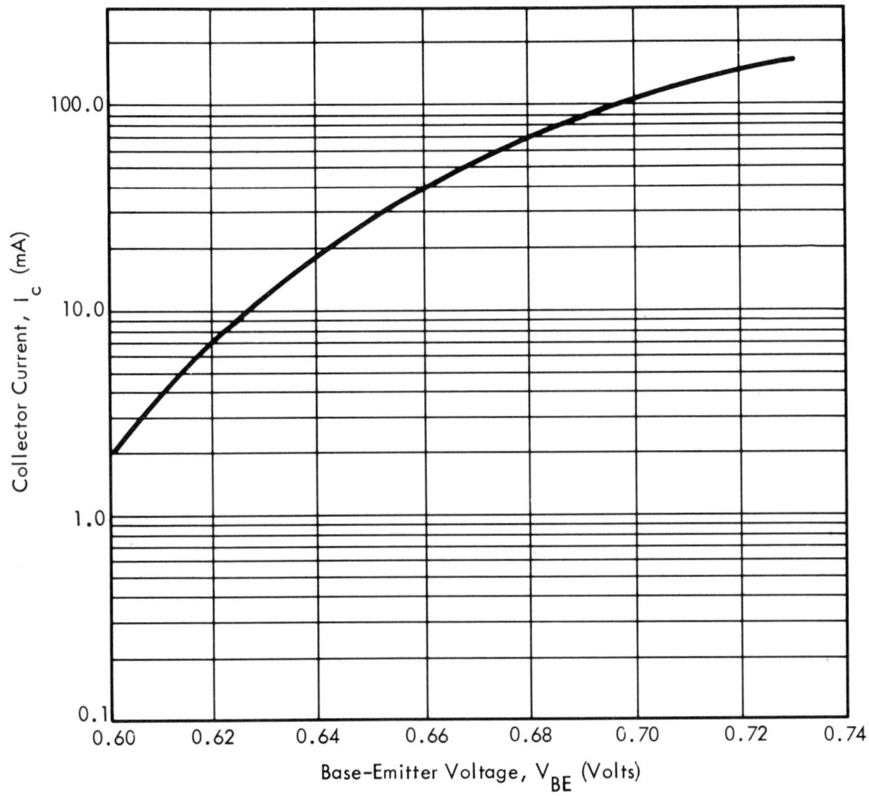

Fig. 1. Transconductance Characteristics y_{FE} for Device SP2158

18. The emitter bypass capacitor C_1 functions to nearly completely bypass the audio frequencies. This substantially removes all degenerative feedback for maximum gain. In this example, the entire R_E is bypassed which yields maximum power gain. The low-end frequency f was selected to be 100 Hz. At this frequency, the gain should be no more than −3 dB of its maximum.

$$C_1 \geq \frac{I_{cp}}{2\pi f K_4} \tag{17}$$

$$\geq \frac{70 \times 10^{-3}\ \text{A}}{2\pi\ (100\ \text{Hz})\ (35 \times 10^{-3})}$$

$$\geq 3200\ \mu\text{F}$$

C_1 as indicated, turned out to be rather large. Often, in practice, only a portion of R_E is bypassed which usually requires a much smaller capacitor. If this technique is employed, less power gain but lower distortion can be expected due to the local feedback. The peak base-emitter voltage is required to compute power gain. V_{BEM} and V_{BEQ} are read from Fig. 1 at I_{CM} and I_{CQ}.

$$V_{bep} = V_{BEM} - V_{BEQ}$$
$$= 0.725\text{ V} - 0.692\text{ V}$$
$$= 0.033\text{ V}$$

19. Power gain of the output stage is the ratio of power output to input.

$$\text{P.G.} = \frac{2P_{o(rms)}}{V_{bep}\,I_{bp}} \tag{18}$$
$$= \frac{2\,(3\text{ W})}{0.033\text{ V}\,(1.75 \times 10^{-3}\text{ A})}$$
$$= 10^5 = 50\text{ dB}$$

The power gain of 50 dB is realizable primarily due to the high voltage capability of the SP-2158 as well as the moderately high current gain. This completes the output stage design. The driver stage requirements can now be defined.

20. For maximum power transfer, the secondary of the driver transformer is matched to the base input impedance Z_b.

$$Z_b = \frac{V_{bep}}{I_{bp}} \tag{19}$$
$$= \frac{0.033\text{ V}}{1.75 \times 10^{-3}\text{ A}} = 18.9\ \Omega$$

21. The secondary impedance of the driver transformer can also be specified as 18.9 ohms. The primary impedance used as the driver collector load $Z_{c(D)}$ should be high for high power gain but not so high as to be unreasonable to manufacture.

22. As a compromise, 20 kilohms is assumed and $I_{cp(D)}$ can be calculated in Eq. (21). Since $Z_{c(D)}$ is assumed, Eq. (20) is omitted.

$$I_{cp(D)} = I_{bp}\sqrt{\frac{Z_b}{\eta Z_{c(D)}}} \tag{21}$$
$$= 1.75 \times 10^{-3}\text{ A}\sqrt{\frac{18.9\ \Omega}{0.80\,(20 \times 10^3\ \Omega)}}$$
$$= 6.02 \times 10^{-5}\text{ A}$$

23. Since $I_{cp(D)}$ can be very small for low driver power requirements, one milliampere is added for the quiescent collector current, $I_{CQ(D)}$. The additional current will ensure, in most transistors suitable for drivers, that the operating point is well within the linear y_{FE} region.

$$I_{CQ(D)} = I_{cp(D)} + 1.0 \times 10^{-3}\text{ A} \tag{22}$$
$$= 6.02 \times 10^{-5}\text{ A} + 1.0 \times 10^{-3}\text{ A}$$
$$= 1.06 \times 10^{-3}\text{ A}$$

$I_{CQ(D)}$ of 1.06 mA is high enough for most small-signal transistors and low enough that dissipation is negligible.

24. The supply voltage for the driver is calculated by assuming a generous allowance for $V_{CE(sat)}$ to ensure that the collector-to-emitter voltage is high enough to allow linear operation while not significantly adding to the power dissipation of the transistor. One-volt $V_{CE(sat)}$ is taken from the data sheet.

$$V_{cc(D)} = \frac{V_{bep}\ I_{bp}}{\eta I_{cp(D)}} + 3V_{CE(sat)(D)} \tag{23}$$

$$= \frac{0.033\ \text{V}\ (1.75 \times 10^{-3}\ \text{A})}{0.80\ (6.02 \times 10^{-5}\ \text{A})} + 3\ (1.0\ \text{V})$$

$$= 4.2\ \text{V}$$

$V_{cc(D)}$ can be derived from the V_{cc} supply by a large decoupling resistor R_D, bypassed with 10 μF at the d-c end of the driver transformer primary. The value of R_D can now be determined.

$$R_D = \frac{V_{cc} - V_{cc(D)}}{I_{CQ}}$$

$$= \frac{110\ \text{V} - 4.2\ \text{V}}{1.06 \times 10^{-3}\ \text{A}}$$

$$= 100\ \text{k}\Omega$$

25. Finally, the power output of the driver is determined:

$$P_{o(rms)(D)} = \frac{I_{cp(D)}^{\ 2}\ Z_{c(D)}\ \eta}{2}$$

$$= \frac{(6.02 \times 10^{-5}\ \text{A})^2\ (20 \times 10^3\ \Omega)\ 0.80}{2} \tag{24}$$

$$= 2.9 \times 10^{-5}\ \text{W}$$

ONE-WATT RC-COUPLED INPUT, TRANSFORMER-COUPLED OUTPUT

The following example illustrates the use of the design procedure for an RC-coupled driven output stage delivering one watt rms power output to a transformer-coupled load. The circuit diagram for this example is Fig. 4 in Chapter 2. Since the output is transformer-coupled, the designer is required to consider only the power output as a starting point. The output transistor ratings establish the required d-c supply voltage and collector load impedance.

1. The first requirement for the output transistor to satisfy is the power dissipation. $P_{(Diss)}$ can be estimated by its relationship to power output in Eq. (1).

$$P_{(Diss)} = 3.12\ P_{o(rms)} \tag{1}$$

$$= 3.12\ (1\ \text{W}) = 3.12\ \text{W}$$

2. The power dissipation required of the output transistor is 3.12 watts at the maximum device temperature in service. The maximum permissible transistor operating temperature T_J may be computed for a specific device by

Eq. (2). The SK-3808 silicon transistor is evaluated here. The $P_{(Rate)}$ term is specified on the data sheet at specified temperature T_2. This power rating must always be higher than $P_{(Diss)}$ calculated in Eq. (1).

$$T_1 = \frac{P_{(Rate)} - P_{(Diss)} + K_1 T_2}{K_1} \tag{2}$$

$$= \frac{5\ W - 3.12\ W + 0.033\ W/°C\ (25°C)}{0.033\ W/°C}$$

$$= 82°C$$

The SK-3808 is capable of dissipating 3.12 watts at 82°C. This transistor will fulfill the power requirement of the application if a heat sink is selected to ensure that the case temperature does not exceed 82°C.

3. The maximum supply voltage V_{cc} is determined by the transistor $V_{(BR)CER}$ rating on the data sheet. It is desirable to select V_{cc} to be as high as possible for maximum power gain but somewhat lower than absolute maximum value determined by the maximum V_{cc} equation for supply voltage tolerance allowance.

$$V_{cc} < \frac{V_{(BR)CER}}{2} < \frac{300\ V}{2} < 150\ V \tag{3}$$

The V_{cc} equation indicates an absolute maximum permissible supply voltage of 150 volts for the SK-3808. To allow for V_{cc} tolerance, 120 volts is selected for this example.

4. For bias stability considerations, ΔV_{BE}, the change in base-to-emitter voltage, must be determined.

$$\Delta V_{BE} = K_2\ (T_1 - T_4) \tag{4}$$

$$= 2.0 \times 10^{-3}\ V/°C\ (82°C - 25°C)$$

$$= 0.114\ V$$

The change in V_{BE} can be expected to be 0.114 volts over the expected transistor temperature range. This quantity is an important consideration in bias network design and consequently, is significant in many of the circuit element determinations to follow.

5. The collector load impedance R_L' can now be determined. This will be the primary impedance of the output transformer. $V_{CE(sat)}$ is specified on the data sheet. The transformer efficiency η is assumed to be 0.80.

$$R_L' = \frac{(V_{cc} - 8\Delta V_{BE} - V_{CE(sat)})^2\ \eta}{2\ P_{o(rms)}} \tag{5}$$

$$= \frac{[120\ V - 8\ (0.114\ V) - 1\ V]^2\ 0.80}{2\ (1\ W)}$$

$$= 5600\ \Omega$$

The primary impedance of the output transformer should be specified as 5.6 kilohms.

6. The peak collector a-c current I_{cp} is computed in Eq. (6):

$$I_{cp} = \frac{2P_{o(rms)}}{(V_{cc} - 8\Delta V_{BE} - V_{CE(sat)})\ \eta} \tag{6}$$

$$= \frac{2\ (1\ W)}{[120\ V - 8\ (0.114\ V) - 1\ V]\ 0.80}$$

$$= 21.2\ mA$$

7. I_{cp} is 21.2 mA for the selected supply voltage and power output. The quiescent collector current I_{CQ} is based on I_{cp}.

$$I_{CQ} = 1.25\ I_{cp} \tag{7}$$

$$= 1.25\ (21.2\ mA) = 26.5\ mA$$

8. The maximum collector current I_{CM} is really the sum of I_{cp} and I_{CQ} or can be expressed in terms of I_{cp}:

$$I_{CM} = 2.25\ I_{cp} \tag{8}$$

$$= 2.25\ (21.2\ mA) = 47.7\ mA$$

9. The I_{CM} of 47.7 mA is within the rating of the SK-3808.

10. The collector cutoff current I_{CBO} is temperature-dependent. Its value at a reference temperature ($T_3 = 25°C$) is specified on the data sheet. Its value at a maximum temperature T_1 is computed in Eq. (9).

$$I_{CBO(T_1)} = I_{CBO(T_3)} 2^{(T_1 - T_3)/K_T} \tag{9}$$

$$= (1\ \mu A)\ 2^{(82°C - 25°C)/12°C}$$

$$= 27\ \mu A$$

11. The base quiescent current is determined for the bias network design:

$$I_{BQ} = \frac{I_{CQ}}{h_{FE(min)}} \tag{10}$$

$$= \frac{26.5\ mA}{15} = 1.77\ mA$$

12. The total current through the biasing resistors I_{BB} is determined in Eq. (11). Its value must always be much larger than the maximum I_{CBO} to ensure temperature stability.

$$I_{BB} = I_{BQ} + 10\ I_{CBO(T_1)} \tag{11}$$

$$= 1.77\ mA + 10\ (0.027\ mA)$$

$$= 2.04\ mA$$

13. The quiescent base-to-emitter voltage V_{BEQ} is determined from the SK-3808 typical y_{FE} curve (Fig. 2) at the quiescent collector current of 26.5 mA:

$$V_{BEQ} = 0.63\ V \tag{12}$$

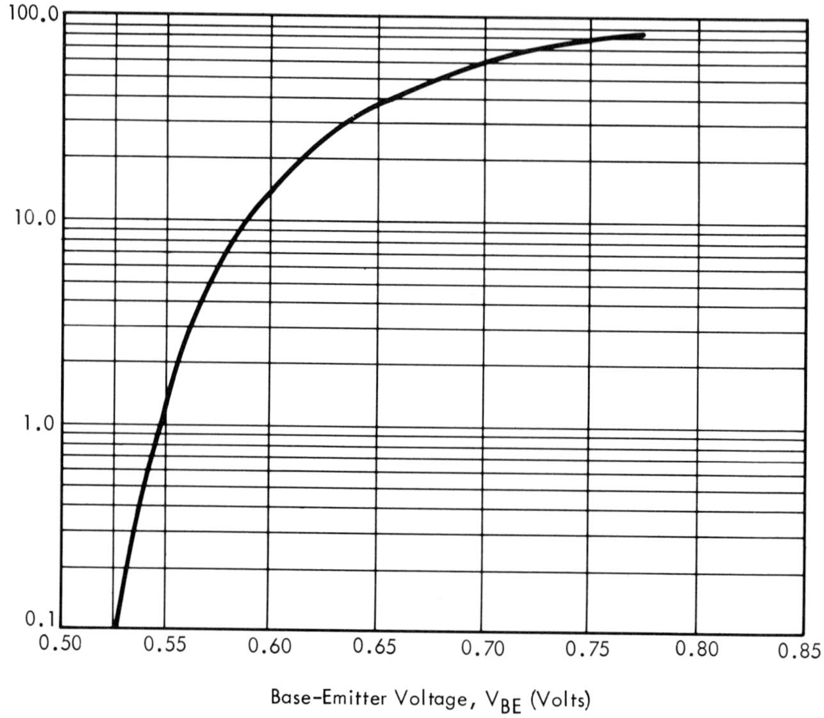

Fig. 2. Transconductance Characteristics y_{FE} for Device SK3808

14. The minimum value of the emitter resistor R_E is computed in Eq. (13). This value is established entirely by bias stability considerations:

$$R_E \geq \frac{100\Delta V_{BE} + V_{BEQ}}{10 I_{CQ}} \tag{13}$$

$$\geq \frac{100\,(0.114\text{V}) + 0.63\text{V}}{10(26.5 \times 10^{-3}\ \text{A})}$$

$$\geq 45\Omega$$

The minimum value of R_E for $\pm 10\%$ quiescent collector current stability is 45 ohms. Larger values may be used if the selected value is also used in all following equations. If R_E is made considerably larger than the value just calculated, the additional voltage loss should be considered in Eqs. (5) and (6).

15. The peak base current I_{bp}, is now determined:

$$I_{bp} = \frac{I_{cp}}{h_{FE(min)}} \tag{14}$$

$$= \frac{21.2\ \text{mA}}{15} = 1.41\ \text{mA}$$

16. The lower resistor R_1 in the bias resistor network is determined:

$$R_1 = \frac{I_{CQ} R_E + V_{BEQ}}{10 I_{CBO(T_1)}} \tag{15}$$

$$= \frac{(26.5 \times 10^{-3}\ \text{A})\ 45\ \Omega + 0.63\ \text{V}}{10(27 \times 10^{-6}\ \text{A})}$$

$$= 6.8\ \text{k}\Omega$$

17. The upper resistor R_2 in the biasing voltage divider is computed:

$$R_2 = \frac{V_{cc} - 10 I_{CBO\ (T_1)}\ R_1}{I_{BB}} \tag{16}$$

$$= \frac{120\text{V} - 10(27 \times 10^{-6}\text{A})\ 6.8 \times 10^{3}\ \Omega}{2.04 \times 10^{-3}\text{A}}$$

$$= 58\ \text{k}\Omega$$

18. The equivalent parallel resistance of the biasing resistors R_o is computed in Eq. (25). R_o appears in parallel with the input of the output transistor.

$$R_o = \frac{R_1 R_2}{R_1 + R_2}$$

$$= \frac{6.8\ \text{k}\Omega\ (58\text{k}\Omega)}{6.8\ \text{k}\Omega + 58\text{k}\Omega}$$

$$= 6.1\text{k}\Omega$$

The input impedance of the output stage R_i is computed. The incremental transconductance y_{fe} is taken from the slope of the y_{FE} curve (Fig. 2) at the quiescent collector current of the output transistor.

$$R_i = \frac{h_{FE}\ R_o}{h_{FE} + y_{fe}\ R_o} \tag{25}$$

$$= \frac{50(6.1 \times 10^{3}\ \Omega)}{50 + 0.47\text{S}(6.1 \times 10^{3}\ \Omega)}$$

$$= 104\ \Omega$$

19. The emitter bypass capacitor C_1 is determined by selecting the minimum frequency f at which the relative response is no more than −3 dB of its maximum. In this example the minimum C_1 frequency was assumed to be 100 Hz.

$$C_1 \geq \frac{I_{cp}}{2\ \pi\ fK_4} \tag{17}$$

$$\geq \frac{21.2 \text{ x } 10^{-3}\ \text{A}}{2\ \pi\ (100\ \text{Hz})\ (35 \text{ x } 10^{-3})}$$

$$\geq 960\ \mu\text{F}$$

20. The power gain P.G. for the output stage is computed from the ratio of output and input impedances taking transformer efficiency and h_{fe} into consideration.

$$P.G. = \frac{h_{fe}^2 R_L' \eta}{R_i} \tag{26}$$

$$= \frac{(15)^2 (5600\ \Omega)\ 0.80}{104\ \Omega}$$

$$= 9600$$

21. The decoupling resistor R_D is established:

$$R_D = \frac{V_{cc} \times 10^3}{40} \tag{27}$$

$$= \frac{120\ V \times 10^3}{40\ A} = 3\ k\Omega$$

22. The maximum driver emitter resistor R_3 is calculated in terms of the decoupling resistor:

$$R_3 \leq 3R_D \tag{28}$$

$$\leq 3\ (3\ k\Omega) \leq 9\ k\Omega$$

23. The peak driver collector-to-emitter voltage must be determined for the driver output requirements. The requirements for $V_{cep(D)}$ are the same as V_{bep}. V_{bep} is determined from the SK-3808 y_{FE} curve (Fig. 2). V_{BEM} is read off at I_{CM} and V_{BEQ} is read off at I_{CQ}.

$$V_{cep(D)} = V_{bep} = V_{BEM} - V_{BEQ} \tag{29}$$

$$= 0.67\ V - 0.63\ V$$

$$= 0.04\ V$$

The maximum driver collector-to-emitter voltage $V_{CEM(D)}$, allowing a 50% overload factor to prevent saturation problems, is calculated in Eq. (30).

$$V_{CEM(D)} = 3(V_{cep(D)} + V_{CE(sat)(D)}) \tag{30}$$

$$= 3\ (0.04\ V + 1\ V)$$

$$= 3.12\ V$$

24. The quiescent driver collector voltage also has an equal overload margin:

$$V_{CEQ(D)} = 2V_{cep(D)} + 3V_{CE(sat)(D)} \tag{31}$$

$$= 2\ (0.04\ V) + 3\ (1\ V)$$

$$= 3.08\ V$$

25. The driver peak collector current $I_{cp(D)}$ is determined which includes both I_{bp} and current shunted by the biasing resistors R_1 and R_2.

$$I_{cp(D)} = 1.8\left(I_{bp} + \frac{V_{BEQ}}{R_o}\right) \tag{32}$$

$$= 1.8\left(1.41\text{ mA} + \frac{0.63\text{V}}{6.1 \times 10^3\ \Omega}\right)$$

$$= 2.73\text{ mA}$$

26. The quiescent driver collector current $I_{CQ(D)}$ is computed in terms of $I_{cp(D)}$.

$$I_{CQ(D)} = 1.25\ I_{cp(D)} \tag{33}$$

$$= 1.25\ (2.73\text{ mA})$$

$$= 3.4\text{ mA}$$

27. The driver-stage collector load resistor R_4 is established by the quiescent voltage and current requirements already determined in Eqs. (31) and (33).

$$R_4 = \frac{V_{cc} - V_{CEQ(D)}}{I_{CQ(D)}} - (R_D + R_3) \tag{34}$$

$$= \frac{120\text{ V} - 3.08\text{ V}}{3.4\text{ mA}} - (3\text{ k}\Omega + 9\text{ k}\Omega)$$

$$= 22.5\text{ k}\Omega$$

28. The minimum value of the coupling capacitor C_c is determined in Eq. (35). The minimum frequency f is chosen as the lowest frequency at which the relative output is no more than three dB down from its higher reference frequency amplitude. The output impedance of the driver is assumed to be approximately the value of the driver load resistor.

$$C_c \geq \frac{1}{2\pi f R_4} \tag{35}$$

$$\geq \frac{1}{2\pi(100)\ (22.5 \times 10^3)}$$

$$\geq 0.071\ \mu\text{F}$$

ONE-HUNDRED-MILLIWATT DIRECT-COUPLED INPUT, DIRECT-COUPLED OUTPUT

This example illustrates the use of the design procedure for a lower-power Class A direct-coupled output amplifier. See circuit diagram, Fig. 6 in Chapter 2. Usually, this design is only practical for very low power outputs due to the high direct-current component required to flow in the speaker. The power content of the d-c component will ordinarily be about double the maximum undistorted sinewave output. This indicates the requirement for a speaker rated at not less than twice the power rating for an a-c coupled speaker with the same audio output capability. At very low power outputs, however, speaker power capability is not a problem and the circuit designer can take advantage of a very-low-cost transformerless design.

1. This design is initiated by assuming only two parameters: power output required and speaker impedance. It will be desirable to select the highest speaker impedance compatible with low-cost availability because power gain is directly proportional to load impedance as shown in Eq. (50). An output of 100 mW undistorted sine wave power $P_{o(rms)}$ and a speaker load impedance R_L of 32 ohms is assumed here.

2. To select the output transistor, consider first the power dissipation required:

$$P_{(Diss)} = 2.5\ P_{o(rms)} \tag{36}$$
$$= 2.5\ (100\ \text{mW}) = 250\ \text{mW}$$

3. The device selected must be capable of dissipating 250 milliwatts at the maximum temperature T_1. The next step is to evaluate a device with a maximum power rating higher than 250 milliwatts and to determine the maximum temperature at which it can be safely operated. The 2N3706 will be evaluated in this example.

$$T_1 = \frac{P_{(Rate)} - P_{(Diss)} + K_1T_2}{K_1} \tag{2}$$
$$= \frac{0.360\ \text{W} - 0.250\ \text{W} + 2.88 \times 10^{-3}\ \text{W/°C}\ (25°\text{C})}{2.88 \times 10^{-3}\ \text{W/°C}}$$
$$= 63°\text{C}$$

The 2N3706 will be within its power rating at ambient temperatures up to 63°C as indicated by T_1 when operated at 250 milliwatts dissipation.

4. In addition to power rating, the device considered must be capable of handling the maximum current I_{CM} required in this application.

$$I_{cp} = \sqrt{\frac{2P_{o(rms)}}{R_L}} \tag{37}$$
$$= \sqrt{\frac{2\ (0.1\ \text{W})}{32\ \Omega}} = 79\ \text{mA}$$

$$I_{CM} = 2.25\ I_{cp} \tag{8}$$
$$= 2.25\ (79\ \text{mA}) = 178\ \text{mA}$$

The 2N3706 is rated at 800 mA and will therefore be well within its current rating at the required 178 mA.

5. Also, the collector-to-emitter voltage estimation must be no more than $V_{BR(CER)}$:

$$V_{(BR)CER} \geq 2\sqrt{2P_{o(rms)}\,R_L} + 4\text{ V} + 2V_{CE(sat)} \tag{39}$$
$$\geq 2\sqrt{2\,(0.1\text{ W})\,32\,\Omega} + 4\text{ V} + 2\,(1\text{ V})$$
$$\geq 11.06\text{ V}$$

The output transistor $V_{(BR)CER}$ must be rated at least 11.06 volts. This requirement is also easily exceeded by the 2N3706. This voltage is an early approximation of the V_{cc} supply required in the circuit and was estimated by adding the maximum assumed voltage drop across R_E and double the saturation voltage to the theoretical minimum a-c voltage necessary to generate the required power output.

6. After having determined that the 2N3706 will meet all the requirements for the output device, the circuit parameters can be determined. The quiescent collector current I_{CQ} for the output stage is calculated by Eq. (7).

$$I_{CQ} = 1.25\,I_{cp} \tag{7}$$
$$= 1.25\,(79\text{ mA}) = 99\text{ mA}$$

7. From I_{CQ}, the base quiescent current I_{BQ} may be determined. The $h_{FE(min)}$ value is taken from the data sheet.

$$I_{BQ} = \frac{I_{CQ}}{h_{FE(min)}} \tag{10}$$
$$= \frac{99\text{ mA}}{30} = 3.3\text{ mA}$$

8. The maximum driver collector current $I_{CM(D)}$ can now be calculated from knowing the output stage drive requirements:

$$I_{CM(D)} = 2.5\,I_{BQ} \tag{40}$$
$$= 2.5\,(3.3\text{ mA}) = 8.2\text{ mA}$$

9. Because a change in quiescent operating current in the driver stage will be reflected as a much larger change in current in the output stage, it is important to give special consideration to driver collector-base current $I_{CBO(D)}$. The worst-case I_{CBO} will occur at maximum ambient temperature T_A The silicon planar 2N3710 has been selected for the driver in this example because of its low I_{CBO} and high h_{FE} characteristics. Equation (41) evaluates driver I_{CBO} with the maximum circuit ambient temperature assumed to be 55°C:

$$I_{CBO(T_A)(D)} = I_{CBO(T_3)}2^{(T_A - T_3)/K_T} \tag{41}$$
$$= (1 \times 10^{-7}\text{ A})\; 2^{(55°C - 25°C)/12°C}$$
$$= 0.57\;\mu\text{A}$$

10. At the maximum ambient temperature, I_{CBO} will be only about one-half microampere. Knowing I_{CBO} in the worst-case situation and determining V_{BE} from the y_{FE} curve for the driver, the driver bias resistor R_3 can be computed. To use the y_{FE} curve, the driver quiescent current $I_{CQ(D)}$ must be determined so $V_{BEQ(D)}$ can be located at this point.

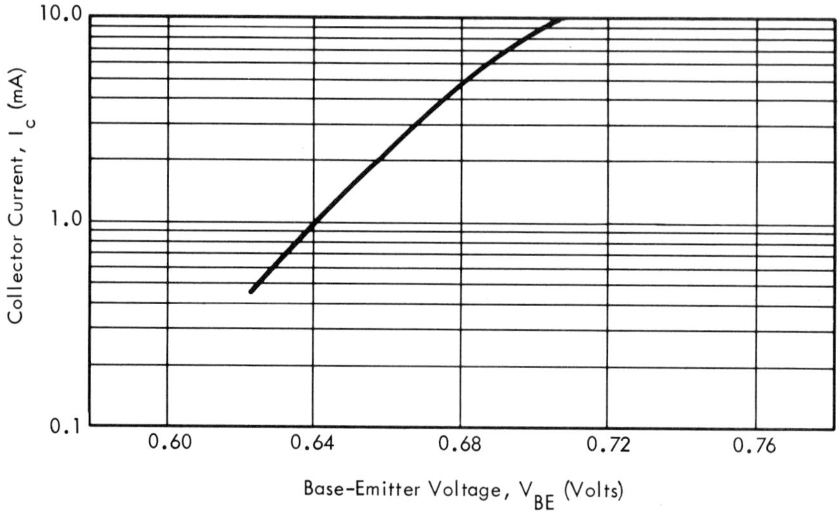

Fig. 3. Transconductance Characteristics y_{FE} for Device 2N3710

$$I_{CQ(D)} = 1.5\ I_{BQ} \qquad (42)$$
$$= 1.5\ (3.3\ \text{mA}) = 4.9\ \text{mA}$$

11. Now, from the y_{FE} curve (Fig. 3) $V_{BEQ(D)}$ can be determined. The quiescent base-emitter voltage is found to be 0.68 volts at a collector current of 4.9 milliamperes.

$$V_{BEQ(D)} = 0.68\ \text{V}$$

12. The driver $V_{BEQ(D)}$ and $I_{CBO(T_A)(D)}$ values needed to calculate R_3 have now been determined. Since any change in driver y_{FE} due to temperature effects is largely compensated for by a similar y_{FE} change in the output stage, quiescent-current stability criteria can be based largely on the value of R_3. Equation (43) determines the maximum value of R_3 allowing $I_{CBO(T_A)(D)}$ to contribute no more than a 5% change in V_{BEQ}. If R_3 calculates to be more than about 20 kilohms, a lower value may be used without greatly reducing power gain of the circuit. Lower values of R_3 will improve the circuit Q-point stability for changes in driver h_{FE} which will permit a wider h_{FE} distribution to be used. If R_3 is selected to be smaller than the value calculated in Eq. (43), that value should also be used to calculate R_E in Eq. (45). Since R_E is a factor in determining V_{cc}, V_B, and R_2 in Eqs. (46), (47) and (48), these quantities will also change. Ordinarily, it is desirable to make R_3 as small as possible down to a minimum of about 1000 ohms. However, this example will use the maximum value calculated in Eq. (43) to illustrate the validity of the procedure.

$$R_3 \leq \frac{V_{BEQ(D)}}{20\ I_{CBO(T_A)(D)}} \tag{43}$$

$$\leq \frac{0.68\ \text{V}}{20\ (0.57 \times 10^{-6}\ \text{A})}$$

$$\leq 59.6\ \text{k}\Omega$$

13. The driver base quiescent current $I_{BQ(D)}$ must be determined to establish a value of R_E. $I_{BQ(D)}$ is calculated in terms of the driver quiescent current:

$$I_{BQ(D)} = \frac{I_{CQ(D)}}{h_{FE(min)(D)}} \tag{44}$$

$$= \frac{4.9 \times 10^{-3}\ \text{A}}{90} = 55\ \mu\text{A}$$

14. The emitter resistor R_E which provides the sensing point for d-c feedback through R_3 can now be computed. This resistor develops a voltage drop proportional to emitter current and is therefore a measure of the collector quiescent current. Its value is rather critical and a 5% tolerance may be justified if it is desired that collector current due to variance in this circuit element be limited to 5%.

$$R_E = \frac{V_{BEQ(D)} + I_{BQ(D)}R_3}{I_{CQ}} \tag{45}$$

$$= \frac{0.68\ \text{V} + (55 \times 10^{-6}\ \text{A})\ 59.6 \times 10^3\ \Omega}{99 \times 10^{-3}\ \text{A}}$$

$$= 40\ \Omega$$

15. This relatively large R_E resulted from using the maximum permissible value of R_3 computed in Eq. (43). All circuit parameters are now known to permit the exact V_{cc} requirement to be determined.

$$V_{cc} = 2\sqrt{2P_{o(rms)}R_L} + I_{CQ}R_E + 2V_{CE(sat)}$$

$$= 2\sqrt{2\ (0.1\ \text{W})\ 32\ \Omega} + 0.099\ \text{A}\ (40\ \Omega) + 2\ (1\ \text{V}) \tag{46}$$

$$= 11.02\ \text{V}$$

16. The driver collector load resistor R_2 can be determined more accurately if the quiescent collector voltage is known. Since the collector is directly coupled to the base of the output stage, determination of this voltage V_B will give the desired Q point. V_{BEQ} is read off the 2N3706 y_{FE} curve (Fig. 4) at I_{CQ} of 99 milliamperes.

$$V_B = I_{CQ}R_E + V_{BEQ} \tag{47}$$

$$= 0.099\ \text{A}\ (40\ \Omega) + 0.88\ \text{V}$$

$$= 4.84\ \text{V}$$

17. Now the driver collector load resistor may be determined:

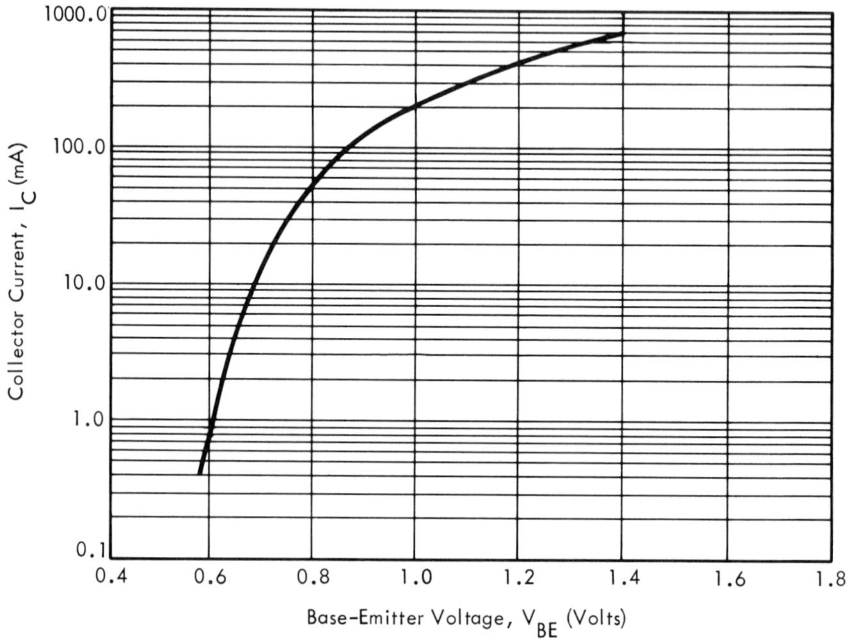

Fig. 4. Transconductance Characteristics y_{FE} for Device 2N3706

$$R_2 = \frac{V_{cc} - V_B}{I_{CM(D)}} \tag{48}$$

$$= \frac{11.02\ V - 4.84\ V}{8.2 \times 10^{-3}\ A}$$

$$= 753\ \Omega$$

18. The emitter bypass capacitor minimum value may now be determined. The minimum frequency was selected at 100 Hz and the output is down three dB at that point.

$$C_1 \geq \frac{I_{cp}}{2\pi f K_4} \tag{17}$$

$$\geq \frac{0.099\ A}{2\ \pi\ (100\ Hz)\ (35 \times 10^{-3})}$$

$$\geq 4500\ \mu F$$

The circuit shown in this example yields maximum power gain by complete bypassing of audio frequency degeneration with C_1. The designer, however, has the option of only partially bypassing R_E which requires a much smaller value of C_1 at the expense of power gain. Another option requiring an even smaller value of capacitance is to eliminate completely the emitter bypass. The elimination of a-c degenerative feedback to the driver is accomplished by splitting R_3 into two series components and bypassing the center junction to ground with a relatively small capacitance. These approaches will yield somewhat lower power gain but are generally more economical.

19. The input impedance of the amplifier $R_{i(D)}$ may now be determined. R_3 was calculated in Eq. (43). The incremental y_{fe} is taken from the slope of the 2N3710 V_{BE} versus I_C curve (Fig. 3) at the driver quiescent current of 4.9 mA.

$$R_{i(D)} = \frac{h_{FE(D)}R_3}{h_{FE(D)} + y_{fe}R_3} \tag{49}$$

$$= \frac{90\ (59.6 \times 10^3\ \Omega)}{90 + 0.15\ \text{S}\ (59.6 \times 10^3\ \Omega)}$$

$$= 600\ \Omega$$

20. Power gain of the entire amplifier may be determined in terms of input impedance and load impedance.

$$PG_{(T)} = \frac{(h_{FE(D)}\ h_{FE})^2\ R_L}{R_{i(D)}} \tag{50}$$

$$= \frac{[90\ (30)]^2\ 32\ \Omega}{600\ \Omega}$$

$$= 3.9 \times 10^5$$

ADDITIONAL CLASS A AMPLIFIERS

This section presents three additional Class A amplifier designs. Component values, device numbers, and performance data are given for each circuit. When available, additional data such as response and distortion curves and transformer winding information are given. All resistors are one-half watt ±10% and values are in ohms unless otherwise noted. All capacitor values are in microfarads unless otherwise noted.

2-WATT AMPLIFIER

Direct-coupled Input
Transformer-coupled Output
Output Device 2N1718

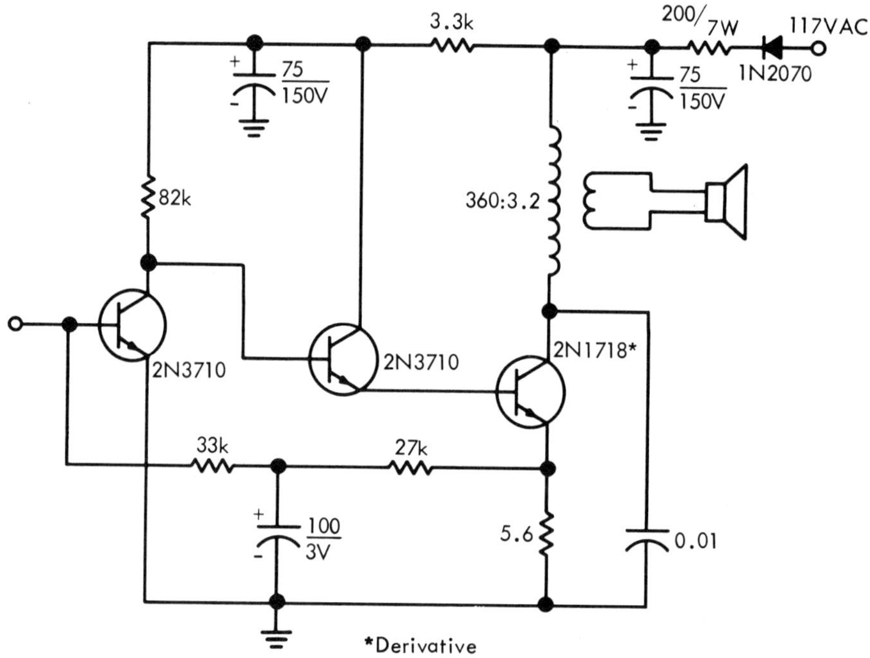

Two-watt Amplifier Schematic

Performance Data

Maximum Power Output: 3.5 W
Maximum Power (10% Distortion): 2.15 W
Sensitivity (100-mW Reference): 56 mV
Frequency Response (1-kHz Reference)
3 dB Low: 135 Hz
3 dB High: 39,000 Hz

Output Transformer

Primary: 360 Ω
Secondary: 3.2 Ω

850-MILLIWATTS-PER-CHANNEL STEREO AMPLIFIER

Diode-coupled Input
Transformer-coupled Output
Output Device 2N497

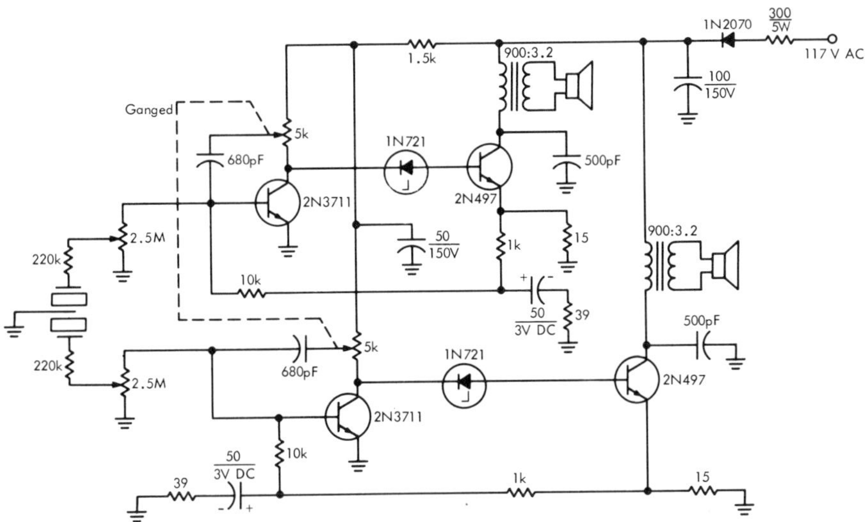

850-mW-per-channel Stereo Phono Amplifier Schematic

Performance Data
Maximum Power Output: 850 mW
Sensitivity (500-mW Reference): 1.25 V
Frequency Response (1-kHz Reference)
3 dB Low: 75 Hz
0.6 dB High: 20,000 Hz

Output Transformers
Bobbin: E-188
Core: EI-21
Stacking: Butt
Winding Order: Secondary First
Primary: 900 Ω, 1100 Turns
#32 Single Soldereze
Random Wound
Secondary: 3.2 Ω, 65 Turns
#21 Single Soldereze
Random Wound

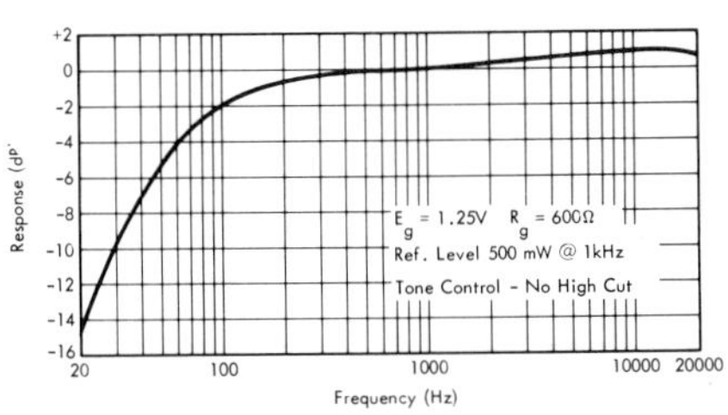

850-mW Stereo Amplifier Response Curve, E_g = 1.25 V

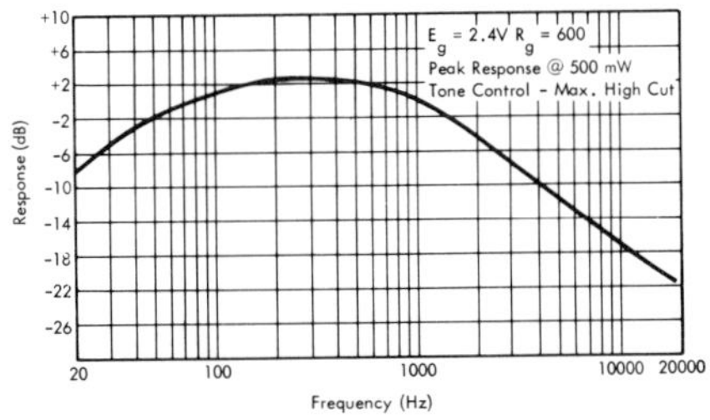

850-mW Stereo Amplifier Response Curve, E_g = 2.4 V

3-WATTS-PER-CHANNEL STEREO AMPLIFIER

Direct-coupled Input
Transformer-coupled Output
Output Device 2N1050

Performance Data
Maximum Power (10% Distortion): 3 W
Distortion (1-W Reference): 0.9%
Frequency Response (1-kHz Reference)
3 dB Low: 80 Hz
3 dB High: 16,500 Hz

Output Transformers
Bobbin: E-188
Core: EI-21
Stacking: Butt with 2 Layers 0.001 Tape on Butt Ends
Winding Order: Secondary First
Primary: 570 Ω, 1130 Turns #32 Single Soldereze Random Wound
Secondary: 8 Ω, 134 Turns #23 Single Soldereze Random Wound

Power Transformer
Bobbin: 5785
Core: EI-100
Stacking: 8 × 8
Winding Order: Secondary First with 2 Layers Mylar Tape
Primary: 1000 Turns #27 Single Soldereze Random Wound
Secondary: 600 Turns #25 Single Soldereze Random Wound

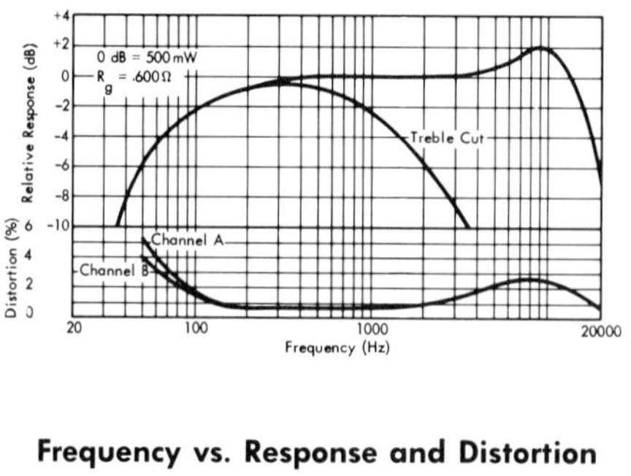

Frequency vs. Response and Distortion

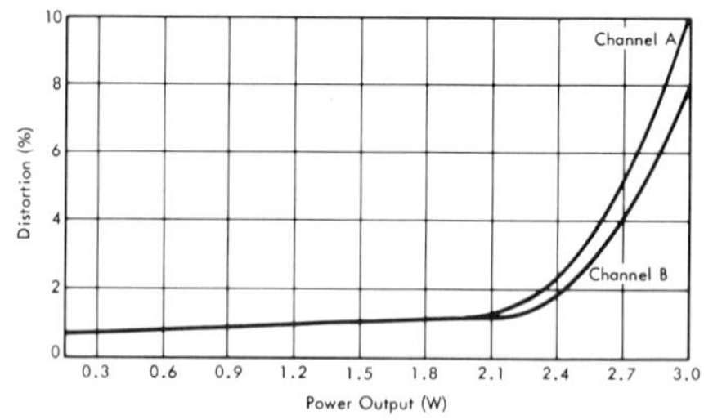

Power Output vs. Distortion

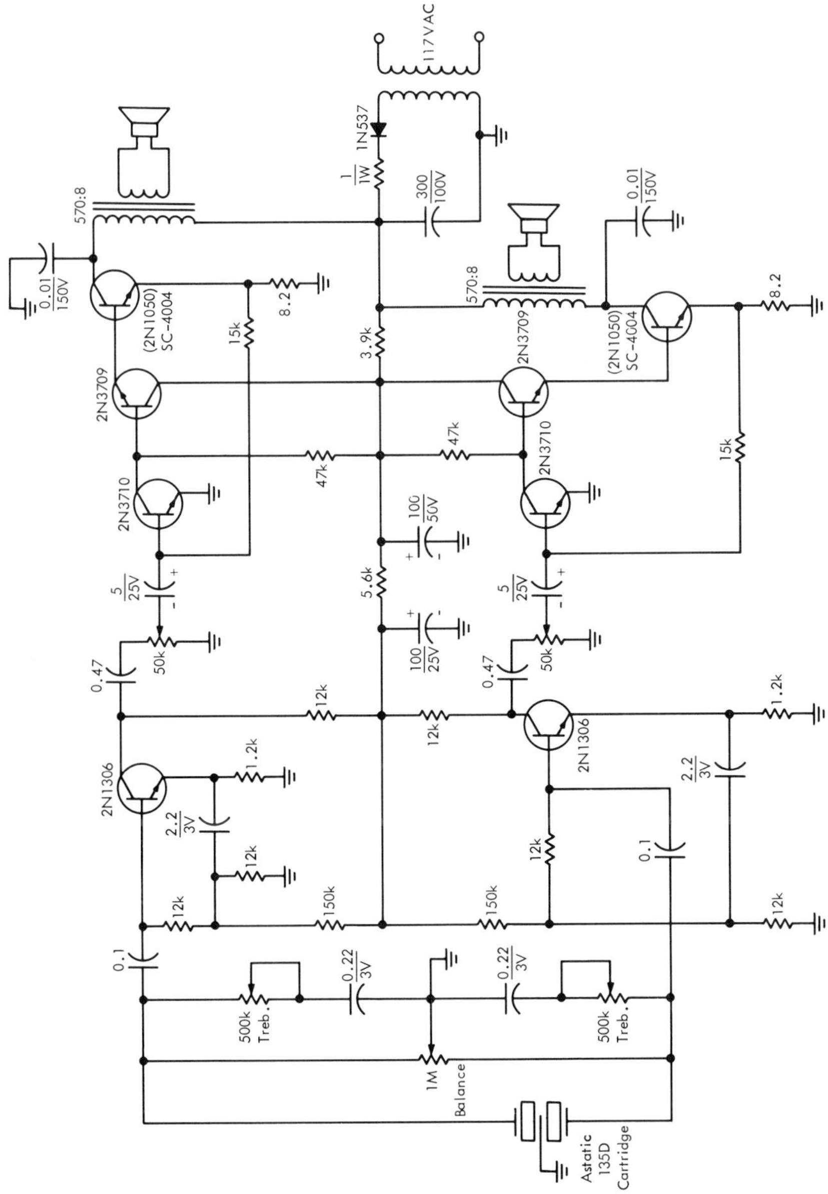

Three-watt-per-channel Stereo Amplifier Schematic

5

Class B Design Examples

The two design examples in this chapter illustrate the respective Class B procedures outlined in Chapter 3. Paragraph and equation numbers correspond to those used in Chapter 3. These examples represent a simple and direct approach to practical circuit design. Although some advantage may be gained by using more complex circuits, the basic design principles remain the same. Practical modifications which will provide performance advantages are suggested in the text. Selection of both the transistors and power output levels was made to show the versatility of the design procedures.

At the end of the chapter, additional Class B circuits are shown complete with component values and performance data.

FIFTEEN-WATT TRANSFORMER-COUPLED INPUT, TRANSFORMER-COUPLED OUTPUT

1. This example illustrates the use of the Class B transformer-driven, transformer-coupled output amplifier procedure. Only the power output of 15 W rms and frequency response of 20 Hz to 20 kHz need be specified for this design. The first step is to select the output transistors. The power dissipation requirements at the maximum temperature anticipated is approximated in Eq. (1).

$$\begin{aligned} P_{(Diss)}/\text{Device} &= 0.312\ P_{o(rms)} \\ &= 0.312\ (15\ W) = 4.7\ W \end{aligned} \tag{1}$$

2. Each transistor must be capable of dissipating 4.7 watts in service at the maximum case temperature T_1. A specific device capable of dissipating more than this amount of power rated at a lower temperature must be evaluated to determine T_1 for that device. The 2N1038 rated at 25°C for 20 watts is considered:

$$T_1 = \frac{P_{(Rate)} - P_{(Diss)} + K_1T_2}{K_1} \tag{2}$$

$$= \frac{20\ W - 4.7\ W + 0.267\ W/°C\ (25°C)}{0.267\ W/°C}$$

$$= 82.5°C$$

The 2N1038 is within its power rating for this application if the case temperature is not allowed to exceed 82.5°C with an appropriate heat sink. This can be done at ambient temperatures below about 55°C so the 2N1038 meets this requirement.

3. Now consider the voltage limitations of the device. Because the load impedance may be chosen to fit the supply voltage used in transformer-coupled output designs, the transistor voltage rating determines the voltage supply V_{cc} used. $V_{(BR)CER}$ and $V_{(BR)CEV}$ specifications from the data sheet should be considered from these relations:

$$V_{(BR)CER} > V_{cc} \tag{3}$$

$$30 > V_{cc}$$

$$V_{(BR)CEV} > 2\ V_{cc} \tag{4}$$

$$40 > 2\ V_{cc}$$

$$20 > V_{cc}$$

V_{cc} for the 2N1038 is primarily limited by the $V_{(BR)CEV}$ specification of 40 volts. This limits V_{cc} to an absolute maximum of 20 volts. 18 volts is chosen for this example to allow for power supply regulation tolerance.

4. The change in the base-to-emitter voltage ΔV_{BE} is an important consideration for bias circuit design. The 2N1038 is a germanium device and the 1.3 millivolts-per-degree C constant is multiplied by the expected temperature change to find ΔV_{BE}.

$$\Delta V_{BE} = K_2\ (T_1 - T_4) \tag{5}$$

$$= 1.3 \times 10^{-3}\ V/°C\ (82.5°C - 25°C)$$

$$= 0.075\ V$$

This 0.075-volt ΔV_{BE} is used in many of the following equations because the bias circuit design is dependent on it.

5. The reflected load impedance R_L', which is the load for each of the output-device collector circuits, is dependent primarily on the V_{cc} already selected and the power output specification. The transformer efficiency factor η is typically about 0.80, the value used in this design. $V_{CE(sat)}$ is the maximum value taken from the 2N1038 data sheet.

$$R_L' = \frac{(V_{cc} - 25\Delta V_{BE} - V_{CE(sat)})^2\ \eta}{2\ P_{o(rms)}} \tag{6}$$

$$= \frac{[18\ V - 25\ (0.075\ V) - 0.25\ V]^2\ 0.80}{2\ (15\ W)}$$

$$= 6.7\ \Omega$$

6. The peak collector current is computed with the same terms.

$$I_{cp} = \frac{2\ P_{o(rms)}}{(V_{cc} - 25\Delta V_{BE} - V_{CE(sat)})\ \eta} \tag{7}$$

$$= \frac{2\ (15\ W)}{[18\ V - 25\ (0.075\ V) - 0.25\ V]\ 0.80}$$

$$= 2.36\ A$$

7. The quiescent collector current is calculated in terms of I_{cp}.

$$I_{CQ} = 0.01\ I_{cp} \tag{8}$$

$$= 0.01\ (2.36\ A) = 23.6\ mA$$

8. The maximum collector current I_{CM} is the sum of I_{cp} and I_{CQ}. I_{CM} must be less than the maximum current rating of the 2N1038 on the data sheet.

$$I_{CM} = 1.01\ I_{cp} \tag{9}$$

$$= 1.01\ (2.36) = 2.38\ A$$

In this circuit I_{CM} is 2.38 amperes which is considerably below the 3-ampere maximum collector current rating.

9. The 2N1038 transistor can be confirmed from the foregoing calculations.

10. The base quiescent current I_{BQ} is determined in terms of I_{CQ}.

$$I_{BQ} = \frac{I_{CQ}}{h_{FE(min)}} \tag{10}$$

$$= \frac{23.6\ mA}{15} = 1.57\ mA$$

11. The collector cutoff current at the maximum operating temperature $I_{CBO(T_1)}$ is calculated based on the data sheet I_{CBO} specification rated at the maximum temperature T_3.

$$I_{CBO(T_1)} = I_{CBO(T_3)} 2^{(T_1 - T_3)/K_T} \tag{11}$$

$$= (125\ \mu A)\ 2^{(82.5°C - 25°C)/8°C}$$

$$= 18.3\ mA$$

12. The base bias current I_{BB} flowing through biasing resistors R_1 and R_2 is calculated.

$$I_{BB} = I_{BQ} + 10\ I_{CBO(T_1)} \tag{12}$$

$$= 1.57\ mA + 10\ (18.3\ mA)$$

$$= 185\ mA$$

13. The change in I_{CBO}, ΔI_{CBO}, is the difference between I_{CBO} at the maximum operating temperature and at the minimum temperature. Its value is used later to determine R_E.

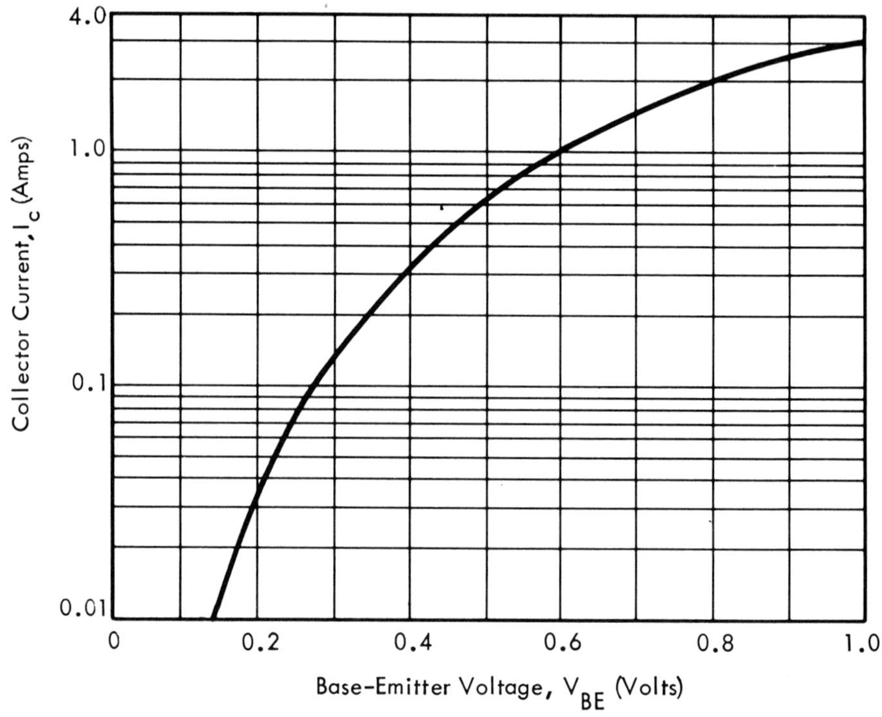

Fig. 1. Transconductance Characteristics y_{FE} for Device 2N1038

$$\Delta I_{CBO} = I_{CBO(T_1)} - I_{CBO(25°C)} \tag{13}$$
$$= 18.3 \text{ mA} - 0.125 \text{ mA}$$
$$= 18.2 \text{ mA}$$

The quiescent base-emitter voltage V_{BEQ} required at room temperature for I_{CQ} is determined from the 2N1038 y_{FE} curve. V_{BEQ} is read as 0.18 V at 23.6 milliamperes from Fig. 1.

$$V_{BEQ} = 0.18 \text{ V}$$

14. The resistor R_1 in the bias voltage divider required to develop V_{BEQ} at I_{BB} is determined.

$$R_1 \leq \frac{5V_{BEQ} + \Delta V_{BE}}{5I_{BB}} \tag{14}$$

$$\leq \frac{5\ (0.18 \text{ V}) + 0.075 \text{ V}}{5\ (0.185 \text{ A})}$$

$$\leq 1.05\ \Omega$$

15. The emitter resistor R_E minimum value is determined.

$$R_E \geq \frac{\Delta V_{BE} + \Delta I_{CBO} R_1}{5 I_{CQ}} \tag{15}$$

$$\geq \frac{0.075\ \text{V} + 0.0182\ \text{A}\ (1.05\ \Omega)}{5\ (0.0236\ \text{A})}$$

$$\geq 0.797\ \Omega$$

16. Total transformer primary impedance is usually specified to the transformer designer. This is the same value as the collector-to-collector load impedance R_{cc}.

$$R_{cc}' = 4R_L' \tag{16}$$

$$= 4\ (6.7\ \Omega) = 26.8\ \Omega$$

17. The largest value resistor R_2 in the base bias voltage divider is designed to supply the proper I_{BB} in Eq. (12).

$$R_2 = \frac{V_{cc}}{I_{BB}} - R_1 \tag{17}$$

$$= \frac{18\ \text{V}}{0.185\ \text{A}} - 1.05\ \Omega$$

$$= 96.5\ \Omega$$

18. The peak base current I_{bp} is calculated in terms of I_{cp} and minimum h_{FE} from the 2N1038 data sheet.

$$I_{bp} = \frac{I_{cp}}{h_{FE(min)}} \tag{18}$$

$$= \frac{2.36\ \text{A}}{15} = 157\ \text{mA}$$

19. The voltage drop across R_1 as a result of I_{bp} is now determined.

$$V_{R1} = I_{bp} R_1 \tag{19}$$

$$= 0.157\ \text{A}\ (1.05\ \Omega) = 165\ \text{mV}$$

20. Also, the voltage drop across the emitter resistor is computed.

$$V_{RE} = I_{cp} R_E \tag{20}$$

$$= 2.36\ \text{A}\ (0.797\ \Omega) = 1.88\ \text{V}$$

21. The maximum base-to-emitter voltage V_{BEM} is determined from Fig. 1, the 2N1038 y_{FE} curve. This voltage is the base-emitter voltage corresponding to the collector current I_{CM}.

$$V_{BEM} = 0.865\ \text{V}$$

22. The total peak voltage V_{ip}, required of each half of the driver transformer, is the sum of the voltage drops just determined.

$$V_{ip} = V_{R1} + V_{RE} + V_{BEM} - V_{BEQ} \tag{21}$$

$$= 0.165\ \text{V} + 1.88\ \text{V} + 0.865\ \text{V} - 0.18\ \text{V}$$

$$= 2.73\ \text{V}$$

23. The output transformer filter capacitor C_0 is determined by Eq. (22). The high-end frequency at which the relative output is to be down no more than 3 dB is selected for f. This capacitor C_0 is simply used to attenuate harmonics resulting from any residual nonlinearity in the output stage and is particularly helpful in reducing crossover distortion.

$$C_0 \leq \frac{1}{2\pi f R_{cc}} \tag{22}$$

$$\leq \frac{1}{2\pi(20 \times 10^3\,\text{Hz})\ 26.8\ \Omega}$$

$$\leq 0.297\ \mu\text{F}$$

The value of 0.297 μF is intended to be the maximum value. If distortion is not troublesome, the designer may elect to reduce the value or eliminate the capacitor.

24. The base input impedance Z_b is the ratio of signal voltage to current.

$$Z_b = \frac{V_{ip}}{I_{bp}} \tag{23}$$

$$= \frac{2.73\ \text{V}}{0.157\ \text{A}} = 17.4\ \Omega$$

The base-to-base input impedance Z_{bb} is significant because the driver transformer total secondary impedance, equal to Z_{bb}, is usually specified for the transformer design.

$$Z_{bb} = 4Z_b \tag{24}$$

$$= 4\ (17.4\ \Omega) = 69.6\ \Omega$$

25. The approximate power gain P.G. is calculated. Minimum, typical or maximum values can be computed by using respective min, typ, or max values of h_{FE} from the data sheet. Approximate minimum power gain is computed in Eq. (25).

$$\text{P.G.} \approx \frac{V_{cc} h_{FE} \eta}{V_{ip}} \tag{25}$$

$$\approx \frac{18\ \text{V}\ (15)\ 0.80}{2.73\ \text{V}}$$

$$\approx 79$$

26. The Class A driver stage can now be designed. The driver is usually decoupled with a single-section RC low-pass filter having also d-c voltage loss. In this example, a driver d-c supply voltage of 10 volts is assumed and the approximate collector load impedance $Z_{c(D)}$, which will be the primary of the driver transformer, is computed.

$$Z_{c(D)} = \frac{(V_{cc(D)} - 3V_{CE(sat)(D)})^2 \eta}{V_{ip} I_{bp}} \tag{26}$$

$$= \frac{[10\ V - 3\ (1\ V)]^2\ 0.80}{2.73\ V\ (0.157\ A)}$$

$$= 91.5\ \Omega$$

27. The peak driver current $I_{cp(D)}$ can now be computed in Eq. (27).

$$I_{cp(D)} = I_{bp} \sqrt{\frac{Z_b}{\eta Z_{c(D)}}} \tag{27}$$

$$= 0.157\ A \sqrt{\frac{17.4\ \Omega}{0.80\ (91.5\ \Omega)}}$$

$$= 76.5\ mA$$

28. The quiescent collector current $I_{CQ(D)}$ is established from $I_{cp(D)}$ plus a small overload current.

$$I_{CQ(D)} = I_{cp(D)} + 1.0 \times 10^{-3} A \tag{28}$$

$$= 76.5\ mA + 1.0\ mA = 77.5\ mA$$

29. The assumed 10-volt $V_{cc(D)}$ is confirmed as follows:

$$V_{cc(D)} = \frac{V_{ip} I_{bp}}{\eta I_{cp(D)}} + 3V_{CE(sat)(D)} \tag{29}$$

$$= \frac{2.73\ V\ (0.157\ A)}{0.80\ (0.0765\ A)} + 3\ (1\ V)$$

$$= 10.0\ V$$

30. Power output of the driver is computed for transformer design information.

$$P_{o(rms)(D)} = \frac{I_{cp(D)}^2\ Z_{c(D)}\ \eta}{2} \tag{30}$$

$$= \frac{(0.0765\ A)^2\ 91.5\ \Omega\ (0.80)}{2}$$

$$= 214\ mW$$

NINETY-FIVE WATT TRANSFORMER-COUPLED INPUT, RC-COUPLED OUTPUT

The following example illustrates the application of the Class B, transformer-driven, capacitor-coupled output amplifier procedure. The only two known parameters required are the required power output $P_{o(rms)}$ of 95 watts and the load impedance R_L of 4 ohms.

1. The first step is to select transistors for the output stage which are capable of fulfilling the requirements. The power dissipation for each output transistor is evaluated in Eq. (31).

$$P_{(Diss)}/\text{Device} = 0.25\ P_{o(rms)} \qquad (31)$$
$$= 0.25\ (95\ W) = 23.7\ W$$

2. Each output transistor must dissipate 23.7 watts at a maximum operating temperature T_1. A specific transistor must be considered which has a power rating of more than 23 watts unless the rating temperature T_2 is higher than T_1. The TI-3031, rated at 150 watts at 25°C case temperature T_2, is considered in Eq. (2).

$$T_1 = \frac{P_{(Rate)} - P_{(Diss)} + K_1 T_2}{K_1} \qquad (2)$$
$$= \frac{150\ W - 23.7\ W + 2\ W/°C\ (25°C)}{2\ W/°C}$$
$$= 88°C$$

The maximum permissible case temperature for the TI-3031 device while dissipating 23.7 watts is 88°C. A heat sink must be used to maintain the case temperature at or below this temperature while dissipating 23.7 watts. This is a reasonably easy task.

3. The change in base-to-emitter voltage ΔV_{BE} is determined over the operating temperature range for the germanium TI-3031 in Eq. (5).

$$\Delta V_{BE} = K_2\ (T_1 - T_4) \qquad (5)$$
$$= 1.3 \times 10^{-3}\ V/°C\ (88°C - 25°C)$$
$$= 0.082\ V$$

4. This base-to-emitter voltage change ΔV_{BE} is important to bias stability. Consequently, bias circuit design is based partly on it and ΔV_{BE} is considered when determining d-c collector supply voltage V_{cc}.

$$V_{cc} = \sqrt{8\ P_{o(rms)} R_L} + 50 \Delta V_{BE} + 2 V_{CE(sat)(max)} \qquad (32)$$
$$= \sqrt{8\ (95\ W)\ 4\ \Omega} + 50\ (0.082\ V) + 2\ (0.7\ V)$$
$$= 60.6\ V$$

5. The V_{cc} required for 95 watts output and a load impedance of 4 ohms is approximately 60 volts using the TI-3031 in this circuit. As a check to ensure that the voltage rating is not exceeded by the circuit requirements, the two following relations must be met or exceeded.

$$V_{(BR)CEV} > V_{cc} \qquad (33)$$
$$65\ V > 60\ V$$

$$V_{(BR)CER} > \frac{V_{cc}}{2} \qquad (34)$$

$$65\ V > \frac{60\ V}{2} > 30\ V$$

Since both $V_{(BR)CER}$ and $V_{(BR)CEV}$ are greater than 60 volts, the TI-3031 will be within its voltage ratings.

6. The maximum collector current of seven amperes should not be exceeded. Equation (35) computes peak collector current.

$$I_{cp} = \sqrt{\frac{2P_{o(rms)}}{R_L}} \qquad (35)$$

$$= \sqrt{\frac{2\ (95\ W)}{4\ \Omega}}$$

$$= 6.9\ A$$

7. The quiescent current is set to be a small portion of I_{cp}.

$$I_{CQ} = 0.01\ I_{cp} \qquad (8)$$

$$= 0.01\ (6.9\ A)$$

$$= 69\ mA$$

8. The maximum collector current I_{CM} must not exceed the maximum current rating for the device considered.

$$I_{CM} = 1.01\ I_{cp}$$

$$= 1.01\ (6.9\ A)$$

$$= 7\ A$$

The maximum current required does not exceed the seven-ampere rating of the TI-3031.

10. The quiescent base current I_{BQ} for later use in the bias network design is computed.

$$I_{BQ} = \frac{I_{CQ}}{h_{FE(min)}} \qquad (10)$$

$$= \frac{69\ mA}{30}$$

$$= 2.3\ mA$$

11. Also, collector cutoff current at the maximum operating temperature $I_{CBO\ (T_1)}$ must be considered in bias circuit design.

$$I_{CBO(T_1)} = I_{CBO(T_3)} 2^{(T_1 - T_3)/K_T} \qquad (11)$$

$$= (10\ mA)\ 2^{(88°C - 70°C)/8°C}$$

$$= 48\ mA$$

12. The total bias current I_{BB} requirement through resistors R_1 and R_2 is now determined.

$$I_{BB} = I_{BQ} + 10\ I_{CBO(T_1)} \qquad (12)$$

$$= 2.3\ mA + 10\ (48\ mA)$$

$$\approx 500\ mA$$

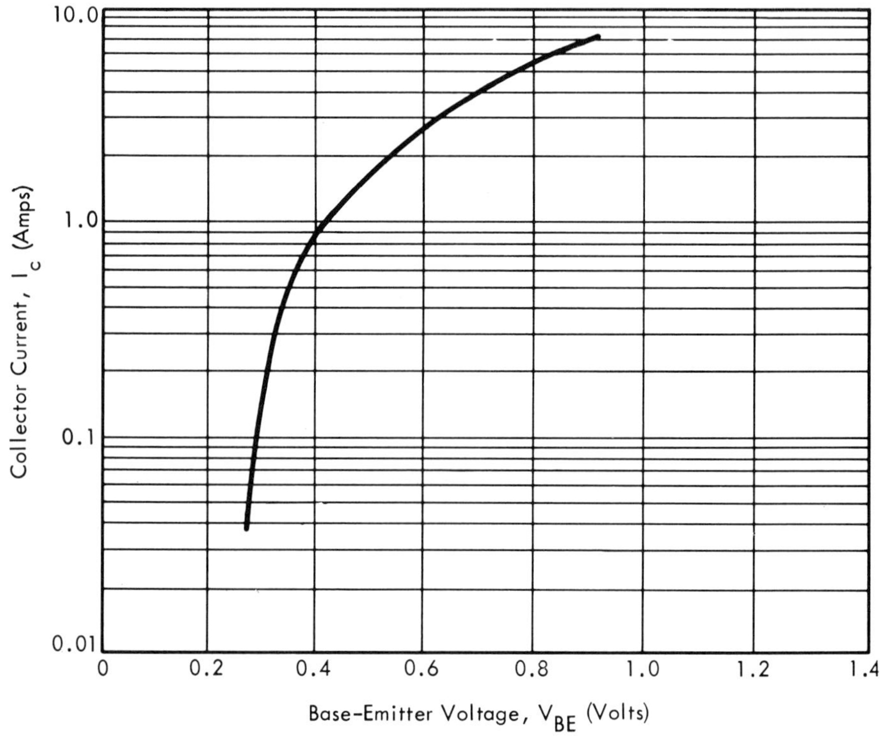

Fig. 2. Transconductance Characteristics y_{FE} for Device TI3031

13. The quiescent base voltage V_{BEQ} requirement is taken from the y_{FE} curve (Fig. 2) at I_{CQ}.

$$V_{BEQ} = 0.28 \text{ V}$$

14. The change in collector cutoff current I_{CBO} is also determined.

$$\Delta I_{CBO} = I_{CBO(T_1)} - I_{CBO(25°C)} \tag{13}$$

$$= 48 \text{ mA} - 1 \text{ mA} = 47 \text{ mA}$$

15. Now, after determining V_{BEQ}, I_{BB}, ΔV_{BE}, V_{cc}, I_{CQ} and ΔI_{CBO}, the biasing circuit can be designed. First, R_1, which develops the proper quiescent biasing voltage, is determined.

$$R_1 \leq \frac{5V_{BEQ} + \Delta V_{BE}}{5I_{BB}} \tag{14}$$

$$\leq \frac{5 \text{ (0.28 V)} + 0.082 \text{ V}}{5 \text{ (0.500 A)}}$$

$$\leq 0.593 \; \Omega$$

16. Now, R_2 is selected to supply the proper value of I_{BB}.

$$R_2 = \frac{V_{cc}}{2I_{BB}} - R_1 \quad (36)$$

$$= \frac{60\ \text{V}}{2\ (0.500\ \text{A})} - 0.59\ \Omega$$

$$= 59.4\ \Omega$$

17. The emitter resistor R_E is selected by considering ΔV_{BE} and ΔI_{CBO} at the quiescent collector current.

$$R_E \geq \frac{\Delta V_{BE} + \Delta I_{CBO} R_1}{5 I_{CQ}} \quad (15)$$

$$\geq \frac{0.082\ \text{V} + (0.047\ \text{A})\ 0.59\ \Omega}{5\ (0.069\ \text{A})}$$

$$\geq 0.318\ \Omega$$

18. The peak base current I_{bp} is determined for the drive circuit requirement.

$$I_{bp} = \frac{I_{cp}}{h_{FE(min)}} \quad (18)$$

$$= \frac{6.9\ \text{A}}{30} = 0.23\ \text{A}$$

19. The voltage drop across R_1, V_{R1} is also determined.

$$V_{R1} = I_{bp} R_1 \quad (19)$$

$$= 0.23\ \text{A}\ (0.59\ \Omega) = 0.136\ \text{V}$$

20. Also, V_{RE} the drop across R_E due to I_{cp} is calculated.

$$V_{RE} = I_{cp} R_E \quad (20)$$

$$= 6.9\ \text{A}\ (0.32\ \Omega) = 2.2\ \text{V}$$

21. Finally, V_{BEM}, the maximum base-to-emitter voltage required to drive the output transistors to their peak currents, is read from the y_{FE} curve in Fig. 2 at I_{CM}.

$$V_{BEM} = 0.90\ \text{V}$$

22. Now the peak input voltage V_{ip} required for each transistor can be determined.

$$V_{ip} = V_{R1} + V_{RE} + V_{BEM} - V_{BEQ} \quad (21)$$

$$= 0.136\ \text{V} + 2.2\ \text{V} + 0.90\ \text{V} - 0.28\ \text{V}$$

$$= 3.52\ \text{V}$$

23. The size of the output coupling capacitor C_c can be determined by assuming a minimum frequency f. Assume low end frequency at −3 dB of 20 Hz.

$$C_c \geq \frac{1}{2\pi f R_L} \tag{37}$$

$$\geq \frac{1}{2\ \pi\ (20\ \text{Hz})\ 4\ \Omega}$$

$$\geq 2000\ \mu\text{F}$$

24. The impedance of each of the driver transformer secondaries Z_b is calculated from the ratio of total voltage to current required.

$$Z_b = \frac{V_{ip}}{I_{bp}} \tag{23}$$

$$= \frac{3.52\ \text{V}}{0.23\ \text{A}} = 15.3\ \Omega$$

25. Finally, the power gain P.G. of the output stage is computed.

$$\text{P.G.} = \frac{V_{cc} h_{FE}}{2V_{ip}} \tag{38}$$

$$= \frac{60\ \text{V}\ (30)}{2\ (3.52\ \text{V})}$$

$$= 256$$

Now that the power gain of 256 is known for the output stage, all requirements for the driver stage can be determined.

26. It is usually desirable to decouple the driver from the output stage power supply by means of an RC decoupling network. In so doing, a voltage loss is encountered and the driver collector supply voltage $V_{cc(D)}$ is somewhat less than V_{cc}. In this example, a $V_{cc(D)}$ of 50 volts is assumed.

The Class A driver load impedance $Z_{c(D)}$ may now be computed. This is the primary impedance of the driver transformer. Transformer efficiency η is assumed to be 0.80.

$$Z_{c(D)} = \frac{(V_{cc(D)} - 3V_{CE(sat)(D)})^2 \eta}{V_{ip} I_{bp}} \tag{26}$$

$$= \frac{[50\ \text{V} - 3\ (1\ \text{V})]^2\ 0.80}{3.52\ \text{V}\ (0.23\ \text{A})}$$

$$= 2.18\ \text{k}\Omega$$

27. The Class A driver peak collector current $I_{cp(D)}$ can now be determined.

$$I_{cp(D)} = I_{bp} \sqrt{\frac{Z_b}{\eta Z_{c(D)}}} \tag{27}$$

$$= 0.23\ \text{A} \sqrt{\frac{15.3\ \Omega}{0.80\ (2180\ \Omega)}}$$

$$= 21.6\ \text{mA}$$

28. The driver quiescent collector current $I_{CQ(D)}$ may be determined. The additional milliampere ensures linear operation particularly when $I_{cp(D)}$ is very small.

$$I_{CQ(D)} = I_{cp(D)} + 1.0 \times 10^{-3} \text{A} \qquad (28)$$
$$= 21.6 \text{ mA} + 1.0 \text{ mA}$$
$$= 22.6 \text{ mA}$$

29. The rms power output of the driver $P_{o(rms)(D)}$ may be computed as a guide to selection of the driver transformer.

$$P_{o(rms)(D)} = \frac{I_{cp(D)}^2 Z_{c(D)} \eta}{2} \qquad (30)$$
$$= \frac{(0.0216 \text{ A})^2 \, 2180 \, \Omega \, (0.80)}{2}$$
$$= 400 \text{ mW}$$

ADDITIONAL CLASS B AMPLIFIERS

Seven additional Class B amplifier designs are shown in this section. Component values, device numbers, and performance data are given. Other data such as response and distortion curves and transformer winding information are given when available. All resistors are one-half watt ±10% and values are in ohms unless otherwise noted. All capacitor values are in microfarads unless otherwise noted.

500-MILLIWATT COMPLEMENTARY-SYMMETRY AMPLIFIER

Direct-coupled Input
RC-coupled Output
Output devices GC1165, GC1442

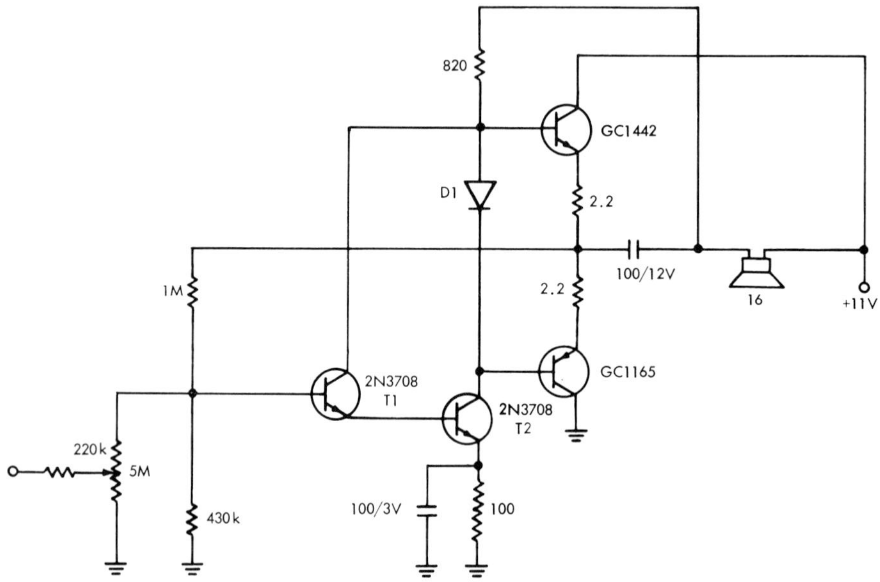

500-mW Complementary Symmetry Amplifier Schematic

Performance Data

Maximum Power Output: 500 mW
Sensitivity (50-mW Reference): 250 mV (220 kΩ input)
Distortion (50-mW Reference): 0.9%
(500-mW Reference): 5.0%
Frequency Response (1 kHz reference)
3 dB Low: 80 Hz
3 dB High: 20 kHz

500-MILLIWATT AMPLIFIER

Transformer-coupled Input
Direct-coupled Output
Output devices (2) 2N1038 without Heatsink

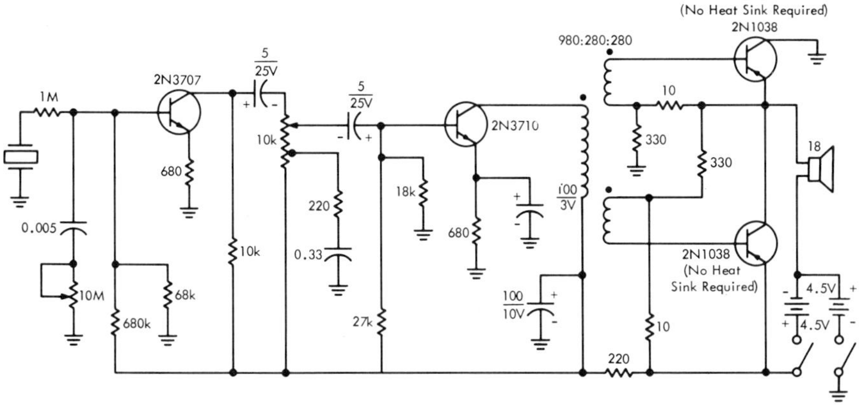

500-mW Amplifier Schematic

Performance Data

Maximum Power Output: 550 mW
Maximum Power (10% Distortion): 530 mW
Distortion (100-mW Reference): 2.9%
Frequency Response (1-kHz Reference)
3 dB Low: 150 Hz
3 dB High: 12,500 Hz

Driver Transformer

Bobbin: 33A
Core: EI-375
Stacking: 1 × 1
Winding Order: Primary First
Primary: 948 Ω, 1750 Turns #36 Single Soldereze, Random Wound
Secondaries: 280 Ω Split, 476 Turns #32 Single Soldereze, Bifilar Random Wound

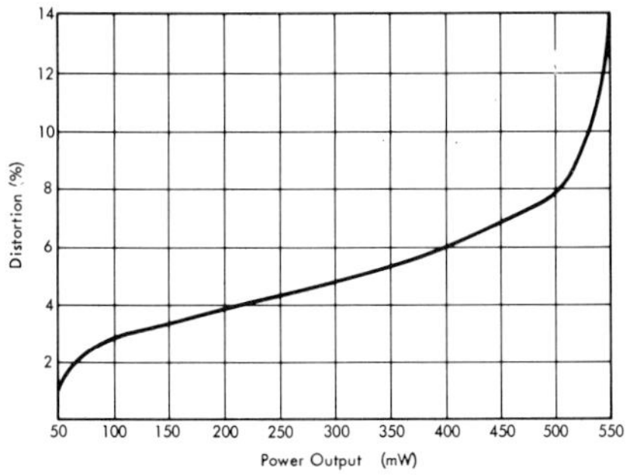

Power Output vs. Distortion

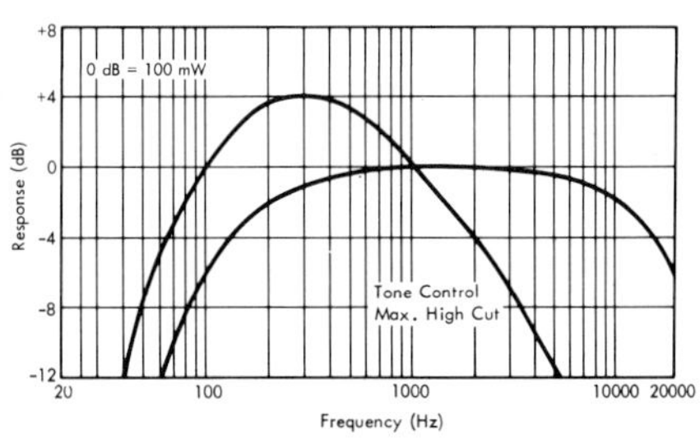

Frequency vs. Response

2-WATT AMPLIFIER

Direct-coupled Input
RC-coupled Output
Output Devices (2) 2N1907

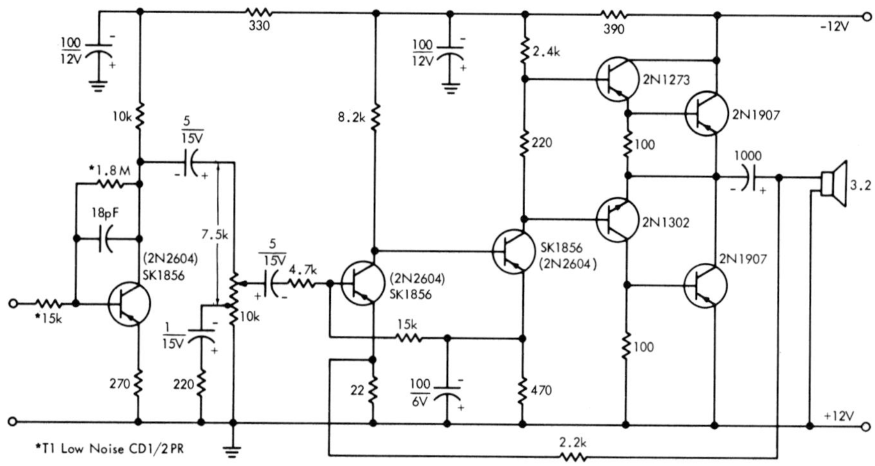

Two-watt Amplifier Schematic

Performance Data

Maximum Power Output: 2.5 W
Sensitivity (50-mW Reference): 0.8 mV
Distortion (2.5-W Reference): 1%
Frequency Response (1-kHz Reference)

3 dB Low: 10 Hz; 3 dB High: 35,000 Hz } 0 dB = 50 mW

3 dB Low: 13 Hz; 3 dB High: 35,000 Hz } 0 dB = 2 W

Distortion Vs. Frequency @ 50 mW: 1.1% @ 40 Hz
0.5% @ 120 Hz
0.7% @ 18,000 Hz

Distortion Vs. Frequency @ 2 W: 0.7% @ 60 Hz
0.6% @ 12,000 Hz

8-WATT AMPLIFIER

Transformer-coupled Input
RC-coupled Output
Output Devices (2) TI3027

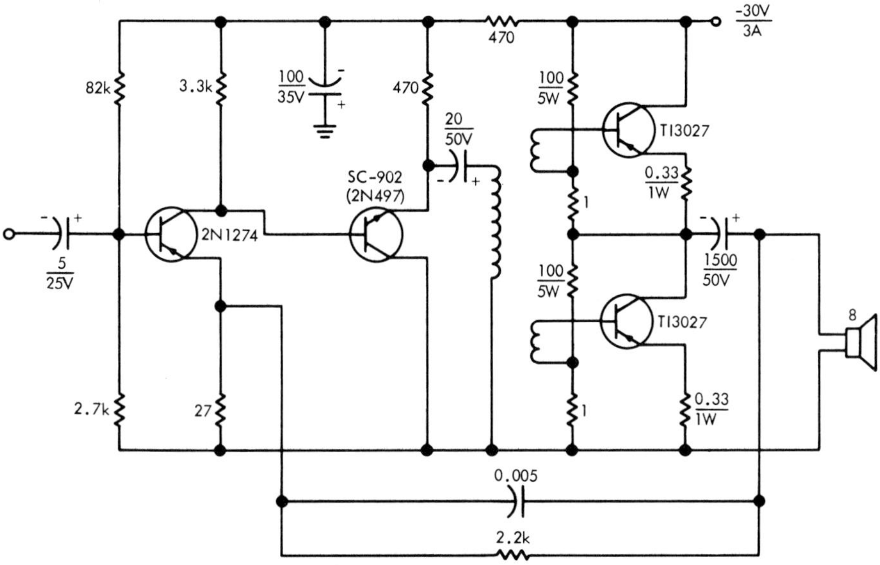

Eight-watt Amplifier Schematic

Performance Data

Maximum Power Output: 12 W
Maximum Power (6% Distortion): 12 W
Distortion (7-W Reference): 0.6%
Frequency Response (1-kHz Reference)
- 0 dB Low: 20 Hz
- 1.5 dB High: 20,000 Hz
- 0 dB Low: 20 Hz
- 0.5 dB High: 10,000 Hz

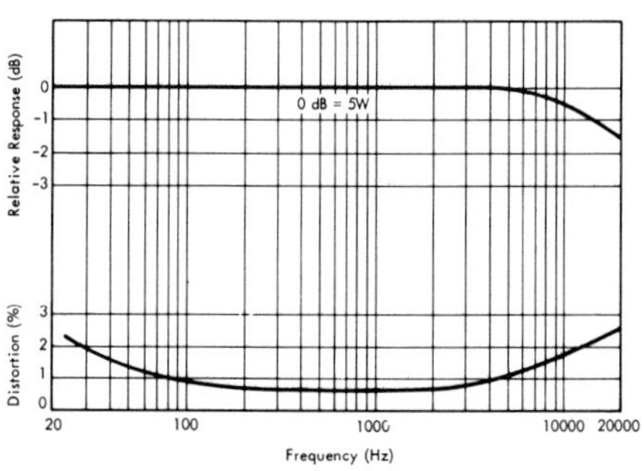

Frequency vs. Response and Distortion

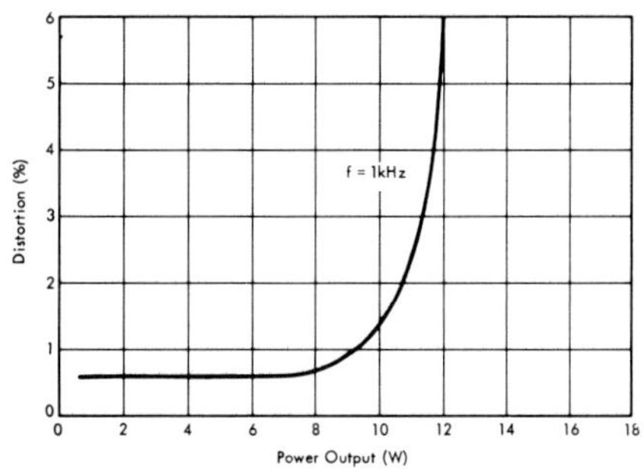

Power Output vs. Distortion

15-WATT AMPLIFIER

Transformer-coupled Input
Transformer-coupled Output
Output devices (2) 2N2984

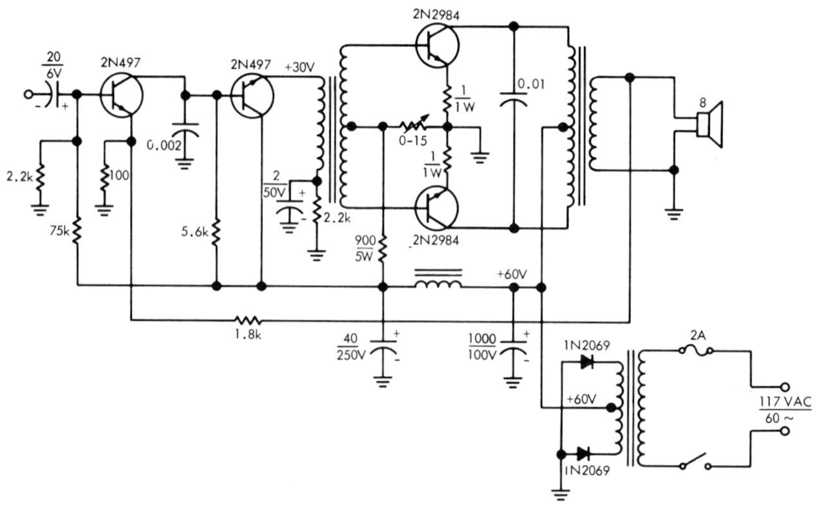

Fifteen-watt Amplifier Schematic

Performance Data

Maximum Power Output: 26 W
Maximum Power (10% Distortion): 20 W
Sensitivity (10-W Reference): 550 mV
Distortion (1-W Reference): 1.5%
Frequency Response (1-kHz Reference)
- 2.25 dB Low: 10 Hz
- 3 dB High: 85,000 Hz
- 0.5 dB Low: 15 Hz
- 0.5 dB High: 70,000 Hz

Driver Transformer

Bobbin: E-188
Core: EI-21
Primary: 1520 Turns #35 Single Soldereze, Random Wound
Secondary: 190 Turns #27 Single Soldereze, Bifilar Random Wound

Output Transformer

Bobbin: 5785
Core: EI-100
Primary: 400 Turns #26 Single Soldereze, Bifilar Random Wound
Secondary: 163 Turns #19 Single Soldereze, Random Wound

15-WATT AMPLIFIER (Continued)

Power Transformer

Bobbin: 5785
Core: EI-100
Primary: 500 Turns #35 Single Soldereze Random Wound
Secondary: 200 Turns #23 Single Soldereze, Bifilar Random Wound

Choke Coil

Bobbin: E-33A
Core: EI-75
Stacking: Butt
Primary: 1000 Turns #27 Single Soldereze, Random Wound

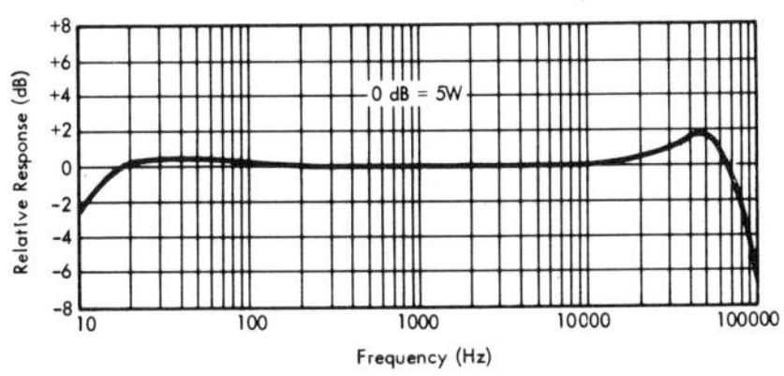

Frequency vs. Response

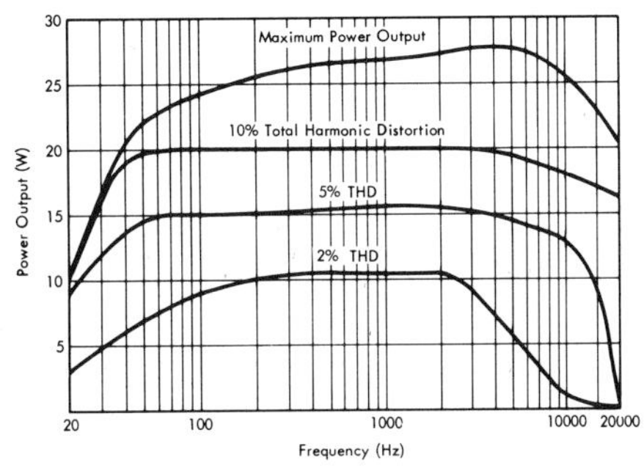

Frequency vs. Power Output

30-WATT AMPLIFIER

Direct-coupled Input
RC-coupled Output
Output devices (2) 2N1724

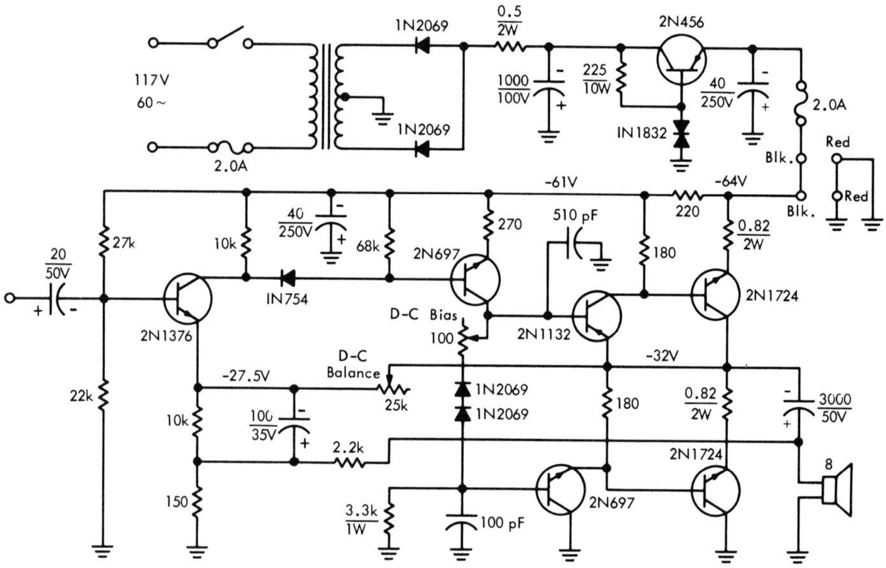

Thirty-watt Power Amplifier Schematic

Performance Data

Maximum Power Output: 30 W
Sensitivity (30-W Reference): 1.5 V
Distortion (30-W Reference): 0.08%
Frequency Response (1-kHz Reference)

1 dB Low: 8 Hz 1 dB High: 100,000 Hz	0 dB = 30 W
1 dB Low: 6 Hz 1 dB High: 430,000 Hz	0 dB = 5 W

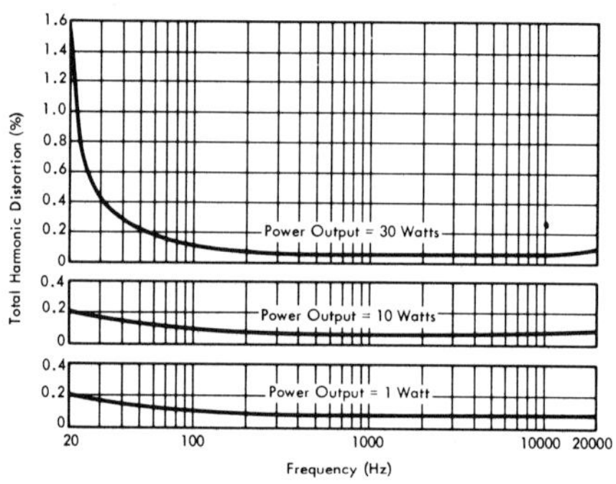

Frequency vs. Distortion

50-WATT AMPLIFIER

Transformer-coupled Input
RC-coupled Output
Output Devices (2) TI3030

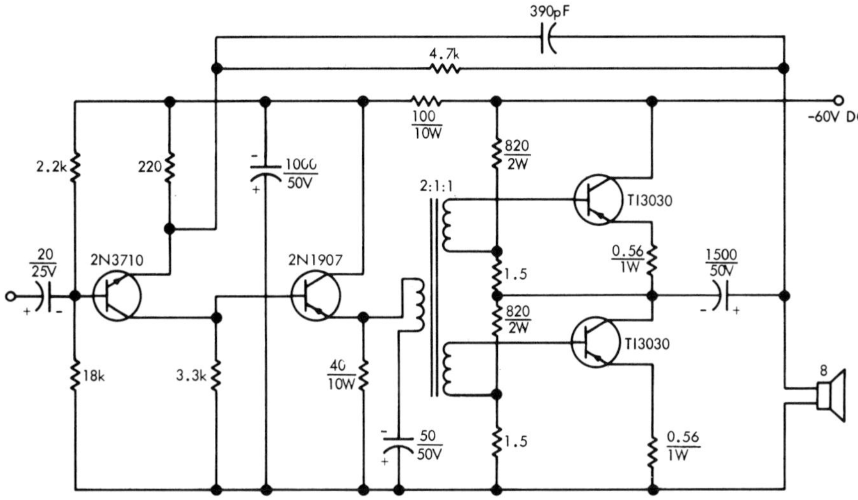

Fifty-watt Amplifier Schematic

Performance Data

Maximum Power Output: 50 W
Maximum Power (1.65% Distortion): 50 W
Sensitivity (50-W Reference): 1.5 V
Distortion (10-W Reference): 0.7%
Frequency Response (1-kHz Reference)
1.6 dB Low: 30 Hz
0.18 dB High: 20,000 Hz

Driver Transformer

Bobbin: E-188
Core: EI-21
Stacking: 1 × 1
Primary: 434 Turns #30 Single Soldereze, Quadfilar Wound
Secondary: 217 Turns #28 Single Soldereze, Quadfilar Wound

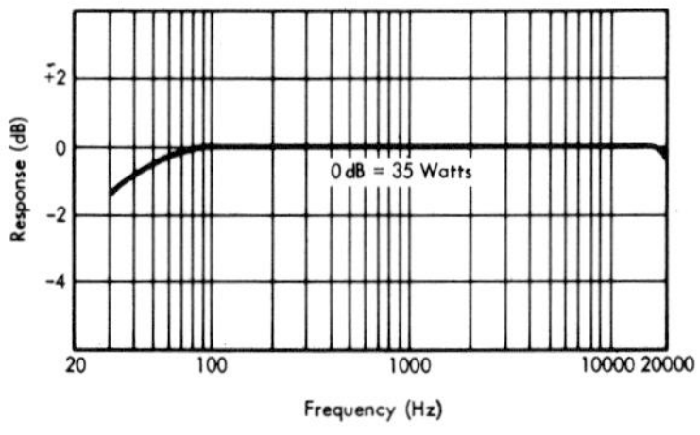

Frequency vs. Response

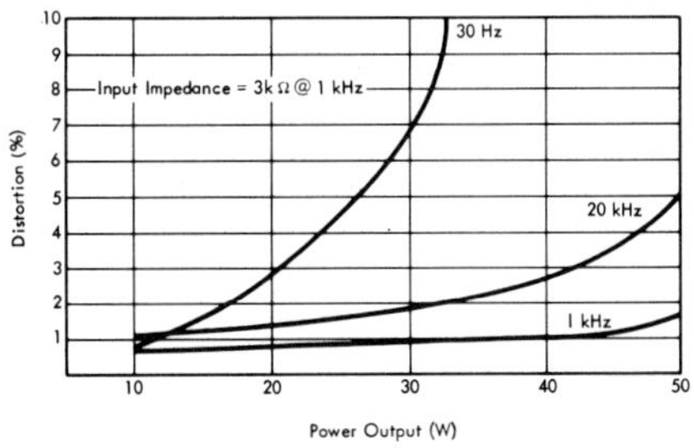

Power Output vs. Distortion

6

Audio Design Equation Derivations

CLASS A (CHAPTER 2)

Equation 1 $P_{(Diss)} = 3.12\ P_{o(rms)}$

Equation 36 $P_{(Diss)} = 2.5\ P_{o(rms)}$

The power output is given in terms of the peak collector parameters.

$$P_{o(rms)} = \frac{V_{cep}\ I_{cp}}{2}$$

The approximation is made:

$$V_{cep} \approx V_{CEQ}$$

Since:

$$I_{CQ} = 1.25\ I_{cp}$$

The Class A stage should be biased for 25% more current than I_{cp}. Hence:

$$P_{o(rms)} = \frac{V_{CEQ}\ I_{CQ}}{1.25\ (2)}$$

$$= 0.40\ V_{CEQ}\ I_{CQ}$$

Maximum device dissipation occurs at zero signal.

$$P_{(Diss)} = V_{CEQ}\ I_{CQ} = 2.5\ P_{o(rms)}$$

For transformer coupled circuits where $\eta = 0.8$:

$$P_{o(rms)} = \frac{V_{CEQ}\ I_{CQ}\ \eta}{1.25\ (2)}$$

$$P_{(Diss)} = \frac{2.50}{\eta}\ P_{o(rms)} = 3.12\ P_{o(rms)}$$

Equation 2 $$T_1 = \frac{P_{(Rate)} - P_{(Diss)} + K_1 T_2}{K_1}$$

$$\Delta \text{ Power} = \Delta \left[\frac{\text{Power}}{^{\circ}\text{C Change}}\right] [\Delta T]$$

Data sheet ratings are used in the equation and the units are kept consistent:

$$P_{(Rate)} \text{ (mW)} - P_{(Diss)} \text{ (mW)} = [K_1 \text{ (mW/}^{\circ}\text{C)}]\ [T_1 \text{ (}^{\circ}\text{C)} - T_2 \text{ (}^{\circ}\text{C)}]$$

The equation is rearranged to solve for the maximum device temperature rating that must not be exceeded while dissipating the required full-load power:

$$T_1 = \frac{P_{(Rate)} - P_{(Diss)} + K_1 T_2}{K_1}$$

Equation 5 $$R_L' = \frac{(V_{cc} - 8\Delta V_{BE} - V_{CE(sat)})^2\ \eta}{2P_{o(rms)}}$$

$$P_{o(rms)} = \frac{(V_{cep})^2\ \eta}{2\ R_L'}$$

$$R_L' = \frac{(V_{cep})^2\ \eta}{2\ P_{o(rms)}}$$

$$V_{cep} = V_{cc} - V_{RE} - V_{CE(sat)}$$

R_E is defined in terms of the temperature-dependent parameters calculating a value for R_E that will not allow more than a 10% change in I_{CQ}:

$$0.1\ I_{CQ}\ R_E = \Delta\ V_{BE} + \Delta\ I_{CBO}\ R_1$$

I_{BB} has been calculated to make the $\Delta I_{CBO} R_1$ contribution to the total V_{BEQ} insignificant, so the equation can be approximated by:

$$R_E = \frac{10\ \Delta\ V_{BE}}{I_{CQ}}$$

Previously defined was:

$$I_{CQ} = 1.25\ I_{cp}$$

Hence:

$$R_E = \frac{8\ \Delta\ V_{BE}}{I_{cp}}$$

and by substitution:

$$V_{RE} = I_{cp}\ R_E = 8\ \Delta\ V_{BE}$$

Then:

$$R_L' = \frac{(V_{cc} - 8\ \Delta V_{BE} - V_{CE(sat)})^2\ \eta}{2P_{o(rms)}}$$

Equation 13

$$R_E \geq \frac{100\ \Delta V_{BE} + V_{BEQ}}{10\ I_{CQ}}$$

R_E is defined as the resistance necessary to achieve quiescent stability over the operating temperature range. The two sources of voltage change due to temperature are $\Delta\ V_{BE}$ and $\Delta\ I_{CBO}\ R_1$. The general form of the equation is:

$$I_{CQ}\ R_E = \Delta\ V_{BE} + \Delta\ I_{CBO}\ R_1$$

The $\Delta\ I_{CBO}$ is contained in the base bias equation. This bias (I_{BB}) has been calculated to prevent the I_{CBO} leakage current from causing greater than a 10% change in the collector quiescent current I_{CQ}. If we then assume that we are to calculate a value of R_E that will maintain I_{CQ} within 10% of its nominal value, the approximate equation will be equal to:

$$I_{CQ}\ R_E = 10\ \Delta\ V_{BE} + \Delta\ I_{CBO}\ R_1$$

From the bias requirement:

$$R_1 = \frac{I_{CQ}R_E + V_{BEQ}}{10\ I_{CBO(T_1)}}$$

Then

$$I_{CQ}R_E = 10\Delta V_{BE} + \Delta I_{CBO}\left[\frac{I_{CQ}R_E + V_{BEQ}}{10\ I_{CBO(T_1)}}\right]$$

The assumption is made

$$I_{CBO(T_1)} \approx \Delta I_{CBO}$$

Then R_E is approximated by

$$R_E \geq \frac{100\ \Delta V_{BE} + V_{BEQ}}{10\ I_{CQ}}$$

Equation 17

$$C_1 \geq \frac{I_{cp}}{2\pi f K_4}$$

The bypass capacitor should be calculated to allow no more than three dB reduction in power at the lowest desired frequency.

At the -3 dB power point:

$$I_{cp} \text{ at } -3 \text{ dB} = 0.707\ I_{cp} \text{ at } 0 \text{ dB}$$

R_E and R_1 are bypassed with the capacitor:

$$R_E + R_1 \gg r_e$$

Since the components in parallel with the capacitor are large in relation to the series element, the two series elements must be assumed equal in order to achieve the 0.707 reduction in collector current.

$$X_c = r_e$$

If the emitter current is assumed almost equal to the collector current, the following relationship is valid:

$$r_e = \frac{K_4}{I_{cp}}$$

where K_4 is a constant dependent upon the transistor material.

K_4 values—

$$\text{silicon} = 35 \times 10^{-3}$$
$$\text{germanium} = 25 \times 10^{-3}$$

Then:

$$X_c = \frac{1}{2\ \pi\ f\ C_1} = \frac{K_4}{I_{cp}}$$

Or:

$$C_1 \geq \frac{I_{cp}}{2\ \pi\ f\ K_4}$$

Equation 23

$$V_{cc(D)} = \frac{V_{bep}\ I_{bp}}{\eta I_{cp(D)}} + 3V_{CE(sat)(D)}$$

The supply voltage for transformer-coupled Class A circuits is given in the following general equation when the voltage drop across R_E is ignored. This is permissible because the collector current in a driver stage is normally very low.

$$V_{cc(D)} = V_{cep(D)} + V_{CE(sat)(D)}$$

The peak load voltage V_{cep} may be expressed as shown below:

$$V_{cep(D)} = \frac{V_{bep}\ I_{bp}}{\eta\ I_{cp(D)}}$$

If a safety factor is assumed for the saturation specification $V_{CE(sat)}$ to allow for the base overdrive in the manufacturer's saturation voltage test, the supply voltage can be calculated by completing the substitution:

$$V_{cc(D)} = \frac{V_{bep}\ I_{bp}}{\eta I_{cp(D)}} + 3V_{CE(sat)(D)}$$

Equation 25

$$R_i = \frac{h_{FE}\ R_o}{h_{FE} + y_{fe}\ R_o}$$

The voltage gain is given by the following equations:

$$A_v = \frac{V_{cep}}{V_{ip}}$$

$$= y_{fe}\ Z_c$$

Hence:

$$V_{ip} = \frac{V_{cep}}{y_{fe} Z_c}$$

The input impedance may be expressed by:

$$h_{ie} = \frac{V_{ip}}{I_{bp}}$$

$$= \frac{h_{FE} V_{cep}}{y_{fe} Z_c I_{cp}}$$

The load collector voltage is the product of the impedance and the collector current:

$$V_{cep} = Z_c I_{cp}$$

Hence:

$$h_{ie} = \frac{h_{FE}}{y_{fe}}$$

The bias resistors R_1 and R_2 represent a parallel load to the a-c signal:

$$R_o = \frac{R_1 R_2}{R_1 + R_2}$$

The input resistance then may be calculated in terms of R_o:

$$R_i = \frac{h_{ie} R_o}{h_{ie} + R_o}$$

$$= \frac{h_{FE} R_o}{h_{FE} + y_{fe} R_o}$$

Equation 30 $$V_{CEM(D)} = 3(V_{cep(D)} + V_{CE(sat)(D)})$$

In Class A amplifier collector circuits:

$$V_{CEM(D)} = V_{CEQ(D)} + V_{cep(D)}$$

For Class A driver operation, it is normal practice to provide for a 50% overload in the maximum voltage to prevent any distortion due to high $V_{CE(sat)}$, R_E drop, or parameter variations.

$$V_{CEM(D)} = 2\ V_{CEQ} + V_{cep(D)}$$

The quiescent voltage is calculated with the assumption that the actual saturation voltage in the circuit will be 3 $V_{CE(sat)}$ due to the base overdrive condition in the manufacturer's saturation voltage test:

$$2\ V_{CEQ} = 2\ V_{cep(D)} + 3\ V_{CE(sat)(D)}$$

Hence:

$$V_{CEM(D)} = (2\ V_{cep(D)} + 3\ V_{CE(sat)(D)}) + V_{cep(D)}$$
$$= 3(V_{cep(D)} + V_{CE(sat)(D)})$$

Equation 46 $V_{cc} = 2\sqrt{2P_{o(rms)}R_L} + I_{CQ}R_E + 2V_{CE(sat)}$

The collector supply voltage in direct-coupled amplifiers may be expressed as the sum of the collector series elements. The peak voltage V_{cep} is about one-half V_{cc} so the supply voltage equation becomes:

$$V_{cc} = 2\ V_{cep} + I_{CQ}\ R_E + V_{CE(sat)}$$

The peak load voltage is found in the following equation:

$$P_{o(rms)} = \frac{(V_{cep})^2}{2\ R_L}$$
$$V_{cep} = \sqrt{2\ P_{o(rms)}\ R_L}$$

V_{cep} is substituted in the V_{cc} equation and the circuit saturation voltage is assumed to be 2 $V_{CE(sat)}$:

$$V_{cc} = 2\sqrt{2\ P_{o(rms)}\ R_L} + I_{CQ}\ R_E + 2\ V_{CE(sat)}$$

Equation 50 $P.G._{(T)} = \dfrac{(h_{FE(D)}h_{FE})^2 R_L}{R_{i(D)}}$

The power gain of the driver:

$$P.G._{(D)} = \frac{P_{o(rms)(D)}}{P_{i(rms)(D)}} = h_{FE(D)}{}^2\ \frac{R_{L(D)}}{R_{i(D)}}$$

The power gain of the power output stage:

$$P.G. = \frac{P_{o(rms)}}{P_{i(rms)}} = h_{FE}{}^2\ \frac{R_L}{R_i}$$

The total power gain is calculated by:

$$P.G._{(T)} = \frac{P_{o(rms)}}{P_{i(rms)(D)}} = \frac{P.G.\ (P_{i(rms)})\ P.G._{(D)}}{P_{o(rms)(D)}}$$

When the driver is delivering power to the power output stage, the driver output power is equal to the base input power of the power output stage. Therefore:

$$P.G._{(T)} = (P.G.)\ (P.G._{(D)})$$
$$= \left(h_{FE}{}^2\ \frac{R_L}{R_i}\right)\left(h_{FE(D)}{}^2\ \frac{R_{L(D)}}{R_{i(D)}}\right)$$

Because the two stages have a common input and output point, the driver output resistance is about equal to the input resistance of the power output stage. Hence:

$$P.G._{(T)} = \frac{(h_{FE(D)} h_{FE})^2 R_L}{R_{i(D)}}$$

CLASS B (CHAPTER 3)

Equation 1 $\quad P_{(Diss)}/\text{Device} = 0.312\ P_{o(rms)}$

Equation 31 $\quad P_{(Diss)}/\text{Device} = 0.25\ P_{o(rms)}$

The instantaneous voltage and current equations:

$$v_c = V_{cc}\ (1 - A \sin \omega t)$$

$$i_c = \frac{V_{cc}}{R_L} A \sin \omega t$$

The power dissipation is given by integrating the product:

$$P_{(Diss)}/\text{Device} = \frac{1}{2\ \pi} \int_0^{\pi} V_{cc}\ (1 - A \sin \omega t)\ \frac{V_{cc}}{R_L} A \sin \omega t\ d\omega t$$

$$= \frac{V_{cc}^2\ A}{R_L\ \pi} \left(1 - \frac{A\ \pi}{4}\right)$$

The maximum power dissipation point is found by taking the derivative of the $P_{(Diss)}$/Device and setting it equal to zero.

$$\frac{d\ P_{(Diss)}/\text{Device}}{d\ A} = \frac{V_{cc}^2}{\pi\ R_L} - \frac{A\ V_{cc}^2}{2\ R_L} = 0$$

Maximum power dissipation occurs when

$$A = 0.636$$

Hence:

$$P_{(Diss)}/\text{Device} = \frac{V_{cc}^2}{\pi^2\ R_L}$$

The power output is related to the same V_{cc} by the equation:

$$P_{o(rms)} = \frac{V_{cc}^2}{2\ R_L}$$

The $P_{o(rms)}$ equation may be substituted into the $P_{(Diss)}$ equation.

$$P_{(Diss)}/\text{Device} = \frac{2\ P_{o(rms)}}{\pi^2} = 0.203\ P_{o(rms)}$$

The small quiescent current will require additional power dissipation capability. Therefore, the total $P_{(Diss)}$/Device for RC-coupled circuits is approximated by

$$P_{(Diss)}/\text{Device} = 0.25\ P_{o(rms)}$$

For transformer-coupled circuits

$$P_{o(rms)} = \frac{V_{cc}^{2}\ \eta}{2\ R_{L}'}$$

The $P_{o(rms)}$ equation may be substituted into the $P_{(Diss)}$ equation.

$$P_{(Diss)}/\text{Device} = \frac{2\ P_{o(rms)}}{\pi^{2}\ \eta} = \frac{0.203}{\eta}\ P_{o(rms)}$$

The small quiescent current will require additional power dissipation capability. Therefore, the total $P_{(Diss)}$/Device is approximated by

$$P_{(Diss)}/\text{Device} = \frac{0.250}{0.80}\ P_{o(rms)} = 0.312\ P_{o(rms)}$$

Equation 2

$$T_1 = \frac{P_{(Rate)} - P_{(Diss)} + K_1\ T_2}{K_1}$$

$$\Delta\ \text{Power} = \Delta\left[\frac{\text{Power}}{{}^{\circ}\text{C Change}}\right]\ [\Delta T]$$

Data sheet ratings are used in the equation and the units are kept constant:

$$P_{(Rate)}\ (\text{mW}) - P_{(Diss)}\ (\text{mW}) = [K_1\ (\text{mW}/{}^{\circ}\text{C})]\ [T_1({}^{\circ}\text{C}) - T_2({}^{\circ}\text{C})]$$

The equation is rearranged to solve for the maximum device temperature rating that must not be exceeded while dissipating the required full-load power:

$$T_1 = \frac{P_{(Rate)} - P_{(Diss)} + K_1\ T_2}{K_1}$$

Equation 14

$$R_1 \leq \frac{5V_{BEQ} + \Delta V_{BE}}{5I_{BB}}$$

The voltage across resistance R_1 is approximated by

$$V_{R1} = R_1\ I_{BB}$$

The voltage V_{R1} is also defined in terms of the power output device:

$$V_{R1} = V_{BEQ} + I_{CQ}\ R_E$$

The resistance R_E for Class B circuits has been defined as

$$R_E = \frac{\Delta V_{BE} + \Delta I_{CBO} R_1}{5 I_{CQ}}$$

The voltage V_{R1} may be calculated by substituting for R_E:

$$V_{R1} = V_{BEQ} + I_{CQ} \left(\frac{\Delta V_{BE} + \Delta I_{CBO} R_1}{5 I_{CQ}} \right)$$

When the first equation for V_{R1} is substituted, R_1 may be defined as

$$R_1 = \frac{5 V_{BEQ} + \Delta V_{BE} + \Delta I_{CBO} R_1}{5 I_{BB}}$$

Solving for R_1:

$$R_1 = \frac{5 V_{BEQ} + \Delta V_{BE}}{5 I_{BB} - \Delta I_{CBO}}$$

$$I_{BB} \gg \Delta I_{CBO}$$

Hence:

$$R_1 \leq \frac{5 V_{BEQ} + \Delta V_{BE}}{5 I_{BB}}$$

Equation 25 $$P.G. = \frac{V_{cc} h_{FE} \eta}{V_{ip}}$$

Equation 38 $$P.G. = \frac{V_{cc}\, h_{FE}}{2 V_{ip}}$$

The general expression for power gain is

$$P.G. = \frac{P_{o(rms)}}{P_{i(rms)}}$$

The power output is

$$P_{o(rms)} = \frac{V_{cep}\, I_{cp}\, \eta}{2}$$

V_{cep} is assumed to be almost equal to V_{cc} for transformer-coupled output Class B circuits:

$$P_{o(rms)} = \frac{V_{cc}\, I_{cp}\, \eta}{2}$$

The total power input to the base is dependent upon the single secondary winding power requirements since each winding is delivering power only half the time:

$$P_{i(rms)} = \frac{V_{ip}\, I_{bp}}{2}$$

For transformer-coupled output Class B circuits, the power gain equation is then

$$\text{P.G.} = \frac{\dfrac{V_{cc}\, I_{cp}\, \eta}{2}}{\dfrac{V_{ip}\, I_{bp}}{2}}$$

$$= \frac{V_{cc}\, h_{FE}\, \eta}{V_{ip}}$$

For capacitor-coupled output Class B circuits, V_{cep} is assumed to be almost equal to $V_{cc}/2$. This assumption makes the power gain equation

$$\text{P.G.} = \frac{V_{cc}\, h_{FE}}{2\, V_{ip}}$$

Equation 32 $\quad V_{cc} = \sqrt{8\, P_{o(rms)} R_L} + 50\Delta V_{BE} + 2V_{CE(sat)}$

The general form of the power output equation is

$$P_{o(rms)} = \frac{V_{cep}^{\ 2}}{2\, R_L}$$

The peak collector-to-emitter voltage is assumed to be equal to $V_{cc}/2$:

$$V_{cep} = \frac{V_{cc}}{2} - V_{RE} - V_{CE(sat)}$$

The voltage V_{RE} must be determined before solving for V_{cc}:

$$V_{RE} = I_{cp}\, R_E$$

The resistance R_E has been defined for Class B operation according to the following equation:

$$R_E = \frac{\Delta\, V_{BE} + \Delta\, I_{CBO}\, R_1}{5\, I_{CQ}}$$

In Class B circuits

$$I_{CQ} = 0.01\, I_{cp}$$

Then:

$$R_E = \frac{100\, (\Delta\, V_{BE} + \Delta\, I_{CBO}\, R_1}{5\, I_{cp}}$$

The resistance R_1 is derived in Eq. 14.

$$R_1 = \frac{5\, V_{BEQ} + \Delta\, V_{BE}}{5\, I_{BB}}$$

The resistance R_1 is substituted into the R_E equation.

$$I_{cp}\ R_E = \frac{100\left[\Delta\ V_{BE} + \Delta\ I_{CBO}\left(\frac{5\ V_{BEQ} + \Delta\ V_{BE}}{5\ I_{BB}}\right)\right]}{5}$$

or

$$V_{RE} = 100\left[\frac{5\ I_{BB}\ \Delta\ V_{BE}}{25\ I_{BB}} + \frac{5\ \Delta\ I_{CBO}\ V_{BEQ}}{25\ I_{BB}}\right]$$

The following approximations are assumed:

$$I_{BB} \geq 10\ \Delta\ I_{CBO}$$

$$V_{BEQ} \leq 2.5\ \Delta\ V_{BE}$$

The R_E equation now assumes the following form after simplifying:

$$V_{RE} = 20\ \Delta\ V_{BE} + 5\ \Delta\ V_{BE} = 25\ \Delta\ V_{BE}$$

The peak collector voltage now becomes

$$V_{cep} = \frac{V_{cc}}{2} - 25\ \Delta\ V_{BE} - V_{CE(sat)}$$

The power output is then derived from substituting the peak collector voltage in the original power equation.

$$P_{o(rms)} = \frac{(V_{cc} - 50\ \Delta\ V_{BE} - 2\ V_{CE(sat)})^2}{8\ R_L}$$

The collector supply voltage may now be obtained by rearranging the power equation:

$$V_{cc} = \sqrt{8\ P_{o(rms)}R_L} + 50\Delta V_{BE} + 2V_{CE(sat)}$$

PART II

AM/FM Design

7

AM IF Amplifier Design

This chapter discusses the design of tuned narrow-bandwidth IF amplifiers using high-frequency transistors and coupling transformers between stages. Gain, stability, bandwidth, and large-signal handling capability for both single- and double-tuned low-frequency amplifiers ($f_{IF} \leq f_T/100$) are considered.

A step-by-step solution to the problem of designing unneutralized IF amplifiers is presented. The method used to achieve this solution has distinct advantages in that the degree of stability is indicated and large-signal handling capability is ensured—advantages lacking in earlier methods using the MAG and mismatch loss concepts.

This simpler method can now be used for low-frequency amplifiers because present high-frequency transistors are so improved that the low-frequency parameters can be used for design calculations (except y_{re}, the feedback parameter). This method is applicable for Texas Instruments Dalmesa®, Amperex PADT®, some mesa transistors, and to a lesser extent, the drift transistors. It is not applicable to mesa transistors with low h_{FE}.

Symbols used throughout the text are defined near the end of the chapter. Boldface numbers are used to designate equations appearing for the first time. Where equations are repeated in design examples, lightface numbers are used.

TRANSISTOR MODEL

A complex model of a transistor and the terminations is shown in Fig. 1. This model may be reduced to the simpler model in Fig. 2 if it can be shown that some of these parameters are negligible in their effect on gain and bandwidth of the amplifier.

The output admittance g_{oe} may be neglected if (1) the emitter current is reasonably low (less than 2.5 milliamps, typically), (2) the low-frequency current gain is not too high (less than 200), and (3) the f_T is greater than 100 f_{IF}.

The parameter R_b' may be neglected if it is less than one-tenth of R_{be}. This

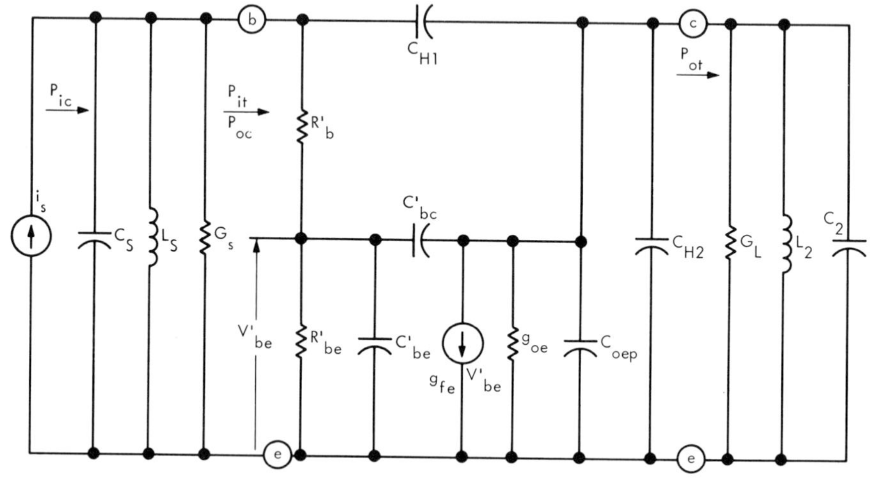

Fig. 1. Complex Model of Transistor and Terminations

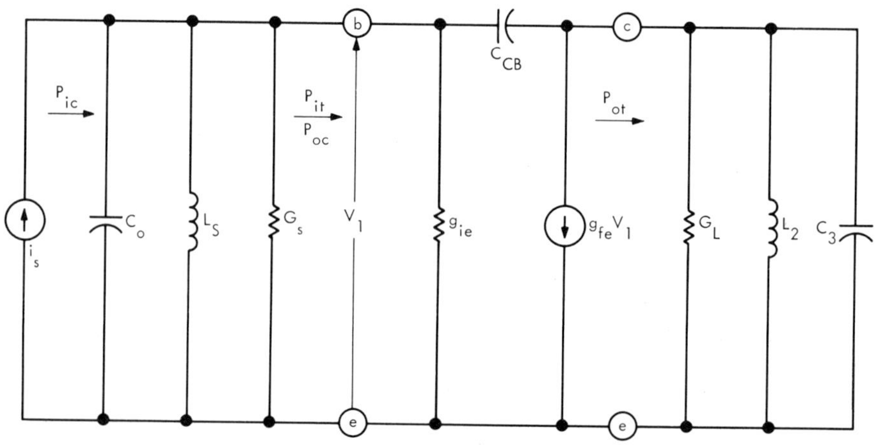

Fig. 2. Simplified Model of Transistor and Terminations

is generally true if R_b' is reasonably low, the beta (h_{fe}) is greater than 20, and the emitter current is less than 2.5 milliamperes. When these conditions are met, g_{ie} is approximately given by

$$g_{ie} = \frac{1}{R_{be}'} \tag{1}$$

Under these same conditions, the output current generator is approximately $y_{fe}V_1$. The complex model may be further simplified by combining capacitances.

Thus: $C_o = C_{be}' + C_S$, $C_{CB} = C_{H1} + C_{bc}'$, and $C_3 = C_{oep} + C_2$. The parameter C_{CB} is approximately equal to the well known C_{OB} for transistors which have the header isolated and ungrounded. If, however, the header is isolated and grounded, $C_{CB} = C_{OB} - C_{H1}$. Most of the transistors previously mentioned are such that $C_{CB} = C_{OB}$.

The power gain of the transistor described by the model in Fig. 2 is

$$\frac{P_{ot}}{P_{it}} = \frac{|y_{fe}|^2}{G_L g_{ie}} \tag{2}$$

Equation (2) assumes no reverse signal transfer (i.e., the effects of C_{CB} are neglected). In the discussion that follows, the terminations are chosen so that the effect of C_{CB} on power gain is small, typically less than one decibel.

SINGLE-TUNED CIRCUITS

Transformer Loss. The transformer (shown simplified as a parallel tuned circuit) has a loss which is a function of the ratio of Q_U to Q_L. The power loss of the transformer is

$$\frac{P_{oc}}{P_{ic}} = 1 - \frac{Q_L}{Q_U} \tag{3}$$

A simple derivation of this equation is shown under Equation Derivations at the end of this chapter.

Stage Gain. The power gain of the stage may be expressed by

$$\frac{P_o}{P_i} = \frac{P_{ot}}{P_{ic}} = \frac{P_{ot}}{P_{it}}\frac{P_{oc}}{P_{ic}}$$

$$= \frac{|y_{fe}|^2}{G_L g_{ie}}\left(1 - \frac{Q_L}{Q_U}\right) \tag{4}$$

If the gain is expressed in decibels,

$$\frac{P_o}{P_i} = 10 \log \frac{|y_{fe}|^2}{G_L g_{ie}} + 10 \log \left(1 - \frac{Q_L}{Q_U}\right) \tag{5}$$

Stable Stage Gain. With the equation for stage gain determined in terms of transistor parameters, load admittance, and coil losses, it is necessary to assure that the amplifier is free from regeneration and the possibility of oscillation. The object is to find the input and output termination that assures stability. Stern[1/] has defined unconditional stability in terms of the two-port admittance parameters by

$$(g_{ie} + G_S)\ (g_{oe} + G_L) > \frac{L + M}{2} \tag{6}$$

If an arbitrary constant σ (stability factor) is added to the inequality of Eq. (6), it may be expressed as an equality.

$$(g_{ie} + G_S)\ (g_{oe} + G_L) = \frac{L + M}{2\sigma} \tag{7}$$

The stability factor σ always has a value less than one if the amplifier is stable.

L and M may be stated in terms of transistor common-emitter admittance parameters as

$$M + jN = y_{re}y_{fe} \tag{8}$$

$$L = |M + jN| = |y_{re}y_{fe}| \tag{9}$$

If L and M are expressed in terms of the parameters of the model in Fig. 2, the stability relationship may be rewritten as

$$R_1R_2 = \frac{2\sigma}{\omega C_{CB}} \frac{0.026}{I_E} \tag{10}$$

where R_1 and R_2 are defined by

$$R_1 = \frac{1}{g_{ie} + G_S}$$

$$R_2 = \frac{1}{g_{oe} + G_L} \approx \frac{1}{G_L}$$

This assumes that y_{fe} is equal to its ideal value of $I_E/0.026$. This equation may again be modified to establish a relationship between the actual termination and a critical termination ($R_1'R_2'$). For a given current condition and a fixed termination, the amplifier is at its least stable condition where C_{CB} is at its maximum value $C_{CB(max)}$. If this critical termination is defined as

$$R_1'R_2' = \frac{0.026}{\omega I_E C_{CB(max)}}, \tag{11}$$

then

$$R_1R_2 = 2\sigma R_1'R_2' \tag{12}$$

This establishes stability with respect to the critical termination for a single stage. When more stages are cascaded, the R_1R_2 product must be reduced by a factor of 2, 2.61, and 3 for 2, 3, and 4 stages respectively.[2] For a two-stage amplifier which is normally used in portable receivers, Eq. (12) is therefore:

$$R_1R_2 = \sigma R_1'R_2' \tag{13}$$

This relationship for stability may be substituted in Eq. (4), and the result is a power gain equation in terms of forward transfer admittance, stability factor, frequency, and maximum collector-base capacitance. The stable power gain equation is

$$\frac{P_o}{P_i} = \frac{|y_{fe}|\sigma}{\omega C_{CB(max)}} \tag{14}$$

The development of this relationship is shown under Equation Derivations.

In choosing a value for σ, the desired symmetry of bandwidth and overall gain must be considered. Tolerable asymmetry results if σ is 0.4 or less. Final choice of the value for σ depends on the desired stage gain. In terms of frequency, $C_{CB(max)}$, y_{fe}, and desired power gain, σ is expressed as

$$\sigma = \frac{|\text{Power Gain}|\ \omega C_{CB(max)}}{|y_{fe}|} \tag{15}$$

Bandwidth. The Q_L is related to the bandwidth by the following equation:

$$BW_{3\ dB}\ \text{Single} = \frac{f_{IF}}{Q_L} \tag{16}$$

If n stages of equal Q_L are used, the overall 3-dB bandwidth is related to single-stage 3-dB bandwidth by

$$\frac{BW_{3\ dB}\ \text{Overall}}{BW_{3\ dB}\ \text{Single}} = (2^{1/n} - 1)^{1/2} \tag{17}$$

After Eq. (16) is substituted in Eq. (17), it may be manipulated to express the Q_L of each transformer in terms of the center frequency, overall 3-dB bandwidth, and number of stages.

$$Q_L = (2^{1/n} - 1)^{1/2} \frac{f_{IF}}{BW_{3\ dB}\ \text{Overall}} \tag{18}$$

If the overall 6-dB bandwidth is given, then Eq. (18) becomes

$$Q_L = (2^{1/n} - 1)^{1/2} \frac{1.56\ f_{IF}}{BW_{6\ dB}\ \text{Overall}} \tag{18a}$$

Another means of selecting the Q based on the desired bandwidth is shown in Fig. 7.

Large-Signal Handling Ability. With large input signals, an IF amplifier must be capable of passing the signal with a minimum of distortion, if the signal is amplitude modulated. With proper design techniques, the signal handling capabilities may be extended considerably before limiting occurs.

If it is assumed that the transistor is linear over the entire operating region, including the large-signal condition, a good rule of thumb which can be used is that no more than a peak voltage of 26 millivolts can be applied to the base of the transistor without distorting the modulation. This can be shown as follows: Let y_{fe} equal its theoretical value for linear operation of $I_E/0.026$. Assume further that the collector load is expressed in terms of the collector-emitter voltage and collector current as $R_L = V_{CE}/I_C$. Then

$$E_o = V_{be} y_{fe} R_L$$

or

$$E_o = V_{be} \left(\frac{I_E}{0.026}\right) \left(\frac{V_{CE}}{I_C}\right)$$

Since $I_C \approx I_E$, E_0 has its maximum values of V_{CE} when $V_{be} = 0.026$. This condition can be summarized by the following statement for the class of amplifier considered here. If the output amplifier or any fixed gain amplifier is designed to handle large signals, the output is limited to a peak value of $I_C R_L$ and the input to 26 millivolts.

If it is desirable to include the effect of R_b' at high input levels, the approximate drop across R_b' may be added to the 26 millivolts. However, if the peak amplitude applied to the base is limited at 26 millivolts (for $R_L \leq V_{CE}/I_C$) by the driving source or previous stage, no limiting can take place. Therefore, this limiting condition (Eq. (19)) is used in the following development.

$$R_L \leq \frac{V_{CE}}{I_C} \tag{19}$$

If the input stage is controlled for automatic gain purposes, its large-signal handling ability must be considered when the current is at some specified minimum I_E'. The input stage should be designed to limit at the same time that the output stage limits. The collector load impedance for the input stage is found by determining the voltage level on the collector when the current is at the chosen minimum and dividing it by I_E'. The peak collector voltage on the gain controlled stage is 26 millivolts multiplied by the turns ratio of the transformer between the stages.

The collector load impedance is approximately

$$R_L = \left(\frac{N2}{N3}\right)\frac{0.026}{I_E'} \tag{20}$$

This can be expressed in terms of the turns ratios and total tuned impedance by using the equations for the input impedance of the transformer and the above relationship. The turns ratios are defined in Fig. 3.

$$\frac{N1}{N2} = \frac{Q_L \omega L_1 I_E'}{0.026}\left(\frac{N3}{N1}\right) \tag{21}$$

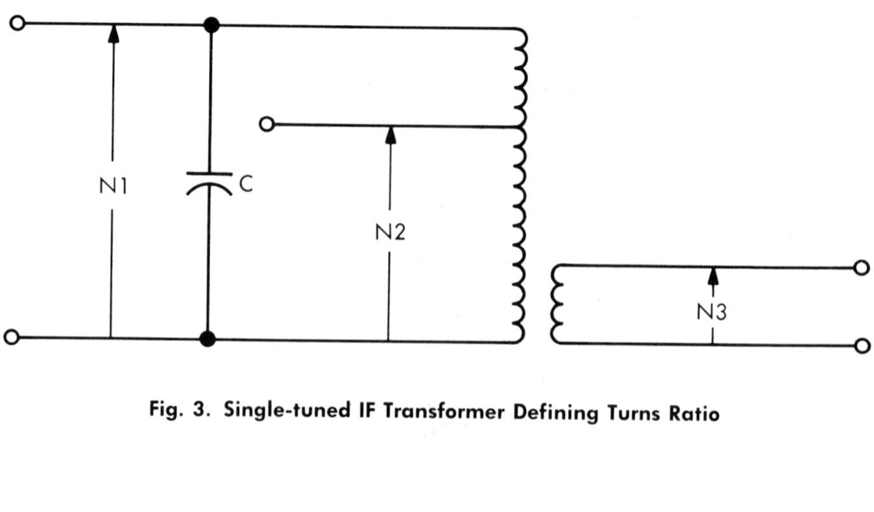

Fig. 3. Single-tuned IF Transformer Defining Turns Ratio

The choice of I_E' depends on how much change in gain is required. This is expressed by

$$I_E' \approx \frac{I_E}{\Delta (P_o/P_i)} \tag{22}$$

Design Method. The method discussed here uses the principles already set forth for the design of a two-stage single-tuned IF amplifier. The first stage is gain-controlled by decreasing the collector current. The amplifier has input, interstage, and output transformers.

In the equations that are developed, the second subscripts on Q, turns N, and inductance L refer to the position of the transformer in the circuit. For example, N11/N31 refers to the N1/N3 ratio of the input transformer.

The starting point is the third or output transformer. The N1/N3 turns ratio is

$$\frac{N13}{N33} = \left[\frac{Q_{U3}\,(Q_{L3}\,\omega L_{13})}{(Q_{U3} - Q_{L3})\,R_D}\right]^{1/2} \tag{23}$$

The load on the output stage is usually a germanium diode whose dynamic impedance, R_D, is approximately 1500 ohms.

The quantity of $Q_{L3}\omega L_{13}$ is shown as a magnitude to emphasize that this applies at the resonant frequency. In many cases, L is dictated by transformer manufacturers for economic reasons. The choice of Q_{U3} and Q_{L3} depends on the bandwidth and amount of loss between the last stage and the detector. In most cases, it is desirable to have high efficiency; therefore, Q_{U3} is chosen as high as economics and physical size permit. A low value of Q_L may also be chosen and thus the output transformer will not contribute to the selectivity.

The N1/N2 ratio for the output transformer is given by

$$\frac{N13}{N23} = \left[\frac{(Q_{L3}\omega L_{13})\,I_C}{0.8\,V_{CE}}\right]^{1/2} \tag{24}$$

The safety factor, 0.8, is included here to prevent a "bottoming" effect described in Reference 3. A sacrifice in maximum power output is made to prevent this effect under large-signal conditions. The choice of V_{CE} depends on the supply voltage and the drop across the emitter resistor. The choice of I_C depends on how much power output is needed to operate the AGC and drive the audio amplifier. The efficiency of the detector must also be considered. A useful equation for selecting I_C, if the coil loss of the output transformer is one dB or less, is

$$I_C \approx \frac{10\,P_{mo}}{V_{CE}} \tag{25}$$

This equation assumes that P_{mo} is the maximum rms power that is delivered to the detector when the carrier is 100 per cent modulated.

The turns ratio N1/N3 for the interstage is

$$\frac{N12}{N32} = \left[\frac{(Q_{L2}\omega L_{12})\,0.8\,V_{CE}}{\sigma_2\,(R_1'R_2')\,I_C}\right]^{1/2} \tag{26}$$

The determination of this turns ratio is based on the desired stability, bandwidth, and stage gain. Eq. (26) is derived by expressing the turns ratio in terms of the tuned impedance reflected to primary and secondary. The quantity, R_1, is found by solving Eq. (13) for R_1 after expressing R_2 in terms of V_{CE} and I_C of the output stage. The quantity $R_1'R_2'$ is found by solving Eq. (11). The stability factor is determined from Eq. (15) and Q_{L2} from Eq. (18).

The N1/N2 turns ratio of interstage is expressed by

$$\frac{N12}{N22} = \frac{I_E'\,(Q_{L2}\omega L_{12})}{0.026}\left(\frac{N32}{N12}\right) \tag{27}$$

Here the turns ratio is chosen for maximum signal handling of the interstage amplifier. This is a simple restatement of Eq. (21) with subscripts added. The minimum current I_E' is given by Eq. (22).

Before calculating turns ratios for the input transformer, inductance and capacitance for the stage must be chosen. In some cases, the same inductance may be chosen for each of the transformers. In this case

$$\frac{N11}{N21} = \left(\frac{Q_{L1}\omega L_{11}}{R_{in}}\right)^{1/2} \tag{28}$$

However, in cases where it is not desirable to tap the primary, the tuned impedance, $Q_{L1}\omega L_{11}$ should equal the desired input impedance of the amplifier.

Because of standard values of capacitance, it may be necessary to alter the desired input impedance slightly. In this case

$$R_{in} = Q_{L1}\omega L_{11} = Q_{L1}/\omega C_1 \tag{29}$$

The N1/N3 turns ratio may now be found by using the following relationship which is developed under Equation Derivations at the end of this chapter. This equation is derived in the same manner as Eq. (26).

$$\frac{N11}{N31} = \frac{0.026}{I_E'}\left[\frac{N12}{N32}\right]\left[\frac{Q_{L1}\omega L_{11}}{Q_{L2}\omega L_2\ \sigma_1(R_1'R_2')}\right]^{1/2} \tag{30}$$

The remaining step is to find Q_{U1} and Q_{U2}. This can be done by using the input admittance of the transistor, the Q_L, and N1/N3 turns ratio. The quantity R_1 is expressed in terms of $Q_L\omega L$ and the turns ratio.

$$R_1 = \frac{Q_L\omega L}{\left(\frac{N1}{N3}\right)^2} \tag{31}$$

Equation (32) is developed in the section on Equation Derivations. Since minimum gain is usually specified, the maximum value of g_{ie} is chosen at the quiescent current level. This value can be obtained from the data sheet or by measuring a sample of the transistor type to be used. Equation (32) then takes the form of

$$Q_U = \frac{Q_L\left[\frac{1}{g_{ie(max)}}\right]}{\frac{1}{g_{ie(max)}} - R_1} \tag{32}$$

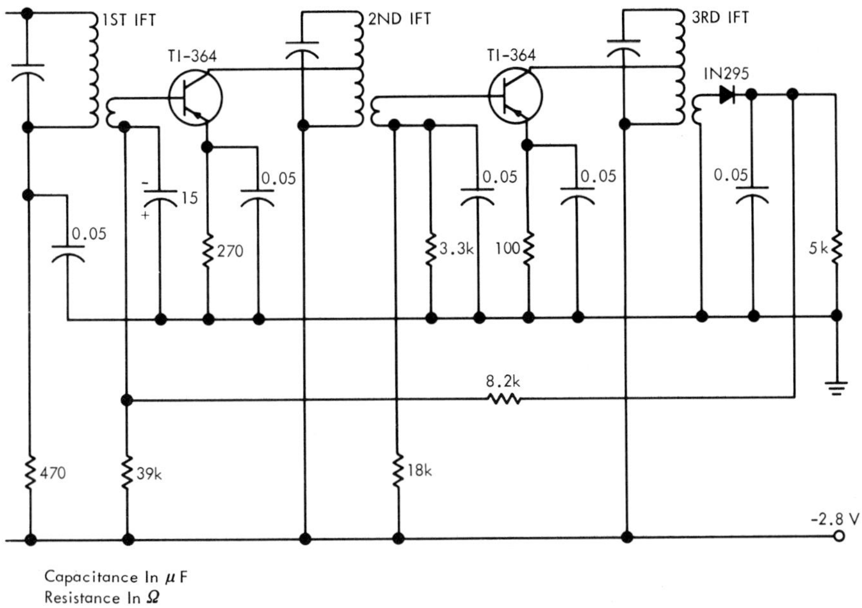

Fig. 4. Typical Single-tuned IF Amplifier Circuit

This completes the information required to design the amplifier. The gain and bandwidth are known. The degree of stability and the ability to handle the maximum input signal have been considered.

Design Example. To demonstrate the use of this method, an example will be shown for the design of a typical IF amplifier for a pocket size portable radio which derives its energy from two small pen light cells.

The desired properties of the amplifier shown in Fig. 4 are:

Total IF gain = 55 dB
Supply Voltage = 2.8 volts
Maximum IF power from output transformer = 0.25 mW
Center frequency = 455 kHz
Total 6 dB bandwidth = 10 kHz
Number of stages = 2
Source impedance = greater than 500 kΩ (neglect its effect on bandwidth)
Input impedance = Approximately 25 kΩ
IF load = 1N295 diode
Change in power gain for AGC = 35 dB

The output IF will be made very efficient and the Q_L will be 15 so it does not contribute greatly to the selectivity. The efficiency is made high so that the detector load will have a high ac/dc ratio for handling large percentages of modulation without diagonal clipping. A one-dB loss will be assumed. Note that one-dB loss = 0.79. From Eq. (3),

$$0.79 = 1 - \frac{Q_L}{Q_U} \tag{3}$$

$$Q_U = \frac{15}{0.21} = 72$$

A standard ½-inch IF transformer was selected which has a tuning capacitance equal to 190 pF. For 455 kHz, the inductance is 640 μH.

$$Q_{L3}\omega L_{13} = 15\ (2\pi)\ (455 \times 10^3)\ (640 \times 10^{-6})$$
$$= 27300$$

Therefore, from Eq. (23)

$$\frac{N13}{N33} = \left[\frac{72\ (27300)}{(72 - 15)\ 1500}\right]^{1/2} = 4.8 \tag{23}$$

Before calculating N13/N23, the I_C must be chosen. The collector current is chosen to give the specified maximum power into the output transformer. From Eq. (25), therefore,

$$I_C = \frac{10\ (0.25 \times 10^{-3})}{2.5} = 10^{-3}\ \text{A} \tag{25}$$

The collector emitter voltage will be approximately 2.5 volts after a small drop across the emitter bias resistor is subtracted from the 2.8-volt supply. Now, from Eq. (24)

$$\frac{N13}{N23} = \left[\frac{(27300)\ 10^{-3}}{0.8\ (2.5)}\right]^{1/2} = 3.69 \tag{24}$$

Before designing the interstage transformer, the stage gain Q_L and type of transistor must be selected. The TI-364 transistor is used since it has a low collector-base capacitance which will provide greater stability for a given power gain. The data sheet specifies that $C_{CB(max)}$ is 2.5 pF at V_{CB} equal to 9 volts. At a V_{CB} equal to approximately 2.5 volts, the $C_{CB(max)}$ is 4.7 pF. Also, for $I_C = 1$ mA, $y_{fe} = 35 \times 10^{-3}$ S. The stage gain is divided equally between the stages. Therefore,

$$\text{Stage gain (dB)} = \frac{\text{Total Gain + Output Transformer Coil Loss}}{\text{No. of Stages}}$$

$$= \frac{55 + 1}{2} = 28\ \text{dB}$$

The stability factor can now be found for each stage using Eq. (15). Note that 28 dB gain = 630.

$$\sigma = \frac{630\ (2\pi)\ (455 \times 10^3)\ (4.7 \times 10^{-12})}{35 \times 10^{-3}} \tag{15}$$

$$= 0.25$$

The critical termination impedance is, from Eq. (11)

$$R_1'R_2' = \frac{0.026}{2\pi\ (455 \times 10^3)\ 10^{-3}\ (4.7 \times 10^{-12})} \qquad (11)$$

$$= 1.92 \times 10^6\ \Omega^2$$

And the Q_L is, from Eq. (18a)

$$Q_L = (2^{1/2} - 1)^{1/2}\ \frac{1.56\ (455 \times 10^3)}{10 \times 10^3} \qquad (18a)$$

$$= 45.5$$

Again selecting a tuning capacitance equal to 190 pF, from Eq. (26)

$$\frac{N12}{N32} = \left[\frac{45.5\ (1820)\ 0.8\ (2.5)}{0.25\ (1.92 \times 10^6)\ 10^{-3}}\right]^{1/2} \qquad (26)$$

$$= 18.5$$

The value of I_E' is dictated by the AGC requirements and no-signal emitter current of the first stage. Since the supply voltage and stage gain are the same for the first stage as the last, the no-signal current will be the same. If $\Delta(P_o/P_i)$ is equal to 35 dB (56), then from Eq. (22)

$$I_E' = \frac{1000\ \mu A}{56} = 17.8\ \mu A, \qquad (22)$$

and using Eq. (27)

$$\frac{N12}{N22} = \frac{(17.8 \times 10^{-6})\ (45.5 \times 1820)}{0.026\ (18.5)} \qquad (27)$$

$$= 3.05$$

Before N1/N3 is calculated for the input transformer, the value of the tuning inductance must be chosen. In this case, it is undesirable to tap down on the input; therefore, an inductance will be chosen that will yield the desired tuned input impedance:

$$L_1 = \frac{25000}{45.5\ (2\pi)\ (455 \times 10^3)}$$

$$= 200\ \mu H$$

and

$$C_1 = \frac{1}{\omega^2 L_1} = \frac{1}{[2\pi\ (455 \times 10^3)]^2\ [200 \times 10^{-6}]} = 620\ pF$$

This is not a standard capacitance, so the next highest standard value (680 pF) is chosen. Now, from Eq. (29)

$$Q_{L1}\omega L_{11} = Q_{L1}/\omega C_1 = 4.5/2\pi\ (455 \times 10^3)(680 \times 10^{-12}) \qquad (29)$$

$$= 23100\ \Omega$$

With $Q_{L1}\omega L_{11}$ chosen, N1/N3 can be calculated, using Eq. (30).

$$\frac{N11}{N31} = \left[\frac{0.026\ (18.5)}{17.8 \times 10^{-6}}\right]\left[\frac{23100}{45.5\ (1820)\ 0.25\ (1.92 \times 10^{6})}\right]^{1/2} \tag{30}$$

$$= 20.6$$

The remaining steps are the calculations of Q_{U1} and Q_{U2}. First find Q_{U1} using Eq. (31) and Eq. (32).

$$R_1 = \frac{23100}{20.6^2} = 55\ \Omega \tag{31}$$

$$Q_{U1} = \frac{45.5\ (1000)}{1000 - 55} = 48 \tag{32}$$

$1/g_{ie(max)}$ for the TI-364 is found on the r_{iep} vs I_C curve of the published data sheet. r_{iep} is the reciprocal of g_{ie}. In this case, the minimum r_{iep} is 1000 ohms at a current of one milliampere.

Now, Q_{U2} may be calculated using Eq. (31) and (32).

$$R_1 = \frac{45.5\ (1820)}{(18.5)^2} = 245\ \Omega \tag{31}$$

$$Q_{U2} = \frac{45.5\ (1000)}{1000 - 245} = 60 \tag{32}$$

A summary of the results is shown in the following table:

	N1/N2	N1/N3	C	Q_U	Q_L	Load Resistance
T_1	1.0	20.6	680 pF	48	45.5	1 kΩ
T_2	3.05	18.5	190 pF	60	45.5	1 kΩ
T_3	3.69	4.8	190 pF	72	15	1.5 kΩ

The amplifier constructed using the foregoing design data showed the following performance when compared with the desired characteristics:

	Desired	Actual
Total Gain	55 dB	56 * dB
Max. Undistorted Output Power at 90% Modulation	0.20 mW	0.22* mW
Center Frequency	455 kHz	455 kHz
6-dB B.W.	10 kHz	9.8 kHz
Input Impedance	25 kΩ	22 kΩ
AGC Range	35 dB	45 ** dB
Max. Input Level Before Overload	—	0.5*** V

* Measured across a 1.5-kΩ load resistor to simulate diode load.
** Measured with 0.22 milliwatts as the reference power output. The input was decreased until the output had fallen 10 dB.
***10% distortion occurs at approximately 1 volt input.

DOUBLE-TUNED CIRCUITS

Double-tuned transformers are needed in an IF amplifier when more adjacent channel rejection is desired or when more selectivity is needed with few transformers. The design of double-tuned transformers is discussed in many texts and articles.[4,5,6,7] Therefore, only the characteristics that are necessary to ensure that the amplifier can handle the large signals and provide adequate gain and bandwidth are discussed. The equations are kept as general as possible so that any type of double-tuned configuration can be used. In the equations that are developed, the second number subscript on Q, turns N and inductance L refer to the relative position of the transformer in the circuit. For example, N12/N32 refers to the N1/N3 ratio of the secondary winding of the interstage transformer, and L_{13} is the inductance of the primary winding of the output transformer. The second or third letter subscript on Q designates either the primary or secondary. Turns ratios for double-tuned transformers are defined in Fig. 5.

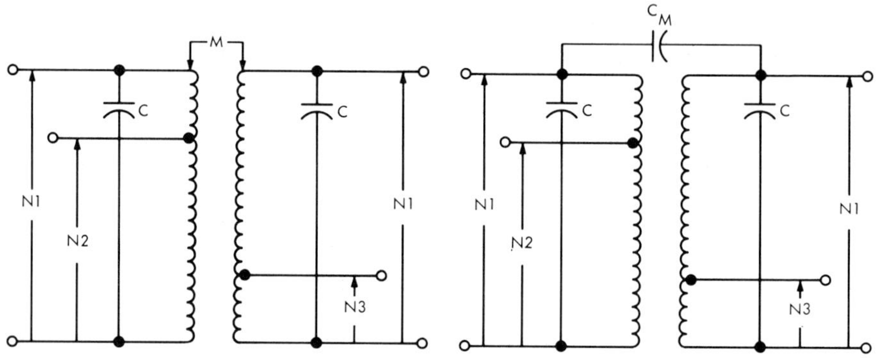

Fig. 5. Two-Types of Double-tuned IF Transformers with Turns Ratios Defined

Stage Power Gain. The single most important figure of merit of an amplifier is its gain. The power gain at the center frequency of a double-tuned stage is expressed by

$$\frac{P_o}{P_i} = \frac{|y_{fe}|^2}{g_{ie}} \left(\frac{a}{1 + a^2}\right)^2 Q_{LP}\omega L_1' \left(1 - \frac{Q_{LS}}{Q_{US}}\right) \tag{33}$$

where the coil loss is given by

$$\frac{P_{oc}}{P_{ic}} = \frac{a^2}{1 + a^2}\left(1 - \frac{Q_{LS}}{Q_{US}}\right)$$

and $Q_{LP}\omega L_1'$ is the uncoupled antiresonant impedance of the primary referred to the collector.

A derivation of the power gain at the center frequency is shown in the section on Equation Derivations at the end of this chapter.

The power gain of the stage in terms of forward transfer admittance, stability factor, frequency, and the maximum collector base capacitance is

$$\frac{P_o}{P_i} = \frac{|y_{fe}|\ 2\sigma a^2}{\omega C_{CB(max)}} \tag{34}$$

This equation for the stage gain for a two-stage amplifier is expressed as

$$\frac{P_o}{P_i} = \frac{|y_{fe}|\ \sigma a^2}{\omega C_{CB(max)}} \tag{35}$$

This equation is analogous to Eq. (14) for the single-tuned circuit. It defines the maximum usable gain if σ is equal to 1. A derivation of the stage power gain and the maximum usable stage gain is shown in the Equation Derivation section.

Parameter Limits for Design. In the single-tuned design procedure, the maximum limit $g_{ie(max)}$ is chosen as the value for the calculation of coil loss and power gain. Using $g_{ie(max)}$ yields the minimum gain case.

For double-tuned coupling transformers, the variance of both the critical coupling coefficient and gain with respect to g_{ie} must be considered. Although optimum flatness and transducer gain are obtained with critical coupling ($a = 1$), transformers in narrow-band double-tuned amplifiers are usually undercoupled. An "a" = 0.9 is usually chosen so that a simple procedure of alignment can be used. Very little gain is lost with this choice.

The critical coefficient of coupling is expressed in general terms of secondary turns ratio, loaded and unloaded antiresonant impedances, and g_{ie} as

$$a = k \left[\frac{Q_{LP}Q_{U3}\,(N1/N3)^2}{Q_{US}\omega L_3 g_{ie} + (N1/N3)^2} \right]^{1/2}$$

The minimum value of g_{ie} and the maximum values of Q_{LP}, Q_{US}, and k are chosen to ensure that the k/k_c ratio is always less than 0.9.

Large-signal Handling Ability. Large-signal handling ability in double-tuned amplifiers is identical to the single-tuned case. The primary N1/N2 ratio of the output transformer is expressed in terms of the primary Q_{LP}, coefficient of coupling, and transistor bias.

$$\frac{N13}{N23} = \left[\frac{(Q_{LP}\omega L_{1P})\, I_C}{(1 + a^2)\, 0.8\, V_{CE}} \right]^{1/2} \tag{36}$$

Note that this expression is similar to Eq. (24).

The termination for the stage which is controlled for AGC purposes is derived in the same manner as the single-tuned case. The collector load impedance is made equal to the ratio of V_{CE} to I_C at the point of minimum current. The turns ratio of this termination is

$$\frac{N1}{N2} = \left[\frac{(Q_{LP}\omega L_1)(Q_{LS}\omega L_2)}{a\,(1 + a^2)} \right]^{1/2} \left[\frac{N3}{N1} \right] \frac{I_E'}{0.026} \tag{37}$$

Note that if it is assumed $Q_{LP}\omega L_1 = Q_{LS}\omega L_2$, and $a = 0.9$, the equation reduces to

$$\frac{N1}{N2} = \frac{Q_L\omega L_1}{1.6} \left[\frac{N3}{N1} \right] \frac{I_E'}{0.026} \tag{38}$$

Design Method. The method discussed in the paragraphs to follow utilizes the principle previously discussed for the design of a two-stage double-tuned IF amplifier. The first stage is gain-controlled by decreasing the collector current (reverse AGC). The amplifier has input, interstage, and output transformers.

It is first assumed that the loaded, uncoupled Q's of the transformers have been determined for the desired overall bandwidth.

The design of an amplifier begins at the third or output transformer. The N1/N3 turns ratio is

$$\frac{N13}{N33} = \left[\frac{Q_{U3S}Q_{L3S}\omega L_{23}}{(Q_{U3S} - Q_{L3S})\, R_D} \right]^{1/2} \tag{39}$$

The load on the output stage is usually a germanium diode in which case R_D is 1500 ohms.

The choice of Q_{U3} and Q_{L3} for the secondary of the output transformer depends on the bandwidth and the amount of loss between the last stage and the detector. The Q_{U3} is chosen as high as economics and physical size will permit. A low value of Q_{L3} may also be chosen and further increase the efficiency.

The N1/N2 ratio of the output transformer is given by

$$\frac{N13}{N23} = \left[\frac{Q_{L3P}\omega L_{23}\ I_C}{(1 + a_3^2)\ 0.8\ V_{CE}}\right]^{1/2} \tag{40}$$

The collector current is found by using Eq. (25) or Eq. (41). If the coil loss is as high as three dB,

$$I_C \approx \frac{16\ P_{mo}}{V_{CE}} \tag{41}$$

This is a reasonable figure for double-tuned since the Q_{LP}/Q_{LS} ratio can be as high as two without making bandwidth calculations too difficult.

The turns ratio N1/N3 for the interstage is

$$\frac{N12}{N32} = \left[\frac{Q_{L2S}\omega L_{22}\ 0.8\ V_{CE}}{(1 + a_2^2)I_C\ \sigma_2\ (R_1'R_2')}\right]^{1/2} \tag{42}$$

Note the similarity between Eq. (42) and Eq. (26). The derivation steps are the same.

The N1/N2 turns ratio of the interstage is

$$\frac{N12}{N22} = \frac{[(Q_{L2P}\omega L_{12})\ (Q_{L2S}\omega L_{22})]^{1/2}}{a_2\ (1 + a_2^2)} \left[\frac{N32}{N12}\right] \frac{I_E'}{0.026} \tag{43}$$

Here the turns ratio is chosen for maximum signal handling of the interstage amplifier. This is a simple restatement of Eq. (37). The minimum current, I_E' is given by Eq. (22).

The N1/N3 turns ratio of the input transformer is found by

$$\frac{N11}{N31} = \frac{N22}{N12}\left[\frac{(Q_{L1S}\omega L_{21})\ (Q_{L2P}\omega L_{12})}{(1 + a_1^2)\ (1 + a_2^2)\ \sigma_1(R_1'R_2')}\right]^{1/2} \tag{44}$$

This equation is derived by solving Eq. (13) for R_1 and equating R_1 as

$$R_1 = \left(\frac{N31}{N11}\right)^2 \frac{Q_{L1S}\omega L_{21}}{1 + a_1^2}$$

then expressing R_2 in Eq. (13) as

$$R_2 = \left(\frac{N22}{N12}\right) \frac{Q_{L2P}\omega L_{12}}{1 + a_2^2}$$

then solve the equation for N11/N31.

The determination of the primary turns ratio depends on the desired load impedance for the frequency converter. When it is undesirable to tap, the tuning capacitance is found by

$$C = \frac{Q_{L1P}}{R_{in}\omega(1 + a_1^2)} \tag{45}$$

The remaining step is to find the unloaded Q of the secondary windings. This can be done by using the input admittance of the transistor, the Q_L, and the N1/N3 turns ratio. The uncoupled tuned impedance at the input of the transistor is expressed as

$$R_{1U} = Q_L \omega L_1 \left(\frac{N3}{N1}\right)^2 \tag{46}$$

This is the same as R_1 in Eq. (31) for the single-tuned. The remaining factor Q_U/Q_L is expressed as

$$\frac{Q_U}{Q_L} = \frac{\dfrac{1}{g_{ie(min)}}}{\dfrac{1}{g_{ie(min)}} - R_{1U}} \tag{47}$$

An alternate IF amplifier consists of two double-tuned transformers and one single-tuned. This complement is desirable because of the simplicity of designing an efficient single-tuned output transformer.

The procedure for designing an amplifier with this combination is very simple if "a" is chosen near unity for the double-tuned transformer.

The single-tuned output IF transformer is designed by using Eq. (23) and (24). The remaining transformers are designed by using the double-tuned method beginning with Eq. (42) and ending with Eq. (47).

Another alternative, which is popular in 117-volt line-operated transistor radios is the use of one double-tuned and one single-tuned transformer. The stable gain should be determined by considering only one stage of IF gain. If "a" is between 0.8 and 1.2, Eq. (25), (15), (23), (24), can be used with less than 2 dB of error. This is less than the expected variance of gain due to interchangeability of transistors.

Design Example. An example of one of the alternatives is given (see Fig. 6). The desired properties of the amplifier are:

1. Total IF gain = 53 dB
2. Maximum IF power into output transformer = 1 mW.
3. Center frequency = 455 kHz
4. Total 6-dB bandwidth = 10 kHz
5. Total 20-dB bandwidth $\leq$ 19 kHz
6. Input impedance $\approx$ 25 kΩ
7. IF load = 1N295 diode detector
8. Change of power gain for AGC = 35 dB
9. Supply voltage = 9 V.

An examination of requirements 4 and 5 indicates that two double-tuned IF transformers are needed if "a" is chosen as 0.9. This is found by using information in Fig. 7. This allows the third or output transformer to be single-tuned and low Q (very efficient). The fact that three transformers (or two stages) are needed is found by the gain requirements. The gain requirement is too great for one stage if any degree of stability is required.

The output-stage current must be determined to satisfy the IF power output requirements. From Eq. (25)

$$I_C = \frac{10 \times 10^{-3}}{8} = 1.2 \times 10^{-3}\ \text{A} \tag{25}$$

A one-volt drop in V_{CE} is allowed for the emitter resistor.

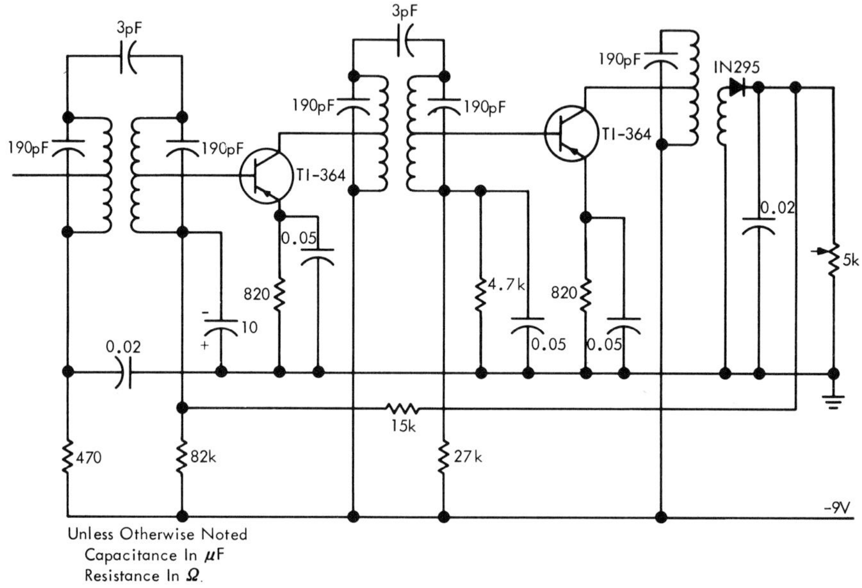

Fig. 6. Typical Double-tuned IF Amplifier Circuit

The Q_U, Q_L, and N1/N3 turns ratio for the single-tuned transformer are identical to those of the output transformer in the single-tuned example. $Q_U = 72$, $Q_L = 15$, and N13/N33 = 4.8. Now, from Eq. (24)

$$\frac{N13}{N23} = \left[\frac{27\ 300\ (1.2 \times 10^{-3})}{0.8\ (8)}\right]^{1/2} = 2.25 \tag{24}$$

The TI-364 data sheet specifies that $C_{CB(max)}$ is 2.5 pF at V_{CB} equal to 9 volts. At V_{CB} equal to 8 volts, the $C_{CB(max)}$ is 2.65 pF. At $I_C = 1.2$ mA, $|y_{fe}| = 42$ mS. The stage gain is divided equally between stages. Therefore,

$$\text{Stage Gain (dB)} = \frac{\text{Total Gain + Output Transformer Coil Loss}}{\text{No. of stages}}$$

$$= \frac{53 + 1}{2} = 27 \text{ dB } (510)$$

If both stages are operated at $I_C = 1.2$ mA and $V_{CE} = 8$ volts, the stability factor by Eq. (35) is

$$\sigma_2 = \frac{510\ (2\pi)\ (455 \times 10^3)\ (2.65 \times 10^{-12})}{0.9^2\ (42 \times 10^{-3})} \tag{15}$$

$$= 0.113$$

The critical termination impedance from Eq. (11) is

$$R_1'R_2' = \frac{0.026}{2\pi\ (455 \times 10^3)\ (1.2 \times 10^{-3})\ (2.65 \times 10^{-12})} \tag{11}$$

$$= 2.86 \times 10^6\ \Omega^2$$

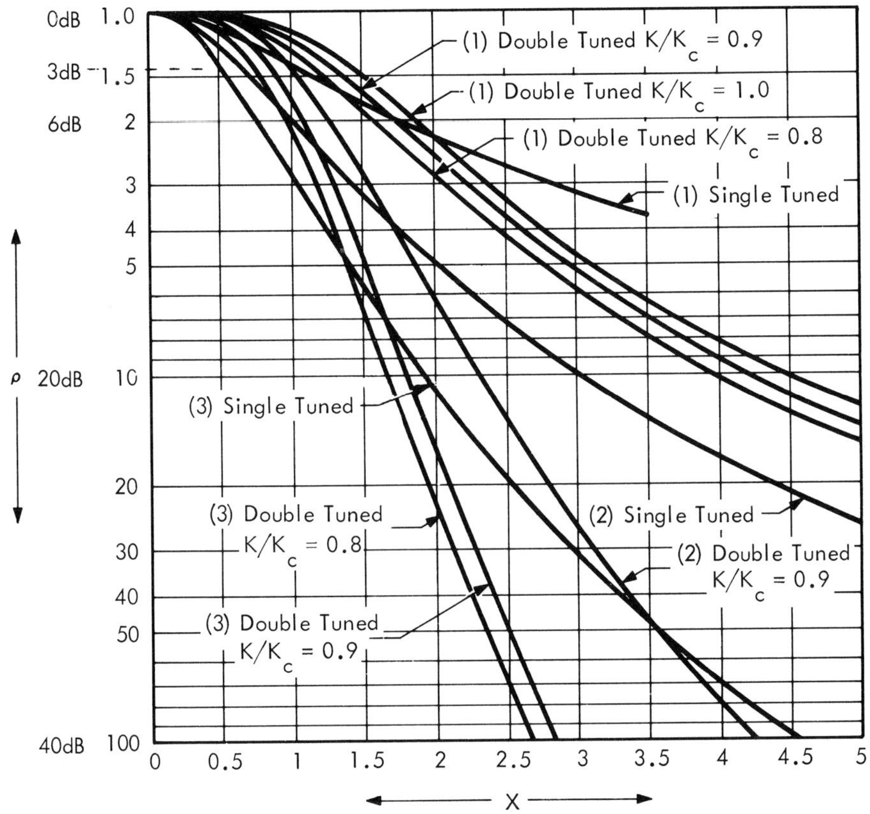

For Single-Tuned Ckts.

$$X = \sqrt{\rho^{2/N} - 1}$$

Double-Tuned Ckts. General

$$X = \sqrt{\sqrt{(1+a^2)^2\,\rho^{2/N} - 4a^2} - (1-a^2)}$$

Where: $a = K/K_c$

$$X = \frac{2\Delta f}{f_o} Q$$

Where: $Q = \frac{1}{K_c} = \sqrt{Q_1 Q_2}$

For Double-Tuned Ckts.

$K/K_c = 1.0$ $\qquad \frac{10x}{2x} = 2.4$

$$X = \sqrt{2}\;\sqrt[4]{\rho^{2/N} - 1}$$

$K/K_c = 0.90$ $\qquad \frac{10x}{2x} = 2.44$

$$X = \sqrt{\sqrt{3.28\rho^{2/N} - 3.24} - 0.19}$$

$K/K_c = 0.80$ $\qquad \frac{10x}{2x} = 2.5$

$$X = \sqrt{\sqrt{2.69\rho^{2/N} - 2.56} - 0.36}$$

Fig. 7. Universal Selectivity Curves for Single- and Double-tuned Transformers

Since the same ½-inch IF transformer material is chosen for both the single-tuned and double-tuned transformers, the tuning capacitance is again 190 pF and the inductance is 640 μH. If it is assumed the two transformers have the same bandpass characteristics, then the uncoupled Q_L for the two transformers from Fig. 7 is

$$Q_L = \frac{1.25\ f_o}{BW_6} = \frac{1.25\ (455 \times 10^3)}{10 \times 10^3} = 57$$

The reactance of 640 μH at 455 kHz is 1840 ohms.

From Eq. (42)

$$\frac{N12}{N32} = \left[\frac{57\ (1.84 \times 10^3)\ 0.8\ (8)}{(1 + 0.81)\ (1.2 \times 10^{-3})\ 0.113\ (2.86 \times 10^6)}\right]^{1/2}$$

$$= 30.5 \qquad (42)$$

The minimum current from Eq. (22) is

$$I_E' = \frac{1200\ \mu A}{56} = 21.5\ \mu A \qquad (22)$$

Since the same change of gain is required for this amplifier as in the single-tuned example, the divisor is again 56.

From Eq. (38)

$$\frac{N12}{N22} = \frac{57\ (1.84 \times 10^3)\ (21.5 \times 10^{-6})}{1.6\ (30.5)\ 0.026} = 1.8 \qquad (38)$$

From Eq. (44)

$$\frac{N11}{N31} = \frac{1}{1.8}\left[\left(\frac{57\ (1.84 \times 10^3)}{1 + 0.81}\right)^2 \frac{1}{0.113\ (2.86 \times 10^6)}\right]^{1/2} = 58 \qquad (44)$$

This is a rather large turns ratio and probably unusable since the coefficient of coupling between the start to tap and the start to overall winding is probably much less than unity. Another method for determining the turns ratio is given later. For the time, assume the coupling is unity.

The input transformer must couple the frequency converter to the first IF amplifier. Its load on the converter is sometimes critical in determining the stability. It is assumed that the primary of the input transformer can be tapped, that $Q_{LP} = Q_{LS}$, and that $L_P = L_S$. The primary turns ratio is

$$\frac{N11}{N21} = \left[\frac{57\ (1.84 \times 10^3)}{(1 + 0.81)\ (25 \times 10^3)}\right]^{1/2} = 1.51$$

To determine the Q_U for the secondary, it is first necessary to find the R_{1U}. Using Eq. (46), R_{1U} for the interstage transformer is

$$R_{1U} = \frac{57\ (1.84 \times 10^3)}{30.5^2} = 110\ \Omega \qquad (46)$$

and using Eq. (47),

$$\frac{Q_{U2}}{Q_{L2}} = \frac{4000}{4000 - 110} = \frac{4000}{3890} = 1.03 \qquad (47)$$

Therefore,

$$Q_{U2} = 57\ (1.03) = 59$$

For the input transformer

$$R_{1U} = \frac{57\ (1.84 \times 10^3)}{58^2} = 30.5\ \Omega \tag{46}$$

$$\frac{Q_{U1}}{Q_{L1}} = \frac{4000}{4000 - 30.5} \approx 1.0 \tag{47}$$

Therefore,

$$Q_{U1} = 57$$

In this case where the turns ratio is rather large, it is necessary to correct for the low coefficient of coupling. This can be done by experimenting with several secondary coils with turns ratios slightly smaller than the calculated value. Take the secondary of the input transformer for example. Two coils should be wound with ratios of approximately 45 and 55. Then, measure the loaded Q on a Q meter with a load of 30 ohms between the start and tap. (The R_{1U} for this transformer). The turns ratio is correct when the loaded Q is one-half the unloaded Q. If this method is employed for smaller turns ratios (less than 25), it is necessary to use a correction factor for finding the required load resistance, (R_g). This factor is

$$R_g = \frac{1}{g_{ie}}\left(\frac{Q_U}{Q_L} - 1\right) \tag{48}$$

For the two transformers in the discussion, the actual turns ratios were N11/N31 = 50, and N12/N32 = 25.

A summary of the results is shown in the following table:

Transformer Data

	N1/N2	N1/N3	C	C_M	Q_{UP}	Q_{US}	Q_{LS}	Load Resistance
T_1	1.51	58	190	3.0	57	57	57	4 kΩ
T_2	1.8	30.5	190	3.0	57	59	57	4 kΩ
T_3	2.25	4.8	190	—	—	72	15	1.5 kΩ

Performance Data

	Desired	Actual
Total Gain*	53 dB	48-52 dB
Maximum Power into Output Transformer (90% Modulation)	1 mW	1.1 mW
Total 6 dB bandwidth	10 kHz	9.2 kHz
Total 20 dB bandwidth	≤ 19 kHz	17 kHz
Input Impedance	25 kΩ	24.2 kΩ
AGC Range**	45 dB	43 dB

*Measured across a 1.5-kΩ load resistor to simulate the diode load. Gain limit transistors were used to make the measurement.
**Measured with 1 milliwatt as the reference power output. The input was decreased until the output had fallen 10 dB.

EQUATION DERIVATIONS

Equation (3) $$\frac{P_{oc}}{P_{ic}} = 1 - \frac{Q_L}{Q_U}$$

A model of a single tuned circuit is shown below.

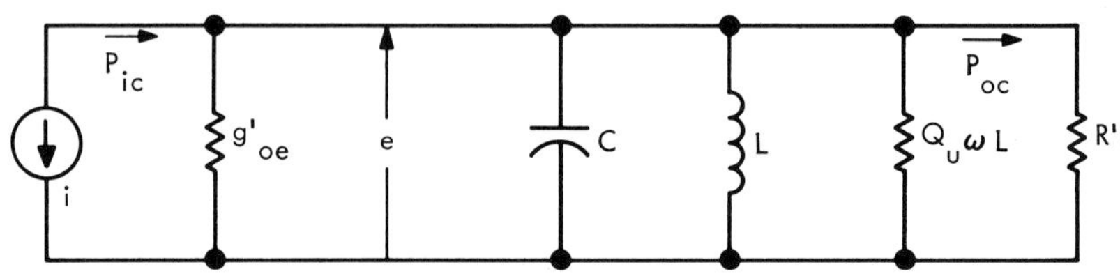

Single-tuned Circuit Model

R_L' is the load on the transformer referred to the full winding by the turns ratio.

$Q_U\omega L$ is the parallel equivalent resistance of the coil loss.

g_{oe}' is the output admittance of the transistor or the load on the primary side referred to the full winding by the turns ratio. For this analysis g_{oe}' is neglected. The ratio of the power delivered to the load to the power delivered into the transformer is given by P_{oc}/P_{ic}.

The power input is expressed by

$$P_{ic} = \frac{e^2}{Q_L\omega L}$$

The power output is expressed by

$$P_{oc} = \frac{e^2}{R_L'}$$

Therefore,

$$\frac{P_{oc}}{P_{ic}} = \frac{Q_L \omega L}{R_L'}$$

R_L' may be expressed in terms of $Q_U\omega L$ and $Q_L\omega L$ by solving the equation for $Q_L\omega L$. The equation for $Q_L\omega L$ is

$$Q_L \omega L = \frac{Q_U \omega L R_L'}{Q_U \omega L + R_L'}$$

Solving for R_L'

$$R_L' = \frac{Q_U Q_L \omega L}{Q_U - Q_L}$$

And by substitution

$$\frac{P_{oc}}{P_{ic}} = 1 - \frac{Q_L}{Q_U} \tag{3}$$

Equation (14)

$$\frac{P_o}{P_i} = |y_{fe}| \frac{\sigma}{\omega C_{CB(max)}}$$

The stage gain from Eq. (4) in the text is

$$\frac{P_o}{P_i} = \frac{|y_{fe}|^2}{G_L g_{ie}} \left(1 - \frac{Q_L}{Q_U}\right) \tag{4}$$

Solve Eq. (10) for R_2 and substitute the result in Eq. (4), noting that $R_2 = 1/G_L$. Now,

$$\frac{P_o}{P_i} = \frac{|y_{fe}|^2\, 2\sigma\, R_1'R_2'}{R_1 g_{ie}} \left(1 - \frac{Q_L}{Q_U}\right)$$

From the derivation of Eq. (32)

$$\frac{Q_U}{Q_L} = \frac{\dfrac{1}{g_{ie}}}{\dfrac{1}{g_{ie}} - R_1}$$

Solving for R_1

$$R_1 = \frac{1}{g_{ie}} \left(1 - \frac{Q_L}{Q_U}\right).$$

And substituting

$$\frac{P_o}{P_i} = |y_{fe}|^2\, 2\sigma\, R_1'R_2'$$

Equation (11) could now be substituted in the equation above, but for the sake of more accuracy, replace $I_E/0.026$ by y_{fe}, then make the substitution.

$$\frac{P_o}{P_i} = \frac{|y_{fe}|\ 2\sigma}{\omega C_{CB(max)}}$$

The difference between this equation and Eq. (14) is a factor of 2. Equation (14) is for the special case of a 2-stage amplifier. The table below shows these equations as they apply for different number of stages.

Number of Stages	P_o/P_i
1	$\frac{\lvert y_{fe}\rvert\ 2\sigma}{\omega C_{CB(max)}}$
2	$\frac{\lvert y_{fe}\rvert\ \sigma}{\omega C_{CB(max)}}$
3	$\frac{0.77\ \lvert y_{fe}\rvert\ \sigma}{\omega C_{CB(max)}}$
4	$\frac{0.67\ \lvert y_{fe}\rvert\ \sigma}{\omega C_{CB(max)}}$

Equation (30) $$\frac{N11}{N31} = \frac{0.026}{I_E'}\left[\frac{N12}{N32}\right]\left[\frac{Q_{L1}\omega L_1}{Q_{L2}\omega L_2 \sigma_1\ (R_1'R_2')}\right]^{1/2}$$

The turns ratio is related by

$$\frac{N11}{N31} = \left(\frac{Q_{L1}\omega L_1}{R_1}\right)^{1/2}$$

R_1 may be stated in terms of stability factor, turns ratio, and tuned impedance of the interstage transformer. From Eq. (13)

$$R_1 = \frac{\sigma_1(R_1'R_2')}{R_2}$$

R_2, if expressed in terms of tuned impedance and turns ratio, is

$$R_2 = \frac{Q_{L2}\omega L_2}{(N12/N22)^2}$$

And, from Eq. (27)

$$\left[\frac{N12}{N22}\right]^2 = \left[\frac{I_E'\ (Q_{L2}\omega L_2)}{0.026}\left(\frac{N32}{N12}\right)\right]^2 \qquad (27)$$

Therefore, if Eq. (27) is substituted in the R_2 equation above

$$R_2 = \frac{1}{Q_{L2}\omega L_2}\left[\frac{0.026}{I_E'}\left(\frac{N12}{N32}\right)\right]^2$$

Substituting this equation in the R_1 equation,

$$R_1 = \frac{\sigma_1 (R_1'R_2')(Q_{L2}\omega L_2)}{\left[\dfrac{0.026}{I_E'}\left(\dfrac{N12}{N32}\right)\right]^2}$$

So that N11/N31 may be expressed as

$$\frac{N11}{N31} = \frac{0.026}{I_E'}\left[\frac{N12}{N32}\right]\left[\frac{Q_{L1}\omega L_1}{Q_{L2}\omega L_2 \sigma_1 (R_1'R_2')}\right]^{1/2} \tag{30}$$

Equation (32)

$$Q_U = \frac{Q_L\left[\dfrac{1}{g_{ie}}\right]}{\dfrac{1}{g_{ie}} - R_1}$$

A useful expression for finding the Q_U when Q_L, input admittance, and R_1 are known can be found by solving the $Q_L\omega L$ equation (shown in the derivation of Eq. (3)) for $Q_U\omega L$.

$$Q_U\omega L = \frac{Q_L\omega L\, R_L'}{R_L' - Q_L\omega L}$$

Then R_L' may be expressed in terms of turns ratio and load admittance which is the input admittance of the transistor.

$$R_L' = \left(\frac{N1}{N3}\right)^2 \frac{1}{g_{ie}}$$

After the substitution and subsequent division of both sides by $Q_L\omega L$,

$$\frac{Q_U}{Q_L} = \frac{(N1/N3)^2 \dfrac{1}{g_{ie}}}{(N1/N3)^2 \dfrac{1}{g_{ie}} - Q_L\omega L}$$

$(N1/N3)^2$ is factored from the numerator and denominator and R_1 is substituted for $Q_L\omega L/(N1/N3)^2$. Then,

$$\frac{Q_U}{Q_L} = \frac{\dfrac{1}{g_{ie}}}{\dfrac{1}{g_{ie}} - R_1}$$

or

$$Q_U = \frac{Q_L\left[\dfrac{1}{g_{ie}}\right]}{\dfrac{1}{g_{ie}} - R_1} \tag{32}$$

Equation (33)

$$\frac{P_o}{P_i} = \frac{|y_{fe}|^2}{g_{ie}}\left(\frac{a}{1+a^2}\right)^2 Q_{LP}\omega L_1 \left(1 - \frac{Q_{LS}}{Q_{US}}\right)$$

Usable Stage Gain. The approximate model of a double-tuned stage is shown below:

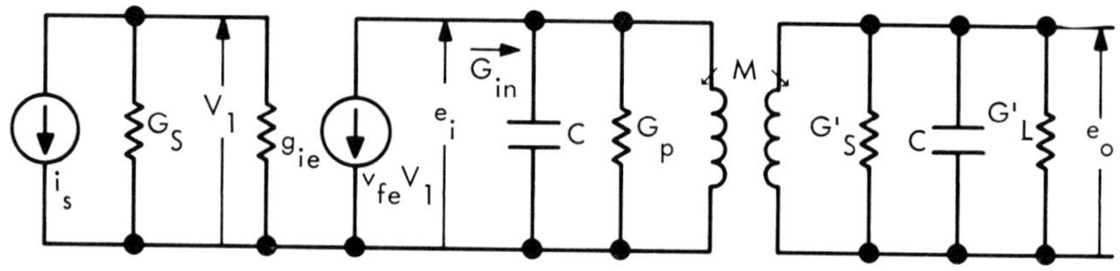

Double-tuned Circuit Model

The power output at the center frequency is

$$P_o = e_o^2\, G_L'$$

The voltage ratio of the transformer at the center frequency is

$$\frac{e_o}{e_1} = j\, a \left(\frac{Q_{LS}\omega L_2}{Q_{LP}\omega L_1}\right)^{1/2}$$

The voltage at the input side of the transformer is expressed in terms of the current source, $y_{fe}V_1$, and the input admittance as

$$e_1 = |y_{fe}|\; V_1 \frac{Q_{LP}\omega L_1}{1 + a^2}$$

The output voltage is

$$e_o = \frac{j\, a\, |y_{fe}|}{1 + a^2} \left(\frac{Q_{LS}\omega L_2}{Q_{LP}\omega L_1}\right)^{1/2} (Q_{LP}\omega L_1)\, V_i$$

The G_L' or load admittance is expressed in terms of the secondary tuned impedance and the uncoupled Q's as

$$G_L' = \frac{\left(1 - \dfrac{Q_{LS}}{Q_{US}}\right)}{Q_{LS}\omega L_2}$$

The power output is given by

$$P_o = |y_{fe}|^2 \left(\frac{a}{1 + a^2}\right)^2 Q_{LP}\omega L_1 \left(1 - \frac{Q_{LS}}{Q_{US}}\right) V_i^2$$

Since the power input is expressed as

$$P_i = V_i^2\, g_{ie}$$

the power gain is

$$\frac{P_o}{P_i} = \frac{|y_{fe}|^2}{g_{ie}} \left(\frac{a}{1 + a^2}\right)^2 Q_{LP}\omega L_1 \left(1 - \frac{Q_{LS}}{Q_{US}}\right) \qquad (33)$$

Equation (34) $$\frac{P_o}{P_i} = \frac{|y_{fe}|\; 2\sigma\, a^2}{\omega C_{CB(max)}}$$

The power gain can be expressed in terms of stability factor and $C_{CB(max)}$ by recognizing first that the collector load admittance can be expressed in terms of the circuit parameters and the critical impedances.

$$\frac{1}{G_L} = \frac{Q_{LP}\omega L_1}{1 + a^2}$$

and

$$\frac{1}{G_L} = \frac{2\sigma R_1'R_2'}{R_1}$$

This power gain is now expressed as

$$\frac{P_o}{P_i} = \frac{|y_{fe}|^2}{g_{ie}}\left(\frac{2\sigma\ R_1'R_2'}{R_1}\right)\left(\frac{a^2}{1 + a^2}\right)\left(1 - \frac{Q_{LS}}{Q_{US}}\right)$$

If it is further recognized that R_1, the shunt impedance of the tuned circuit at the input of the transistor, is

$$R_1 = \frac{1 - \dfrac{Q_{LS}}{Q_{US}}}{g_{ie}\ (1 + a^2)}$$

The power gain is

$$\frac{P_o}{P_i} = |y_{fe}|^2\ (2\sigma\ R_1'R_2')\ a^2.$$

This is further expressed as

$$\frac{P_o}{P_i} = \frac{|y_{fe}|\ 2\sigma\ a^2}{\omega C_{CB(max)}} \tag{34}$$

since

$$R_1'R_2' = \frac{1}{|y_{fe}|\ \omega C_{CB(max)}}$$

The power gain expression for double-tuned circuits is identical to the single tuned with the exception of the critical coefficient ratio. The same power gain can be realized for the same collector current, frequency, σ, and $C_{CB(max)}$ if "a" is unity. For "a" less than unity, less gain is achieved for the same conditions.

SYMBOLS

a	Ratio of the actual to the critical coefficient of coupling (k/k_c)
C_{bc}'	Internal collector-base capacitance
C_{be}'	Internal emitter-base capacitance
C_H	Capacitance contributed by header or can
C_{CB}	Total collector-base capacitance measured with emitter open

C_{oep}	Parallel equivalent common-emitter output capacitance measured with the input short-circuited
C_s	Source capacitance
f_{IF}	Intermediate frequency
f_T	Frequency at which the common-emitter forward current gain is equal to unity
G_L	Collector load conductance
G_S	Source conductance
g_{fe}	Real part of y_{fe}
g_{ie}	Real part of y_{ie}
g_{oe}	Real part of y_{oe}
I_E	The d-c emitter current
I_E'	The minimum permissible value of I_E
k	Coefficient of coupling
k_c	Coefficient of coupling for critical coupling
L	General term for magnitude of $M + jN$ in Eq. (6), (7), and (9)
L_1	Total primary inductance
L_2	Total secondary inductance
L_1'	Primary inductance referred to tap
L_S	Inductance of source
N	Number of turns on winding of transformer; imaginary component of $y_{re}y_{fe}$ term in Eq. (8) and (9); number of identical transformers in Fig. 7
n	Number of stages
P_i	Power input to amplifier stage
P_{ic}	Power input to input coil and equal to the input power to the stage ($P_{ic} = P_i$)
P_{it}	Power input to transistor and equal to the output power of the coil ($P_{it} = P_{oc}$)
P_{mo}	Maximum undistorted RF power output of the final stage considering a 100% amplitude-modulated signal
P_o	Power output of the amplifier stage
P_{oc}	Power output of the input coil and equal to the power input to the transistor ($P_{oc} = P_{it}$)
P_{ot}	Power output of the transistor and equal to the power output of the stage ($P_{ot} = P_o$)
Q_L	Loaded Q of the coil
Q_{LP}	Loaded uncoupled Q of the primary
Q_U	Unloaded Q of the coil
Q_{UP}	Unloaded uncoupled Q of the primary
R_b'	Base spreading resistance
R_{be}'	Resistance between emitter connection and intrinsic base
R_{in}	Input resistance of the input transformer at the resonant frequency
R_1	Defined in Eq. (10)
R_2	Defined in Eq. (10) and equal to collector load resistance
$R_1'R_2'$	Critical product of R_1R_2
V_1	Voltage at the input of the transistor model

REFERENCES

1. Stern, A. P., Stability and Power Gain of Tuned Transistors, *Proc. IRE,* Vol. 45, No. 3, p. 336, March, 1957.
2. Holmes, D. D., and T. O. Stanley, Stability Considerations in Transistor Intermediate Frequency Amplifiers, *Transistors I,* p. 405, RCA Laboratories, Princeton, N. J., 1956.
3. Parasitic Oscillations in I.F. Stages and Frequency Changers of AM Receivers, *Electronic Applications,* Eindhoven, Netherlands, Vol. 20, No. 2, p. 41, 1959-1960.
4. Aiken, C. B., Two-mesh Tuned Coupled Circuit Filters, *Proc. IRE,* Vol. 25, No. 2, p. 230, Feb., 1937.
5. Langford-Smith, F., "Radiotron Designers Handbook," pp. 412-426, 943, 1025-1036, 1065, Radio Corporation of America.
6. Transistor I. F. Amplifiers, *Electronic Applications,* Eindhoven, Netherlands, Vol. 21, No. 4, 1960-1961.
7. Transistor I. F. Amplifiers, *Electronic Applications,* Eindhoven, Netherlands, Vol. 22, Nos. 1, 2, and 3, 1961-1962.

8

FM Tuner Design

This chapter analyzes some of the design and performance problems associated with FM tuners and shows how these problems may be minimized or eliminated by using Texas Instruments TI-400 series germanium epitaxial mesa transistors. The problems discussed include noise performance, spurious responses, and AGC. The relationship of transistor parameter specifications to these problems is analyzed.

NOISE PERFORMANCE

Tuner noise performance is associated primarily with the noise figure and gain of the RF stage. The noise figure can be very good, but if sufficient RF gain is not attained, the overall tuner noise figure is degraded, because the mixer stage also affects noise figure. Maximum available gain, feedback capacity, noise figure, and input and output impedances are the primary specifications that should be considered when selecting an RF transistor. The limits of these parameters are just as important as their absolute magnitudes.

The RF transistor noise performance and its circuit gain are functions of the driving source impedance. For most transistors, the source impedance for optimum noise figure is not the same as for maximum gain. However, this is not true for the TI-400. The optimum source impedance for best noise performance at 100 MHz is a matched condition. Matching the TI-400 input impedance yields both minimum noise figure and optimum gain in the RF stage. IHFM (Institute of High Fidelity Manufacturers) sensitivities in the order of 1.0 to 2.0 μV are attainable using the TI-400 as an RF stage and the TI-401 as the mixer stage, if the IF stages and FM detector are properly designed.

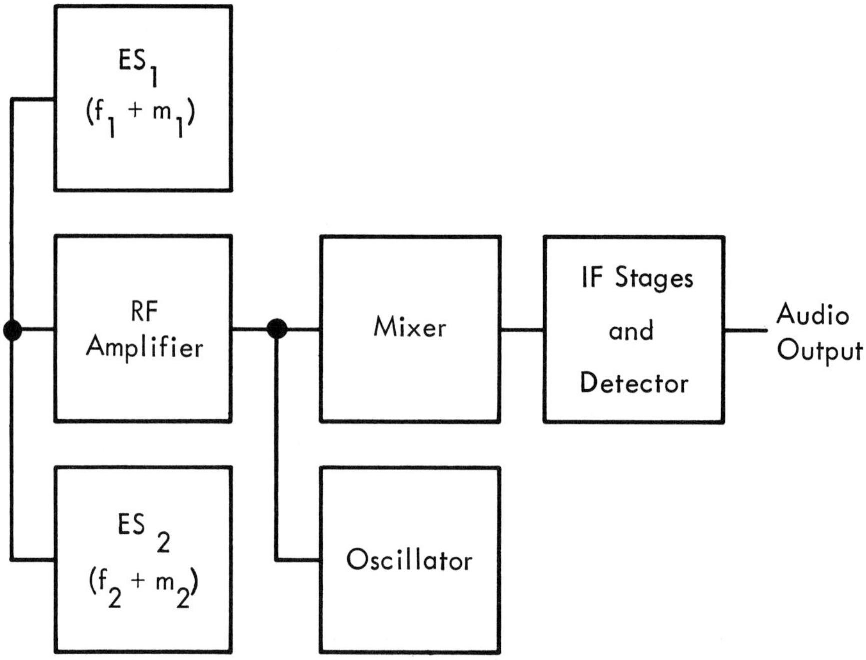

Fig. 1. Wanted and Unwanted Signal Input to the Receiver

SPURIOUS RESPONSE

There are a number of undesirable responses which can occur in an FM tuner. These responses are often referred to as cross modulation distortion, intermodulation, and spurious responses. Before attempting to identify these undesirable responses that can occur in an FM tuner, it would be well to define cross modulation, intermodulation, and spurious responses.

Cross modulation occurs when a receiver is tuned to a small (wanted) signal, and a large (unwanted) signal causes interference. The large signal is usually in the range of selectivity of the RF stage. Figure 1 illustrates the wanted and unwanted signal input to the receiver. ES_1 is the small (wanted) signal and consists of carrier frequency f_1 and modulation frequency m_1. ES_2 is the large (unwanted) signal and consists of carrier frequency f_2 and modulation frequency m_2. When ES_2 is made large with respect to ES_1, an actual transfer of modulation m_2 from f_2 to f_1 is accomplished due to the non-linear input of the RF amplifier. Cross modulation takes place and the output waveform contains $f_1 + m_1 + m_2$ instead of the desired $f_1 + m_1$ only.

According to Terman[1] and Wheeler[2], cross modulation of a desired FM signal by an undesired signal in an adjacent band occurs only to the extent that the undesired signal possesses amplitude modulation and the receiver is incidentally sensitive to amplitude modulation. This means the m_2 must be of an amplitude-modulated and not a frequency-modulated nature. Limiters and detectors in FM receivers are used to prohibit amplitude variations from being present in the receiver output. Cross modulation resulting from the presence of ES_2 does not exist when ES_2 possesses FM only. If the FM receiver is designed properly for good AM rejection, cross modulation is not a problem.

Intermodulation occurs when two or more different frequencies are impressed across the non-linear transfer characteristic of an amplifier. Intermodulation can occur in an FM tuner and produce an output. The resultant frequencies, present after the intermodulation process, must be harmonically related to the center frequency of the tuner, the FM oscillator frequency, such as to produce the IF frequency or its subharmonics; or the resultant frequencies must be subharmonically related to the IF frequency.

Assume the RF amplifier is tuned to a desired frequency of f_d, and ES_1 and ES_2 are displaced by a difference frequency of f_d. If insufficient selectivity is present on the input side of the RF amplifier, both ES_1 and ES_2 are accepted in this stage and intermodulation produces their difference frequency of f_d. This is one example of intermodulation of the RF amplifier which produces a cross talk between m_1 and m_2 in the output.

Another example of intermodulation of the RF stage is if ES_1 and ES_2 are separated by a frequency such as to produce a difference frequency of f_d + IF/2. Thus, f_d + IF/2 would be present on the base of the mixer with the oscillator frequency. Mixer action would take place and generate a difference frequency with its harmonics. This could yield an IF output.

Another example of this type of interference which could exist is when ES_1 and ES_2 are both in the pass band of the tuner selective circuits and are separated by a frequency IF/2. ES_1 and ES_2 would be present on the input of the mixer. The second harmonic of this difference frequency would produce an output. This is caused by poor selectivity and not intermodulation of the RF stage. The FM oscillator may be removed under these conditions and the output still observed.

Intermodulation differs from cross modulation in that intermodulation is a form of mixing between the two carriers which produce either a signal harmonically related to the frequency to which the tuner is tuned, or the IF frequency; and cross modulation is the actual transfer of modulation from one carrier to another. The carriers producing this cross modulation do not have to combine to produce the harmonic relationship needed for intermodulation.

Spurious response in an FM tuner is a result of its inability to reject undesirable frequencies that enter the tuner and beat with the oscillator or its harmonics to produce an IF frequency. The method used to measure the spurious response of an FM tuner and the analysis of the results is very important in determining the actual source of these undesired responses. Particular attention should be paid to the FM generator used to detect spurious response of a tuner. Many commercial FM generators do not supply the generator oscillator fundamental at the output terminals. In most cases, frequency doubler circuits are used to supply an output. This arrangement yields poor generator spurious response rejection.

To examine spurious response in FM tuners, Texas Instruments constructed an FM receiver using TI-400 series epitaxial mesa transistors in the RF and mixer stages. Figure 2 shows the spectrum of responses accepted by the tuner. The tuner was tuned to 100 MHz and, with a 1-μV signal applied, the audio output was recorded. The generator frequency was then varied

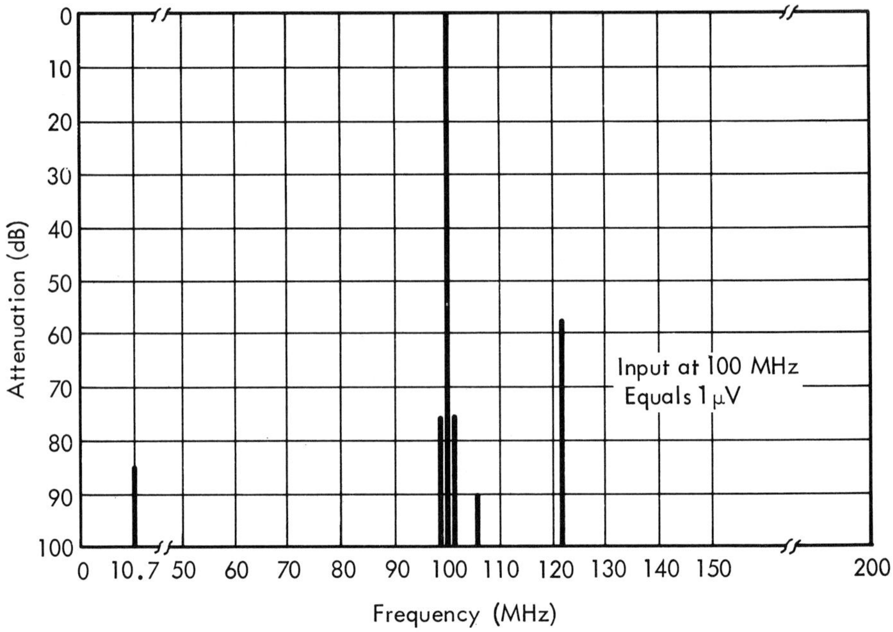

Fig. 2. Spurious Response Spectrum for an FM Tuner

between 10.7 MHz and 216 MHz. All responses (not caused by generator spurious responses) were recorded that were less than 100 dB down from the 1-μV input level. The response at 100 MHz is the tuner frequency and those on both sides of tuner frequency are the upper and lower side responses of the detector. The spurious responses that are of most importance are those that exist at 10.7, 121.4, and 105.35 MHz.

The response at 10.7 MHz is 85 dB down and can be attenuated to a greater extent by use of good shielding of the tuner section of the receiver.

The image frequency response at 121.4 MHz is 60 dB down. It can be further reduced by increasing selectivity in the tuner section. More selectivity causes gain reduction, susceptibility to tracking error, and tuner bandpass narrowness.

The response which exists at 105.35 MHz has two sources and is a result of harmonic generation in the mixer in the following example 1 and by the mixer and FM receiver oscillator in example 2. This response will exist when the tuner is tuned below a strong station transmitting on a frequency of one-half the intermediate frequency above the tuner frequency.

Example 1. This is by far the most important and is due to second harmonic generation in the mixer producing the IF frequency. When the tuner is tuned to 100 MHz, the oscillator is delivering 100 MHz plus the IF frequency to the mixer and the transmitter is delivering 100 MHz plus one-half the IF frequency to the mixer. The only attenuation to this transmitter frequency is the antenna and RF selectivity. The difference frequency of one-half IF is produced in the mixer. Second harmonic generation takes place in the mixer, and an output is observed. One way of reducing this undesired response is to increase selectivity in the antenna and RF tank. Increased selectivity

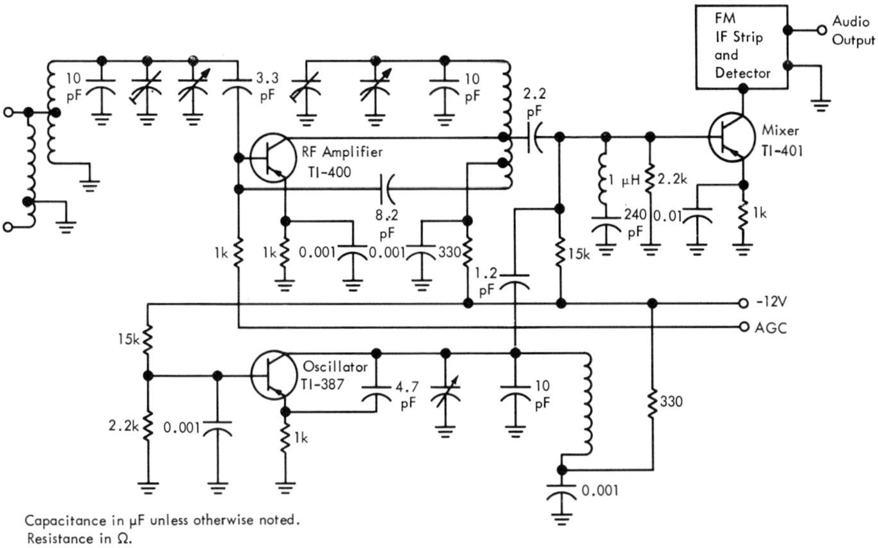

Fig. 3. FM Receiver Schematic, 88-108 MHz

means more insertion loss and reduced tuner gain. A solution to this undesirable response is to select and specify higher-gain transistors at the RF and IF frequencies for use in the FM tuner. By using high-gain transistors, a greater degree of selectivity and overall tuner stability can be achieved with a respectable amount of tuner gain still possible. This selectivity should be taken in the RF collector circuit and not the antenna tank because a high degree of selectivity in the antenna tank results in poorer noiše performance.

Figure 3 demonstrates how proper selectivity is taken in a TI-designed FM tuner using the TI-400 and TI-401 as RF amplifier and mixer. The antenna tank is approximately three MHz wide with an unloaded Q of 180. The TI-400 is driven under matched conditions on the input side. This results in optimum noise performance as well as power gain. The RF collector tank is designed to provide a high degree of selectivity. The unloaded Q of the coil is 180, and the loaded Q approximately 100. The simplest way to achieve this degree of selectivity is by tapping the coil for both the RF transistor collector and mixer base. The IF strip consists of four stages of IF amplification using the Dalmesa® TI-388 and TI-389 transistors.

Example 2. The second type of harmonic generation causes an undesirable response when the transmitter is operating at the tuner frequency plus one-half the IF frequency. The transmitter frequency appears at the base of the mixer where its second harmonic is produced to combine with the existing second harmonic of the oscillator. The difference frequency produced is the IF frequency. This type of harmonic generation usually only accounts for a small percentage of the undesirable response. If this type of harmonic generation is found to be objectionable, the oscillator injection can be cleaned up by the use of low-pass filters between the oscillator output and mixer input.

A degradation of spurious response rejection will result if the RF stage is improperly neutralized, even though an attempt has been made to provide ample selectivity in the tuner section. For this reason, proper neutralization of the RF stage is very important. This problem of neutralization can become formidable if proper selection of the RF transistor is not accomplished. Narrow limits should be placed on the effective feedback capacity at the operating point of the RF transistor so that once an RF stage has been designed, the degree of neutralization will not change appreciably when using the full spread of RF transistors.

AUTOMATIC GAIN CONTROL

There are two types of AGC generally used in FM receivers. These are reverse and forward AGC. Both of these types have their advantages and disadvantages.

Reverse AGC. Reverse AGC is decreasing overall transistor gain by decreasing the emitter current. This is usually done by controlling the base voltage of the RF transistor. Reverse AGC is inexpensive and easy to obtain. It also provides increasing selectivity under increasing signal conditions. With reverse AGC, both input and output impedance of the transistor increase as the emitter current decreases. This results in higher loaded antenna and RF collector tank impedances. Under large-signal conditions, when 0.1 to 0.3 volt of signal is present on the base of the RF transistor, it is possible for the RF stage to operate in a condition other than Class A, due to emitter-base-diode rectification. Because the transistor is still operating in the low-current region, the input impedance is greater than the driving source impedance and a major portion of the antenna voltage is present across the transistor input terminals which further aggravates this condition. Further increase in antenna voltage will result in Class C operation of the RF transistor, overload of the mixer, oscillator pulling, and very poor AM rejection. A partial solution to this problem of large input signal levels is to increase the mismatch between the driving source impedance and transistor input. This would solve one problem but create another. The additional problem would be degradation in FM tuner noise performance at low levels of input.

Forward AGC. Forward AGC is the technique which is used to decrease gain by increasing the current in the RF stage. In this form of AGC, the reduction in tuner gain is dependent on the decreasing power gain of the RF transistor which is caused by the increasing collector current and consequently the decreasing V_{CE}. Figure 4 shows the tuner power gain reduction versus V_{CE} for the TI-400 series transistor. The collector resistor used was 330 ohms and the emitter resistor used was 1.0 kilohm. A fairly linear reduction in gain is shown.

Forward AGC is superior to reverse AGC as far as tuner power gain reduction is concerned. In reverse AGC, the input and output impedances increase, and most of the antenna voltage is developed across the input impedance of the RF stage. In forward AGC, the converse of this is true. When the current is increased in a transistor, the input and output impedances drop to extremely low values. Thus, under large-signal conditions, when the transistor is saturated, only a small portion of the antenna voltage is across the transistor input, and a high insertion loss is obtained on both the input and output sides of the RF stage. Under this condition, large antenna voltages, up to three volts, can be present without oscillator pulling. The Boonton 202H FM-AM Signal Generator was used in conjunction with the Boonton 230A Power Amplifier to

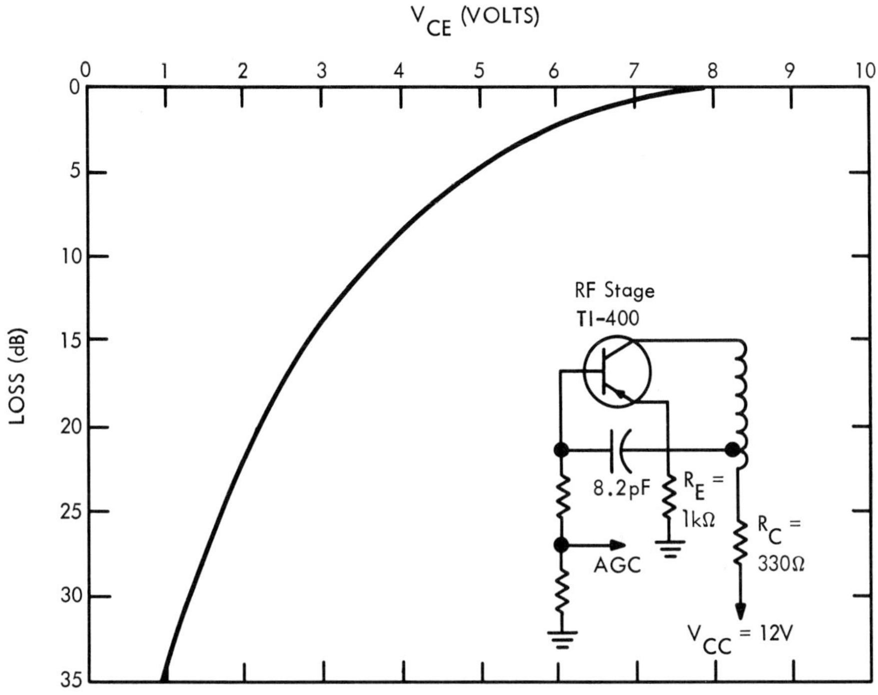

Fig. 4. Forward AGC Characteristics for an FM Tuner Using TI400

feed signals ranging from 5 μV to 5 V into the TI-constructed FM receiver. No degradation in AM rejection or oscillator frequency stability was observed. A forward AGC amplifier was used in conjunction with the receiver for this test.

Three disadvantages of forward AGC compared to reverse AGC are: (1) increased cost, (2) reduced tuner selectivity, and (3) detuning of RF stage with increasing signal. The increased cost would be associated with the possible requirement for a forward AGC amplifier. The decreased tuner selectivity and detuning of the RF stage are of no major consequence as long as the AGC is delayed until the receiver is well into limiting under fairly strong signal conditions.

This analysis of reverse and forward AGC demonstrates that there are applications where either type would be preferable. The primary differences that exist are signal handling and selectivity characteristics. Reverse AGC is recommended for portable applications where an external antenna is normally not required or used. Forward AGC is recommended for installations such as home stereo consoles which may use an external antenna. Both types of AGC may be effectively applied in the same receiver. An example of this would be forward AGC on the RF stage and reverse AGC on the first IF stage. The forward AGC would be used for handling large antenna signals and the reverse AGC used to maintain IF bandwidth under strong signal conditions.

SPECIFICATIONS FOR FM TUNER TRANSISTORS

It has already been shown that transistors play the major role in determining noise performance, interference effects, and AGC characteristics in FM tuners. Texas Instruments has developed the TI-400 and TI-401 epitaxial mesa transistor series specifically for FM tuner applications. These transistors have been parametrically specified at the a-c operating points and frequencies which are normally used in FM tuners. Each of the specifications has been placed on the transistors for particular reasons. Table 1 shows the complete set of specifications on the TI-400 FM RF amplifier and the TI-401 FM mixer. These parameters and their relationship to tuner problems are discussed in following paragraphs. The d-c parameters such as breakdown voltages and dissipation are considered to be sufficient and are fairly well self-explanatory; therefore, major emphasis is placed on the design parameters.

The a-c common emitter forward current transfer ratios h_{fe} of both the TI-400 and TI-401 are specified at their tuner operating points. A minimum and maximum h_{fe} limit is placed on both transistors. This ensures small gain variation and places input impedance restrictions on the transistors. The input impedance follows h_{fe}. High forward transconductance is also a characteristic of these transistors. The minimum h_{fe} assures high gain which is a prime requirement for good spurious response rejection. The imposed restrictions on the input impedance assures a continuance of matched load conditions on the input side of the transistors. The input impedance variations have less than a three-to-one spread at the transistor's specified operating points. These specifications ensure that once a tuner has been placed into production, negligible variation in gain or noise figure will be observed in tuners off the production line.

The d-c forward current transfer ratio h_{FE} is of importance with respect to AGC. A minimum specification is placed on both transistors. This limit is made higher on the TI-400 because the RF amplifier is normally gain controlled. Both of these transistors have a typical h_{FE} of 250. This high h_{FE} means that only a small amount of power is required to forward or reverse AGC the transistor.

The common emitter feedback capacity C_{re} is specified with minimum, typical, and maximum values. All three of these specifications are required for neutralization reasons. The spread must be as narrow as possible since the spread and not the absolute magnitude is one of the prime determining factors in establishing the maximum usable gain that can be achieved. The wider the spread, the more insertion loss must be taken to ensure stability. Improper neutralization can cause a lack of image rejection and spurious response. This is another reason for the tight spread requirements of C_{re}. The parameter C_{re} is therefore a determining factor in both gain and spurious response.

The maximum and typical specifications are placed on the $r_b'C_c$ product of both these devices. This parameter is a product of the base spreading resistance and the intrinsic collector capacitance. $r_b'C_c$ can be used in conjunction with the f_T of the transistor to determine f_{max}, the maximum frequency of oscillation. f_T is the frequency in megahertz where $h_{fe} = 1$. The typical specification for f_T is also shown on the data sheet. The maximum frequency of oscillation occurs where the power gain is equal to 1.

$$f_{max} = \sqrt{\frac{\alpha_o f_T}{8\pi r_b' C_c}}$$

Since $\alpha_o \approx 1.0$ for these transistors,

$$f_{max} \approx 200 \sqrt{\frac{f_T}{r_b' C_c}}$$

f_T and $r_b'C_c$ are in megahertz and picoseconds respectively.

f_{max} for the TI-400 is typically 2200 MHz.

The maximum available gain of the epitaxial mesa follows a 5 dB-per-octave slope from 10 MHz to f_{max}. The maximum available gain can be calculated at any frequency f in this 6 dB-per-octave region.

$$\text{Power gain at } f = \frac{f_{max}^2}{f^2}.$$

Both f_T and $r_b'C_c$ should be specified when selecting a transistor for use in FM tuners. The $r_b'C_c$ of a device is not only inversely proportional to the power gain, but is also inversely proportional to the transistor output impedance. By placing a maximum specification on $r_b'C_c$, the output impedance is held to a minimum limit. When specifying the $r_b'C_c$ product and C_{re} of a device, another limit has actually been set. The value of r_b' can be approximated. The intrinsic collector capacitance is directly proportional to C_{re}. r_b' in these devices is approximately equal to 25 ohms. r_b' is directly proportional to the device noise figure.

The FM RF amplifier should have a maximum noise figure specification. This specification need be placed on the RF amplifier only if this stage yields sufficient gain, as does the TI-400. It is advisable to specify this noise figure at a frequency higher than that at which the amplifier is to be operated. The noise figure on the TI-400 is specified at 200 MHz. This ensures superior noise performance at 100 MHz. The typical NF on the TI-400 is 3.5 dB at 200 MHz with a maximum specification of 5.5 dB. The high-frequency noise corner is at 400 MHz.

The major performance problems in an FM tuner are noise, gain, AGC, and spurious responses. Proper transistor selection can minimize these problems. Transistor parameter specifications, such as those on the TI-400 and TI-401 are required to ensure the production of component quality FM tuners.

REFERENCES

1. Terman, F. E., Sc.D, "Radio Engineers' Handbook," p. 673, McGraw-Hill Book Co. Inc., New York and London, 1943.
2. Wheeler, H. A., Two Signal Cross Modulation in a Frequency Modulated Receiver, *Proc. IRE,* Vol. 28, p. 537, December, 1940.

Table 1. Specifications for Texas Instruments High Frequency Epitaxial Mesa Series

		TI-400	TI-401
h_{fe}	$V_{CE} = -6$	12 dB min.	
	$I_E = -2$ mA	16 dB max.	
	f = 100 MHz		
h_{fe}	$V_{CE} = -6$ V		30 dB min.
	$I_E = -2$ mA		35 dB max.
	f = 10 MHz		
h_{FE}	$V_{CE} = -6$ V	40 min.	20 min.
	$I_E = -2$ mA		
C_{re}	$V_{CE} = -6$ V	0.3 pF min. 0.7 pF typ.	0.7 pF typ.
	$I_E = 0$ mA	1.0 pF max.	1.0 pF max.
$r_b'C_c$	$V_{CE} = -6$ V	15 ps max.	15 ps max.
	$I_E = -2$ mA	4.5 ps typ.	
	header nulled		
*N.F.	$V_{CE} = -6$ V	5.5 dB max.	
	$I_E = -2$ mA	3.5 dB typ.	
	Rg = 75 Ω		
	f = 200 MHz		
$f_{T(h_{fe}=1)}$	$V_{CE} = -6$ V	500 MHz typ.	400 MHz typ.
	$I_E = -2$ mA		
y_{fe}	$V_{CE} = -6$ V		70,000 μS typ.
	$I_E = -2$ mA		
	f = 10 MHz		

*Parameter is measured with a temperature-limited noise diode.

9

FM IF Amplifier Design

A straightforward procedure for the design of double-tuned IF amplifiers used in FM monophonic and FM stereophonic receivers is presented in this chapter. The method considers bandwidth, phase response, power gain, and stability.

AMPLITUDE RESPONSE

Frequency modulation produces an infinite series of sideband frequency pairs. However, the number of significant pairs (those with an amplitude of at least one percent of the unmodulated carrier's amplitude) depends upon the ratio of maximum signal deviation Δf_{max} to the highest modulating frequency f_m. Empirically, all significant sidebands will be passed when

$$BW = 2\left(\frac{\Delta f_{max}}{f_m} + 2\right) f_m \tag{1}$$

For broadcast-band FM monophonic transmission $\Delta f_{max} = 75$ kHz and $f_m = 15$ kHz so that the required receiver bandwidth from Eq. (1) is

$$BW = 2\left(\frac{75}{15} + 2\right) 15 = 210 \text{ kHz}$$

The higher modulating frequencies of FM stereophonic broadcasting, however, impose more stringent requirements on the phase and amplitude response of the FM receiver than are encountered with monophonic FM broadcasting. A 6-dB bandwidth of 240 to 260 kHz is generally accepted as being sufficient for high-quality multiplex reception. With this bandwidth, double-tuned IF transformers with coupling factors near unity must be used to provide adequate selectivity to suppress signals in adjacent channels.

PHASE RESPONSE

Tuned amplifiers exhibit a frequency-dependent phase shift between the secondary current at resonance and the secondary current at frequencies off resonance. At resonance, the secondary current is in phase with the secondary voltage. Above resonance, the tuned secondary is capacitive causing the secondary current to lead the secondary voltage. Hence, at frequencies above resonance, the secondary current has a leading phase angle with respect to the secondary current at resonance. Conversely, at frequencies below resonance, the secondary current lags with respect to the secondary current at resonance. If the phase does not vary linearly with frequency within the IF passband, time-delay distortion of the FM signal occurs. The phase shift between the secondary current at resonance and the secondary current at Δf Hz off resonance may be expressed as

$$\theta = \tan^{-1}\left[\frac{2X}{1 - X^2 + \alpha^2}\right] \tag{2}$$

where

$$X = \frac{2\Delta f Q_L}{f_o} = \left[\{(1-\alpha^2)^2 + (\rho^{2/N} - 1)(1+\alpha^2)^2\}^{1/2} - (1-\alpha^2)\right]^{1/2} \tag{3}$$

$\alpha = \dfrac{k}{k_c}$ = coupling factor (ratio of actual coefficient of coupling to the critical coefficient of coupling)

$k_c = \dfrac{1}{Q_L}$

ρ = attenuation at Δf Hz off resonance

N = number of identical transformers

f_o = resonant frequency

$Q_L = \sqrt{Q_{LUP}Q_{LUS}}$ (in which Q_{LUP} and Q_{LUS} are the loaded uncoupled Q's of the primary and secondary windings) **(4)**

As particular examples with the requirement that a 6-dB bandwidth of 260 kHz must be maintained in all cases (accomplished by letting Q_L vary), the phase shift versus frequency curves for a three-stage amplifier with coupling factor as a running parameter are shown in Fig. 1. Since it is somewhat difficult to tell by inspection of Fig. 1 which curve gives the most linear phase characteristic, the rate of change of phase with respect to frequency has been plotted versus frequency in Fig. 2. From Fig. 2 a coupling factor of 0.7 is seen to yield the most linear phase versus frequency characteristic. However, since power gain is proportional to the square of the coupling factor, a coupling factor of 0.8 to 0.85 is considered to give an adequate compromise between gain and phase shift.

For a required bandwidth, the necessary effective loaded uncoupled coil quality factor Q_L is determined from Eq. (5)

$$Q_L = \frac{Xf_o}{2\Delta f} \tag{5}$$

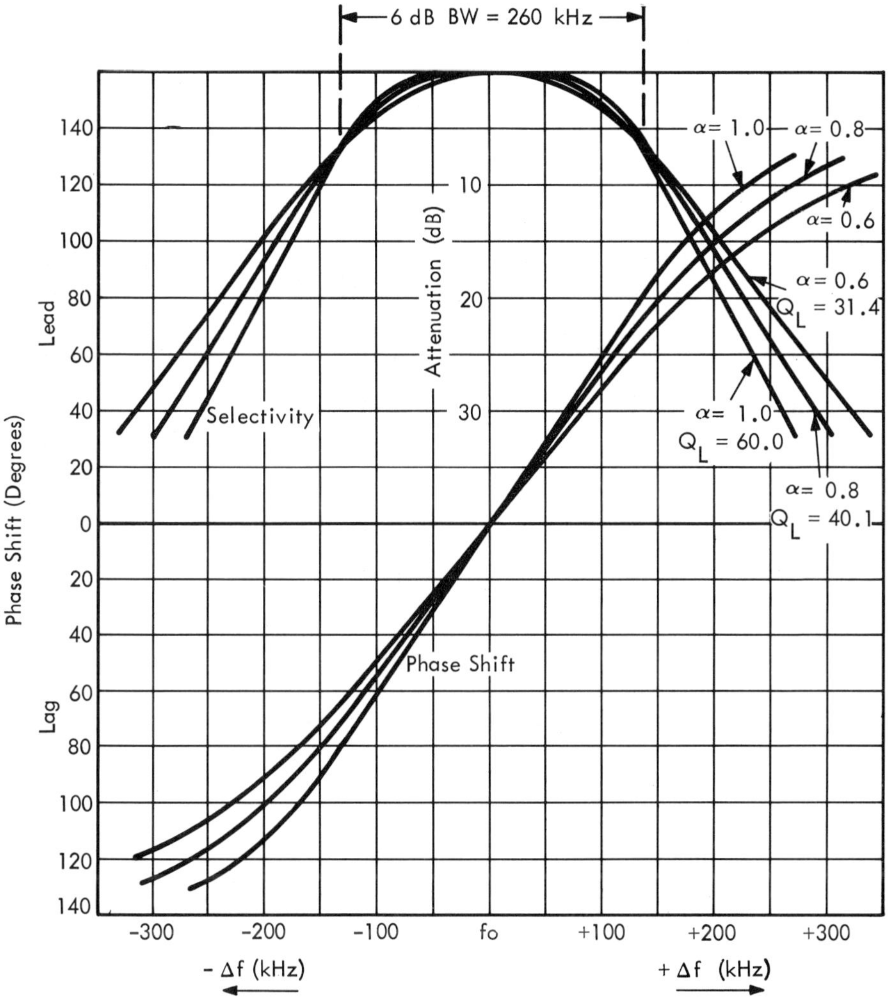

Fig. 1. Phase Shift and Attenuation vs. Frequency for a Three-stage 10.7-MHz IF Amplifier

where

X, f_0, and Δf are as previously defined. Note that 2Δf is equal to the bandwidth at any given attenuation corresponding to the ρ (times down) used in calculating X. The factor X is a dimensionless numeric not to be confused with inductive or capacitive reactance.

To simplify calculations, computer tabulated values of X and θ are given in Tables 1 through 9. To illustrate use of the tables in finding the required Q_L of an amplifier with a given number of stages, desired coupling factor, and required overall bandwidth let:

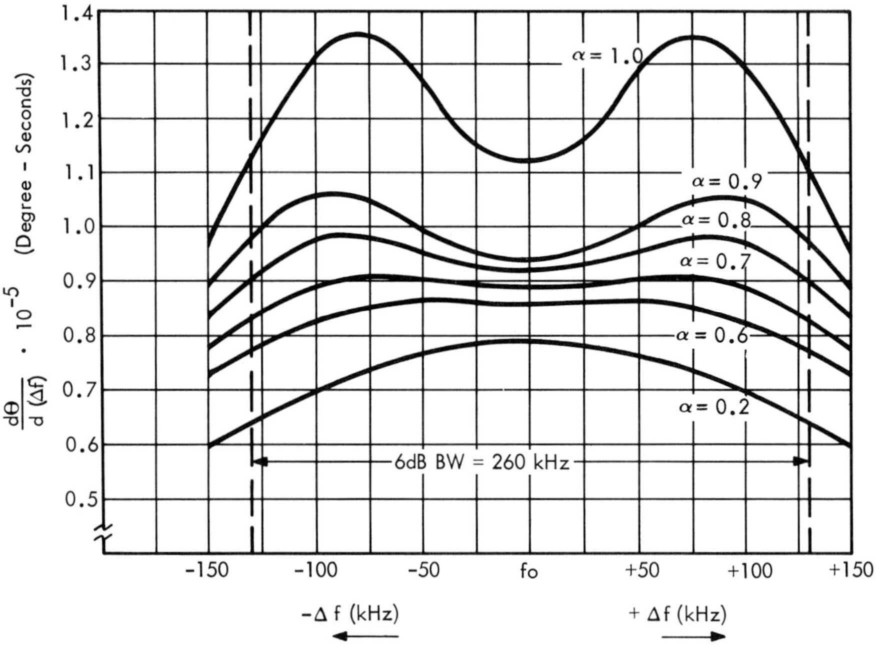

Fig. 2. Rate of Change of Phase with Respect to Frequency vs. Frequency for a Three-stage 10.7-MHz IF Amplifier

$f_o = 10.7$ mHz
$N = 4$ stages
$\alpha = 0.85$
6-dB BW = 260 kHz

From Table 6 find X = 0.9302 so that

$$Q_L = \frac{Xf_o}{2\Delta f} = \frac{0.9302\ (10.7 \times 10^6)}{260 \times 10^3} = 38.3$$

Once Q_L is determined, the selectivity curve may be plotted if desired. For instance, using the Q_L above, the overall 40-dB bandwidth is found simply by rearranging Eq. (5) and using the proper value of X.

$$\text{40-dB BW} = 2\Delta f = \frac{Xf_o}{Q_L} = \frac{2.2130\ (10.7 \times 10^6)}{38.3}$$

$$= 618 \text{ kHz}$$

Table 1

ALPHA(α) = 0.400

RHO(ρ)	DB DOWN	N = 1 X	N = 1 THETA(θ)	N = 2 X	N = 2 THETA(θ)	N = 3 X	N = 3 THETA(θ)	N = 4 X	N = 4 THETA(θ)
1.029	0.24831	0.2142	21.0314	0.1514	14.9094	0.1236	12.1839	0.1070	10.5561
1.059	0.49792	0.3037	29.6338	0.2145	21.0587	0.1751	17.2236	0.1516	14.9288
1.122	0.99986	0.4320	41.5931	0.3043	29.6924	0.2482	24.3241	0.2149	21.1007
1.414	3.00899	0.7694	69.7361	0.5326	50.5580	0.4327	41.6559	0.3739	36.2453
2.000	6.02060	1.1566	94.3932	0.7695	69.7499	0.6190	57.8886	0.5327	50.5685
4.000	12.04120	1.9315	123.6413	1.1566	94.3932	0.9047	79.3127	0.7695	69.7499
7.000	16.90196	2.6908	138.4892	1.4570	108.2865	1.1090	91.8021	0.9321	81.1228
10.000	20.00000	3.2760	145.6094	1.6553	115.5127	1.2360	98.4595	1.0302	87.2589
20.000	26.02060	4.7272	155.9512	2.0702	127.0499	1.4851	109.3908	1.2161	97.4718
40.000	32.04120	6.7493	163.0871	2.5402	136.1725	1.7452	118.3800	1.4019	106.0269
70.000	36.90196	8.9642	167.2445	2.9719	142.2345	1.9681	124.5797	1.5552	112.0300
100.000	40.00000	10.7311	169.3378	3.2760	145.6094	2.1177	128.1300	1.6553	115.5127
1000.000	60.00000	34.0464	176.6348	5.9861	160.9511	3.2760	145.6094	2.3736	133.3027
10000.000	80.00000	107.6994	178.9360	10.7311	169.3378	4.9131	156.8473	3.2760	145.6094

Table 2

ALPHA(α) = 0.500

RHO(ρ)	DB DOWN	N = 1		N = 2		N = 3		N = 4	
		X	THETA(θ)	X	THETA(θ)	X	THETA(θ)	X	THETA(θ)
1.029	0.24831	0.2428	22.1852	0.1721	15.7520	0.1406	12.8796	0.1219	11.1619
1.059	0.49792	0.3426	31.1706	0.2432	22.2139	0.1988	18.1872	0.1723	15.7725
1.122	0.99986	0.4831	43.5416	0.3433	31.2315	0.2810	25.6325	0.2436	22.2579
1.414	3.00899	0.8411	72.1226	0.5914	52.7219	0.4838	43.6062	0.4199	38.0275
2.000	6.02060	1.2415	96.6912	0.8413	72.1364	0.6831	60.1733	0.5915	52.7326
4.000	12.04120	2.0369	125.4363	1.2415	96.6912	0.9818	81.7080	0.8413	72.1364
7.000	16.90196	2.8183	139.8961	1.5500	110.3925	1.1925	94.1256	1.0101	83.5132
10.000	20.00000	3.4220	146.8033	1.7534	117.4845	1.3231	100.7108	1.1114	89.6186
20.000	26.02060	4.9224	156.8094	2.1794	128.7630	1.5787	111.4778	1.3027	99.7352
40.000	32.04120	7.0171	163.6990	2.6631	137.6452	1.8456	120.2925	1.4934	108.1702
70.000	36.90196	9.3137	167.7086	3.1081	143.5314	2.0745	126.3526	1.6506	114.0693
100.000	40.00000	11.1466	169.7266	3.4220	146.8033	2.2283	129.8162	1.7534	117.4845
1000.000	60.00000	35.3447	176.7581	6.2262	161.6377	3.4220	146.8033	2.4915	134.8542
10000.000	80.00000	111.8000	178.9750	11.1466	169.7266	5.1149	157.6752	3.4220	146.8033

Table 3

ALPHA(α) = 0.600

RHO(ρ)	DB DOWN	N = 1		N = 2		N = 3		N = 4	
		X	THETA(θ)	X	THETA(θ)	X	THETA(θ)	X	THETA(θ)
1.029	0.24831	0.2829	23.8459	0.2015	16.9878	0.1650	13.9067	0.1431	12.0596
1.059	0.49792	0.3955	33.3135	0.2832	23.8763	0.2324	19.5911	0.2018	17.0097
1.122	0.99986	0.5501	46.1334	0.3963	33.3772	0.3262	27.4941	0.2838	23.9230
1.414	3.00899	0.9288	75.0125	0.6664	55.5050	0.5509	46.1997	0.4811	40.4487
2.000	6.02060	1.3420	99.3305	0.9290	75.0263	0.7636	63.0390	0.6666	55.5158
4.000	12.04120	2.1601	127.4249	1.3420	99.3305	1.0745	84.5411	0.9290	75.0263
7.000	16.90196	2.9672	141.4386	1.6590	112.7632	1.2916	96.8060	1.1038	86.3293
10.000	20.00000	3.5926	148.1076	1.8681	119.6859	1.4258	103.2796	1.2082	92.3640
20.000	26.02060	5.1511	157.7435	2.3071	130.6556	1.6885	113.8238	1.4049	102.3218
40.000	32.04120	7.3312	164.3637	2.8066	139.2619	1.9631	122.4215	1.6009	110.5898
70.000	36.90196	9.7238	168.2126	3.2673	144.9504	2.1989	128.3152	1.7625	116.3547
100.000	40.00000	11.6342	170.1488	3.5926	148.1076	2.3574	131.6776	1.8681	119.6859
1000.000	60.00000	36.8695	176.8919	6.5077	162.3841	3.5926	148.1076	2.6293	136.5607
10000.000	80.00000	116.6163	179.0174	11.6342	170.1488	5.3512	158.5760	3.5926	148.1076

Table 4

ALPHA(α) = 0.700

RHO(ρ)	DB DOWN	N = 1		N = 2		N = 3		N = 4	
		X	THETA(θ)	X	THETA(θ)	X	THETA(θ)	X	THETA(θ)
1.029	0.24831	0.3392	26.2652	0.2442	18.8525	0.2007	15.4779	0.1744	13.4426
1.059	0.49792	0.4668	36.2727	0.3397	26.2977	0.2806	21.6828	0.2445	18.8765
1.122	0.99986	0.6354	49.4744	0.4676	36.3392	0.3888	30.1514	0.3403	26.3478
1.414	3.00899	1.0317	78.3466	0.7589	58.9416	0.6362	49.5419	0.5609	43.6626
2.000	6.02060	1.4563	102.2182	1.0319	78.3602	0.8606	66.4745	0.7590	58.9524
4.000	12.04120	2.2985	129.5277	1.4563	102.2182	1.1817	87.7314	1.0319	78.3602
7.000	16.90196	3.1346	143.0535	1.7819	115.3079	1.4045	99.7513	1.2118	89.4877
10.000	20.00000	3.7847	149.4683	1.9972	122.0307	1.5423	106.0728	1.3189	95.4050
20.000	26.02060	5.4090	158.7143	2.4505	132.6515	1.8123	116.3389	1.5208	105.1384
40.000	32.04120	7.6859	165.0535	2.9680	140.9567	2.0951	124.6829	1.7221	113.1941
70.000	36.90196	10.1873	168.7351	3.4464	146.4332	2.3386	130.3888	1.8884	118.7976
100.000	40.00000	12.1854	170.5863	3.7847	149.4683	2.5026	133.6390	1.9972	122.0307
1000.000	60.00000	38.5939	177.0305	6.8255	163.1589	3.7847	149.4683	2.7841	138.3526
10000.000	80.00000	122.0635	179.0612	12.1854	170.5863	5.6179	159.5120	3.7847	149.4683

Table 5

ALPHA(α) = 0.800

RHO(ρ)	DB DOWN	N = 1		N = 2		N = 3		N = 4	
		X	THETA(θ)	X	THETA(θ)	X	THETA(θ)	X	THETA(θ)
1.029	0.24831	0.4201	29.8635	0.3092	21.8248	0.2566	18.0581	0.2242	15.7521
1.059	0.49792	0.5617	40.3049	0.4206	29.8981	0.3523	24.9305	0.3096	21.8513
1.122	0.99986	0.7408	53.6279	0.5626	40.3731	0.4761	33.9658	0.4214	29.9514
1.414	3.00899	1.1483	82.0323	0.8689	63.0095	0.7417	53.6952	0.6625	47.8080
2.000	6.02060	1.5822	105.2600	1.1485	82.0455	0.9734	70.4176	0.8690	63.0202
4.000	12.04120	2.4499	131.6764	1.5822	105.2600	1.3014	91.1805	1.1485	82.0455
7.000	16.90196	3.3178	144.6888	1.9166	117.9435	1.5292	102.8655	1.3321	92.8899
10.000	20.00000	3.9952	150.8419	2.1385	124.4424	1.6704	108.9988	1.4416	98.6437
20.000	26.02060	5.6922	159.6910	2.6072	134.6863	1.9479	118.9409	1.6483	108.0927
40.000	32.04120	8.0759	165.7463	3.1445	142.6751	2.2396	127.0033	1.8551	115.8978
70.000	36.90196	10.6971	169.2595	3.6425	147.9321	2.4913	132.5064	2.0263	121.3183
100.000	40.00000	12.7919	171.0253	3.9952	150.8419	2.6612	135.6371	2.1385	124.4424
1000.000	60.00000	40.4925	177.1695	7.1747	163.9374	3.9952	150.8419	2.9534	140.1722
10000.000	80.00000	128.0611	179.1052	12.7919	171.0253	5.9107	160.4534	3.9952	150.8419

Table 6

ALPHA(α) = 0.850

RHO(ρ)	DB DOWN	N = 1		N = 2		N = 3		N = 4	
		X	THETA(θ)	X	THETA(θ)	X	THETA(θ)	X	THETA(θ)
1.029	0.24831	0.4734	32.2861	0.3554	24.0023	0.2978	20.0296	0.2617	17.5621
1.059	0.49792	0.6196	42.7911	0.4739	32.3214	0.4017	27.2304	0.3558	24.0300
1.122	0.99986	0.8011	56.0007	0.6205	42.8592	0.5316	36.4387	0.4746	32.3755
1.414	3.00899	1.2111	83.9711	0.9301	65.2511	0.8020	56.0671	0.7220	50.2459
2.000	6.02060	1.6489	106.8110	1.2113	83.9842	1.0351	72.5443	0.9302	65.2616
4.000	12.04120	2.5297	132.7506	1.6489	106.8110	1.3653	92.9684	1.2113	83.9842
7.000	16.90196	3.4146	145.5014	1.9877	119.2727	1.5953	104.4573	1.3963	94.6494
10.000	20.00000	4.1065	151.5229	2.2130	125.6533	1.7382	110.4854	1.5068	100.3069
20.000	26.02060	5.8421	160.1740	2.6899	135.7019	2.0194	120.2522	1.7158	109.5950
40.000	32.04120	8.2826	166.0885	3.2378	143.5296	2.3158	128.1665	1.9253	117.2635
70.000	36.90196	10.9675	169.5184	3.7461	148.6759	2.5719	133.5645	2.0990	122.5866
100.000	40.00000	13.1135	171.2420	4.1065	151.5229	2.7450	136.6340	2.2130	125.6533
1000.000	60.00000	41.4997	177.2381	7.3598	164.3220	4.1065	151.5229	3.0428	141.0781
10000.000	80.00000	131.2430	179.1269	13.1135	171.2420	6.0658	160.9189	4.1065	151.5229

Table 7

ALPHA(α) = 0.900

RHO(ρ)	DB DOWN	N = 1		N = 2		N = 3		N = 4	
		X	THETA(θ)	X	THETA(θ)	X	THETA(θ)	X	THETA(θ)
1.029	0.24831	0.5370	35.2174	0.4148	26.8639	0.3534	22.7519	0.3139	20.1463
1.059	0.49792	0.6847	45.5981	0.5375	35.2525	0.4633	30.1472	0.4153	26.8923
1.122	0.99986	0.8662	58.5473	0.6857	45.6650	0.5961	39.3381	0.5383	35.3064
1.414	3.00899	1.2765	85.9534	0.9950	67.6050	0.8671	58.6124	0.7872	52.9112
2.000	6.02060	1.7178	108.3687	1.2767	85.9662	1.1001	74.7499	0.9951	67.6154
4.000	12.04120	2.6120	133.8172	1.7178	108.3687	1.4316	94.7813	1.2767	85.9662
7.000	16.90196	3.5145	146.3052	2.0610	120.5993	1.6637	106.0583	1.4628	96.4311
10.000	20.00000	4.2214	152.1957	2.2898	126.8586	1.8081	111.9755	1.5743	101.9841
20.000	26.02060	5.9971	160.6504	2.7752	136.7093	2.0932	121.5604	1.7855	111.1016
40.000	32.04120	8.4963	166.4258	3.3339	144.3753	2.3943	129.3232	1.9977	118.6279
70.000	36.90196	11.2471	169.7736	3.8531	149.4112	2.6550	134.6149	2.1740	123.8506
100.000	40.00000	13.4462	171.4555	4.2214	152.1957	2.8313	137.6225	2.2898	126.8586
1000.000	60.00000	42.5419	177.3057	7.5511	164.7013	4.2214	152.1957	3.1350	141.9752
10000.000	80.00000	134.5355	179.1482	13.4462	171.4555	6.2261	161.3780	4.2214	152.1957

Table 8

ALPHA(α) = 0.950

RHO(ρ)	DB DOWN	N = 1 X	N = 1 THETA(θ)	N = 2 X	N = 2 THETA(θ)	N = 3 X	N = 3 THETA(θ)	N = 4 X	N = 4 THETA(θ)
1.029	0.24831	0.6117	38.6768	0.4908	30.5692	0.4288	26.5166	0.3882	23.9048
1.059	0.49792	0.7568	48.6986	0.6122	38.7107	0.5389	33.7682	0.4912	30.5969
1.122	0.99986	0.9356	61.2383	0.7577	48.7632	0.6698	42.6551	0.6130	38.7628
1.414	3.00899	1.3442	87.9637	1.0632	70.0470	0.9365	61.3014	0.8577	55.7726
2.000	6.02060	1.7886	109.9246	1.3444	87.9763	1.1678	77.0140	1.0634	70.0570
4.000	12.04120	2.6966	134.8716	1.7886	109.9246	1.5000	96.6069	1.3444	87.9763
7.000	16.90196	3.6171	147.0971	2.1363	121.9170	1.7339	107.6593	1.5314	98.2232
10.000	20.00000	4.3396	152.8577	2.3687	128.0529	1.8800	113.4612	1.6437	103.6653
20.000	26.02060	6.1567	161.1186	2.8628	137.7043	2.1690	122.8592	1.8571	112.6042
40.000	32.04120	8.7166	166.7570	3.4328	145.2089	2.4749	130.4683	2.0721	119.9840
70.000	36.90196	11.5353	170.0240	3.9631	150.1352	2.7404	135.6529	2.2510	125.1044
100.000	40.00000	13.7892	171.6651	4.3396	152.8577	2.9200	138.5986	2.3687	128.0529
1000.000	60.00000	43.6165	177.3720	7.7482	165.0737	4.3396	152.8577	3.2297	142.8601
10000.000	80.00000	137.9308	179.1692	13.7892	171.6651	6.3911	161.8291	4.3396	152.8577

Table 9

ALPHA(α) = 1.000

RHO(ρ)	DB DOWN	N = 1		N = 2		N = 3		N = 4	
		X	THETA(θ)	X	THETA(θ)	X	THETA(θ)	X	THETA(θ)
1.029	0.24831	0.6965	42.6012	0.5836	35.1217	0.5267	31.4473	0.4899	29.1028
1.059	0.49792	0.8349	52.0360	0.6970	42.6328	0.6283	38.0535	0.5840	35.1470
1.122	0.99986	1.0088	64.0373	0.8358	52.0974	0.7516	46.3252	0.6977	42.6814
1.414	3.00899	1.4140	89.9875	1.1344	72.5505	1.0097	64.0981	0.9326	58.7856
2.000	6.02060	1.8612	111.4709	1.4142	89.9998	1.2381	79.3158	1.1345	72.5602
4.000	12.04120	2.7832	135.9099	1.8612	111.4709	1.5702	98.4340	1.4142	89.9998
7.000	16.90196	3.7224	147.8747	2.2134	123.2202	1.8060	109.2521	1.6018	100.0151
10.000	20.00000	4.4609	153.5069	2.4495	129.2316	1.9536	114.9354	1.7149	105.3412
20.000	26.02060	6.3206	161.5770	2.9526	138.6835	2.2466	124.1433	1.9305	114.0958
40.000	32.04120	8.9429	167.0811	3.5341	146.0277	2.5575	131.5976	2.1483	121.3262
70.000	36.90196	11.8316	170.2690	4.0759	150.8455	2.8278	136.6751	2.3299	126.3430
100.000	40.00000	14.1418	171.8701	4.4609	153.5069	3.0108	139.5590	2.4495	129.2316
1000.000	60.00000	44.7213	177.4368	7.9507	165.4383	4.4609	153.5069	3.3268	143.7296
10000.000	80.00000	141.4214	179.1897	14.1418	171.8701	6.5607	162.2708	4.4609	153.5069

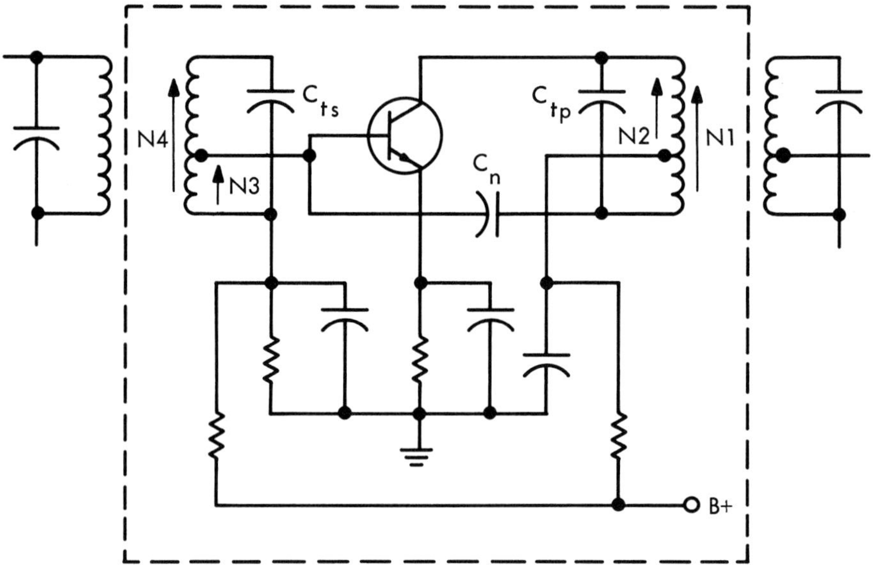

Fig. 3. Typical Neutralized IF Amplifier Stage

DESIGN EQUATIONS

Both neutralized and unneutralized examples are considered. Typical neutralized and unneutralized stages (with coil turns ratios defined) are shown in Fig. 3 and 4. In this treatment, a "stage" is defined as containing the secondary of the transformer on the transistor's input and the primary of the transformer on the transistor's output. With standard IF frequencies and with commonly used transistors having f_T's in excess of 200 MHz, the phase angle associated with the common-emitter forward transfer admittance may be neglected so that

$$y_{fe} = g_{fe} + jb_{fe} \approx g_{fe}$$

The power gain obtained when reverse feedback is neglected and the input and output impedances of the transistor are complex conjugately matched by the generator and the load is referred to as the Maximum Available Gain (MAG) which, for the common emitter configuration, is expressed as

$$\text{MAG} = \frac{|y_{fe}|^2 \, r_{iep} \, r_{oep}}{4} \tag{6}$$

The transistor parameters are defined by the equivalent circuit shown in Fig. 5. Since feedback has been neglected, MAG can never be obtained in practice. Thus, MAG should be used only as a figure of merit.

The minimum feedback capacitance under a conjugate matched condition to cause oscillation is shown in a later section on Equation Derivations to be

$$\Delta C_{fb} = \frac{8}{\omega y_{fe} \, r_{iep} \, r_{oep}} \tag{7}$$

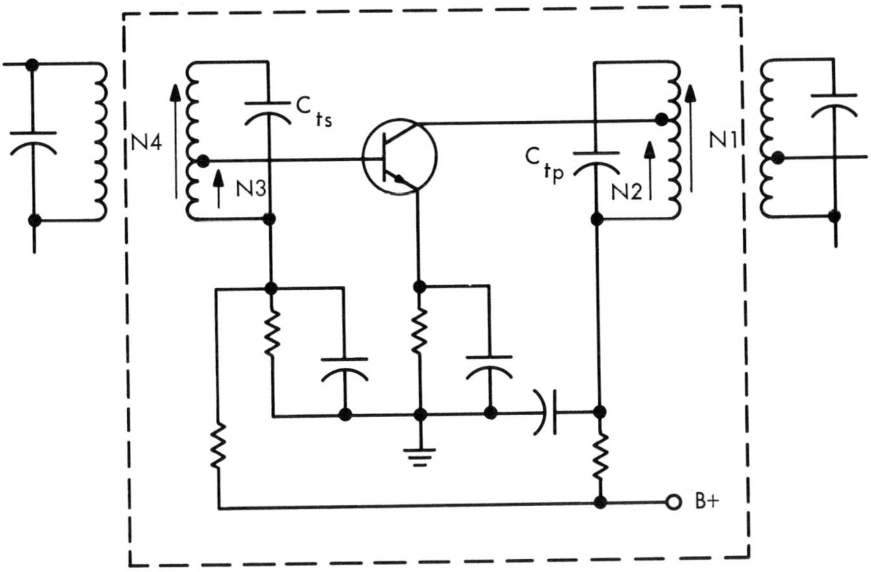

Fig. 4. Typical Unneutralized IF Amplifier Stage

The actual feedback capacitance is defined as C_{fdbk}. To eliminate objectionable skew in the selectivity curve and to maintain stability, the allowable C_{fdbk} must be increased by a skew factor K_s. Practical experiments show that a skew factor K_s of 4 gives excellent results for high-quality FM monophonic and FM stereophonic receivers.

For more than a one-stage amplifier, however, an additional factor K_n must be introduced to ensure stable operation. A stability factor now may be defined as

$$S \equiv K_n K_s \frac{C_{fdbk}}{\Delta C_{fb}} \tag{8}$$

or

$$S = \frac{K_n K_s \ \omega \ C_{fdbk} \ y_{fe} \ r_{iep} \ r_{oep}}{8} \tag{9}$$

where

$$\begin{aligned} K_n &= 1.00 \text{ for a one-stage amplifier} \\ &= 2.00 \text{ for a two-stage amplifier} \\ &= 2.61 \text{ for a three-stage amplifier} \\ &= 3.00 \text{ for a four-stage amplifier.} \end{aligned}$$

Therefore,

One-stage: $S = (0.50) \ \omega \ C_{fdbk} \ y_{fe} \ r_{iep} \ r_{oep}$ **(9a)**

Two-stage: $S = (1.00) \ \omega \ C_{fdbk} \ y_{fe} \ r_{iep} \ r_{oep}$ **(9b)**

Three-stage: $S = (1.31) \ \omega \ C_{fdbk} \ y_{fe} \ r_{iep} \ r_{oep}$ **(9c)**

Four-stage: $S = (1.50) \ \omega \ C_{fdbk} \ y_{fe} \ r_{iep} \ r_{oep}$ **(9d)**

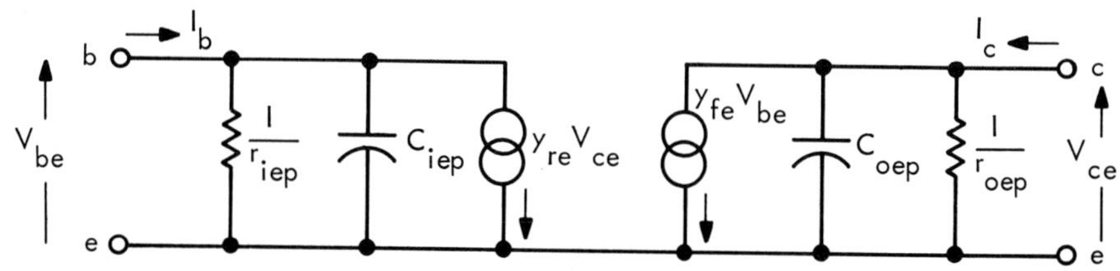

$$I_b = y_{ie}V_{be} + y_{re}V_{ce}$$

$$I_c = y_{fe}V_{be} + y_{oe}V_{ce}$$

$$y_{ie} = \frac{I_b}{V_{be}}\bigg|_{V_{ce}=0} = \frac{1}{r_{iep}} + j\omega C_{iep} \qquad y_{fe} = \frac{I_c}{V_{be}}\bigg|_{V_{ce}=0}$$

$$y_{re} = \frac{I_b}{V_{ce}}\bigg|_{V_{be}=0} \qquad y_{oe} = \frac{I_c}{V_{ce}}\bigg|_{V_{be}=0} = \frac{1}{r_{oep}} + j\omega C_{oep}$$

Fig. 5. Common-emitter Equivalent Circuit Using Short-circuit y-Parameters

The relationship between the stability factor and the coil insertion loss may be shown to be

$$\text{Insertion Loss} = \text{IL} = \text{S}\frac{1}{\alpha^2} \tag{10}$$

or, using the definition of S,

$$\text{IL} = \frac{K_nK_s\ \omega\ C_{fdbk}\ y_{fe}\ r_{iep}\ r_{oep}}{8}\left(\frac{1}{\alpha^2}\right) \tag{11}$$

The maximum usable power gain per stage (MUG) is expressed simply as

$$\text{MUG} = \frac{\text{MAG}}{\text{IL}} \tag{12}$$

Finally, substituting Eq. (6) for MAG and Eq. (11) for IL, the maximum usable power gain per stage is:

$$\text{MUG} = \frac{2}{K_nK_s}\left|\frac{y_{fe}}{\omega\ C_{fdbk}}\right|\alpha^2 \tag{13}$$

Employing previously noted values for K_n and K_s:

$$\text{One-stage: MUG} = 0.50\left|\frac{y_{fe}}{\omega C_{fdbk}}\right|\alpha^2 \tag{13a}$$

$$\text{Two-stage: MUG} = 0.25 \left| \frac{y_{fe}}{\omega C_{fdbk}} \right| \alpha^2 \tag{13b}$$

$$\text{Three-stage: MUG} = 0.19 \left| \frac{y_{fe}}{\omega C_{fdbk}} \right| \alpha^2 \tag{13c}$$

$$\text{Four-stage: MUG} = 0.17 \left| \frac{y_{fe}}{\omega C_{fdbk}} \right| \alpha^2 \tag{13d}$$

The total amplifier gain in decibels then is

$$\text{Total Gain} = (\text{Number of Stages})\ (10 \log \text{MUG}) \tag{14}$$

For an unneutralized design, the value of C_{fdbk} to be used in the equations should be the maximum value of collector-to-base capacitance from a distribution of transistors. For a neutralized design, both the tolerance on the neutralizing capacitor and the maximum variation from average of the transistor's feedback capacitance must be considered.

The value of C_{fdbk} for a neutralized design is

$$C_{fdbk} = (\text{tolerance on neutralizing capacitor})(C_n) + \text{Maximum } C_{cb} \text{ variation from average} \tag{15}$$

where

C_n = neutralizing capacitor value.

However, as shown later,

$$C_n \approx C_{cb\ avg}$$

so that very little error is introduced if in the equations

$$C_{fdbk} = (\text{tolerance on neutralizing capacitor})\ (C_{cb\ avg}) + \text{maximum } C_{cb} \text{ variation from average} \tag{16}$$

As an example, let $C_{cb} = 1.0 \pm 0.2$ pF

For an unneutralized design: $C_{fdbk} = 1.2$ pF
For a neutralized design using a 1.0 pF 10% neutralizing capacitor:

$$C_{fdbk} = 0.1\ (1.0) + 0.2 = 0.3 \text{ pF}$$

At this point, it is beneficial to define an input stability factor IS and an output stability factor OS such that

$$(\text{IS})\ (\text{OS}) = \text{S} \tag{17}$$

As shown in the section on Equation Derivations,

$$Q_{UUP} = \frac{2(\text{OS})\ Q_{LUP}}{2(\text{OS}) - (1 + \alpha^2)} \tag{18}$$

where

Q_{UUP} = unloaded uncoupled Q of the primary winding.

Similarly,

$$Q_{UUS} = \frac{2(IS)\ Q_{LUS}}{2(IS) - (1 + \alpha^2)} \qquad (19)$$

where

Q_{UUS} = unloaded uncoupled Q of the secondary winding.

For transistors having MAG values near 50 dB, practical coil winding considerations usually require that $|10 \log (OS) - 10 \log (IS)| \leqq 6$ dB since large differences between the input and output stability may require unreasonably large turns ratios. Also, by rearranging Equations (18) and (19) so that

$$OS = \left[\frac{Q_{UUP}}{Q_{UUP} - Q_{LUP}}\right]\left[\frac{1 + \alpha^2}{2}\right] \qquad (20)$$

$$IS = \left[\frac{Q_{UUS}}{Q_{UUS} - Q_{LUS}}\right]\left[\frac{1 + \alpha^2}{2}\right] \qquad (21)$$

it becomes evident that a high value for OS or for IS requires a small difference in Q_U and Q_L which causes difficulty in adequately controlling the coil loss.

Using the basic relationships that

$$Q_{LUP}X_{cp} = \frac{(Q_{UUP}X_{cp})(r_{oep})(N1/N2)^2}{(Q_{UUP}X_{cp}) + (r_{oep})(N1/N2)^2}$$

$$Q_{LUS}X_{cs} = \frac{(Q_{UUS}X_{cs})(r_{iep})(N4/N3)^2}{(Q_{UUS}X_{cs}) + (r_{iep})(N4/N3)^2}$$

where

$X_{cp} = \frac{1}{\omega C_{tp}}$ = capacitive reactance of primary tuning capacitor at resonance

$X_{cs} = \frac{1}{\omega C_{ts}}$ = capacitive reactance of secondary tuning capacitor at resonance

then

$$\frac{N1}{N2} = \left[\frac{Q_{UUP}Q_{LUP}X_{cp}}{(Q_{UUP} - Q_{LUP})\ r_{oep}}\right]^{1/2} \qquad (22)$$

$$\frac{N4}{N3} = \left[\frac{Q_{UUS}Q_{LUS}X_{cs}}{(Q_{UUS} - Q_{LUS})\ r_{iep}}\right]^{1/2} \qquad (23)$$

Because the loaded Q is determined from bandwidth requirements and the unloaded Q is determined from stability considerations, tuning capacitance is the only variable which can be manipulated to give a reasonable value for the turns ratio. Note that the primary and secondary tuning capacitances need not be of the same value. The designer should keep in mind, however, that as the tuning capacitor value increases, the total number of coil turns decreases and it is impractical to tap less than one turn.

With reference to Figure 3 for a neutralized design, the following equation must be satisfied:

$$C_n = \frac{1}{\frac{N1}{N2} - 1} C_{cb\ avg} \tag{24}$$

Common values for the neutralizing capacitor are 0.82, 1.0, 1.2, 1.5, 1.8, 2.2, 3.3, and 4.7 pF. In a composite AM/FM IF amplifier C_n should be restricted to small values because it adds to C_{cb} during AM operation.

From the equations just developed, a straightforward design procedure emerges.

DESIGN PROCEDURES

Both neutralized unneutralized procedures are presented. Known values are given:

1. Transistor specifications at a particular bias point and frequency.
   r_{iep}
   r_{oep}
   y_{fe}
   C_{cb} (average and maximum)
2. Center frequency f_o.
3. Desired coupling factor α.
4. Required bandwidth BW.
5. Number of stages.

Neutralized Procedure

1. Calculate C_{fdbk} from Eq. (16).
2. Calculate maximum usable stage gain MUG from Eq. (13).
3. Calculate X from Equation (3) or Tables 1 through 9.
4. Calculate required effective loaded uncoupled Q_L from Eq. (5).
5. Calculate required stability factor S from Eq. (9).
6. Arbitrarily choose values for OS and IS so that 10 log (OS) + 10 log (IS) = 10 log S.
7. Calculate Q_{UUP} and Q_{UUS} from Eq. (18) and (19).
8. Choose a standard value of secondary tuning capacitance C_{ts} to give a reasonable value (< 20) for N4/N3 from Eq. (23). If a standard value of C_{ts} cannot be chosen to yield a reasonable N4/N3, a new set of input and output stability factors must be chosen and steps (7) and (8) repeated. All transformer secondaries are now designed.
9. Choose a standard value of primary tuning capacitance C_{tp} to give reasonable values for both the primary turns ratio and the neutralizing capacitor. The primary turns ratio N1/N2 is calculated from Eq. (22) and the neutralizing capacitor value C_n then is calculated from Eq. (24). If a standard value of C_{tp} cannot be chosen to yield reasonable N1/N2 and C_n, a new set of input and output stability factors must be chosen and steps (7), and (8), and (9) repeated. Except for the AGC stage, all transformer primaries are now designed.

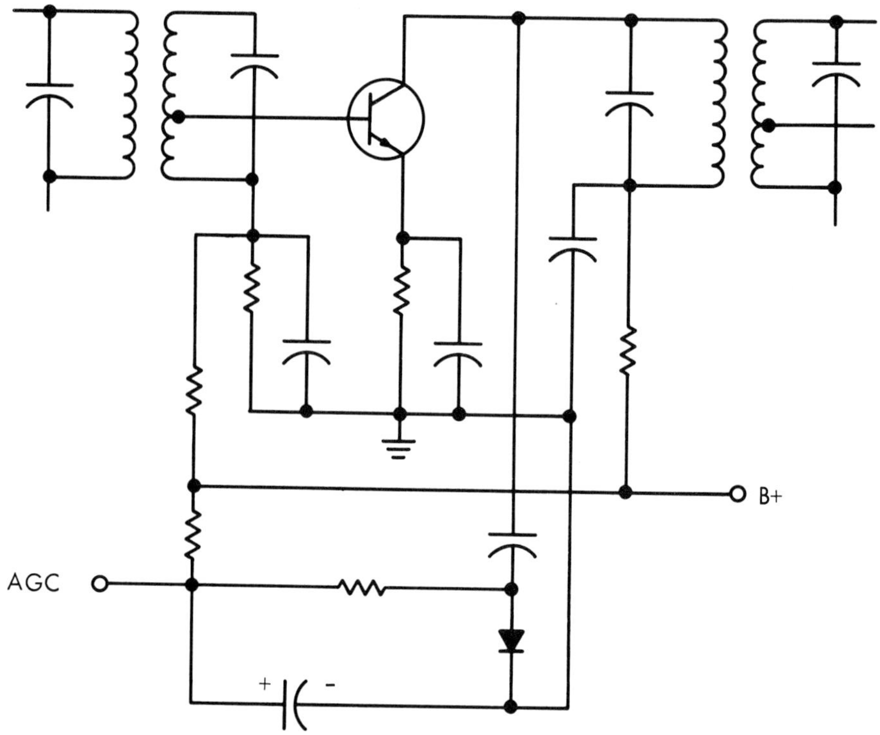

Fig. 6. Typical AGC Stage

10. The loading effect of the AGC network must be considered if the proper bandwidth and stability are to be achieved. The AGC voltage is taken from the first or second stage on a two-stage IF amplifier and is usually taken from the second stage on both three- and four-stage IF amplifiers. A typical AGC stage is shown in Fig. 6. The Q_{LUP} to be used for the AGC stage is the same as found in step (4). The only precaution is that the Q_{LUP} should be measured with a resistor R_{agc} simulating the equivalent resistance for the AGC network in parallel with the resistor R_g simulating the transistor's r_{oep}. Also, the Q_{UUP} to be used is the same as found in step (7) with the precaution that is should be measured with R_{agc} loading the coil. The quality factor of the AGC primary with no loading resistors is readily determined from the relationship

$$\underset{(\text{coil} + R_{agc})}{Q_{UUP}X_{cp}} = \frac{[R_{agc}(N1/N2)^2]\ Q_{UUP}X_{cp}}{R_{agc}(N1/N2)^2 + Q_{UUP}X_{cp}}$$

Solving for Q'_{UUP} gives

$$\underset{(\text{coil only})}{Q'_{UUP}} = \frac{Q_{UUP}R_{agc}(N1/N2)^2}{R_{agc}(N1/N2)^2 - Q_{UUP}X_{cp}} \quad \textbf{(25)}$$

Unneutralized Procedure

The unneutralized design is accomplished in the same manner as the neutralized design with only two exceptions:

1. Neutralizing capacitor C_n is omitted, therefore $C_{fdbk} = C_{cb(max)}$
2. It is usually desirable from a cost standpoint not to tap the primary. This makes $N1/N2 = 1$ in all previous equations.

With $N1/N2 = 1$, the primary tuning capacitance is no longer selected arbitrarily but may be explicitly expressed by rearranging Eq. (22) giving

$$C_{tp} = \frac{Q_{UUP}Q_{LUP}}{2\pi f_o(Q_{UUP} - Q_{LUP})\, r_{oep}} \qquad \textbf{(26)}$$

If a suitable value of C_{tp} is not obtained, a new set of input and output stability factors must be chosen and the design repeated. Alternatively, the designer may wish to initially choose a standard value for C_{tp}, rearrange Eq. (26) to obtain the necessary Q_{UUP}, calculate OS from Eq. (20), and then obtain IS from Eq. (17).

Tapping the primary may become necessary if the values for transistor r_{oep} or MUG are sufficiently low. This necessity will become apparent when calculation of Eq. (26) yields unreasonably large values of C_{tp}. Equation (22) should then be used to determine N1/N2 for satisfactory values of C_{tp}.

DESIGN EXAMPLES

Four-Stage Neutralized 10.7-MHz IF Amplifier

Transistor: PNP Epitaxial Diffused-Base Mesa Germanium TIXM204

Given:

1. TIXM204 parameters at $f = 10$ MHz, $V_{ce} = -6$ V, $I_e = -2$ mA.
   - a. $r_{iep} = 600\ \Omega$
   - b. $r_{oep} = 65{,}000\ \Omega$
   - c. $y_{fe} = 70$ mS
   - *d. $C_{cb} = 1.25 \pm 0.45$ pF
2. Center frequency $f_o = 10.7$ MHz
3. Coupling factor $\alpha = 0.85$
4. Required 6-dB overall bandwidth = 260 kHz
5. Number of stages = 4

Procedure:

1. From Eq. (16) and using a 10% neutralizing capacitor,

$$C_{fdbk} = (\text{tolerance})(C_{cb\ avg}) + C_{cb}\ \text{variation}$$
$$= (0.1)(1.25) + 0.45 = 0.575\ \text{pF}$$

2. Using Eq. (13d),

$$\text{MUG} = 0.17 \left| \frac{y_{fe}}{\omega C_{fdbk}} \right| \alpha^2$$
$$= 0.17 \left[\frac{70 \times 10^{-3}}{2\pi(10.7 \times 10^6)(0.575 \times 10^{-12})} \right] 0.85^2$$
$$= 223 = 23.5\ \text{dB}$$

Total Power Gain = 4(23.5) = 94.0 dB.

---

*C_{cb} is measured at $f = 1$ MHz, $V_{cb} = -6$ V, $I_e = 0$ mA.

3. From Table 6 or Eq. (3) with $N = 4$, $\alpha = 0.85$, $\rho = 2(6\text{ dB})$
   $X = 0.9302$
4. Using Eq. (5)
   $$Q_L = \frac{X_{fo}}{2\Delta f} = \frac{0.9302\,(10.7 \times 10^6)}{260 \times 10^3} = 38.3$$
   and
   $Q_L = \sqrt{Q_{LUP}Q_{LUS}}$
   Let $Q_{LUS} = Q_{LUP}$
   Hence $Q_{LUS} = Q_{LUP} = 38.3$
5. From Eq. (9d)
   $S = 1.5\,\omega\, C_{fdbk}\, y_{fe}\, r_{iep}\, r_{oep}$
   $S = 1.5(2\pi\ 10.7 \times 10^6)(0.575 \times 10^{-12})(70 \times 10^{-3})(600)\ 65 \times 10^3$
   $= 158 = 22.0\text{ dB}$
6. As defined in Eq. (17)
   $(OS)\,(IS) = S$
   $10 \log (OS) + 10 \log (IS) = 10 \log S$
   Let $OS = 12.4\text{ dB} = 17.4$ numeric
   $IS = 9.6\text{ dB} = 9.09$ numeric
7. Using Eq. (18)
   $$Q_{UUP} = \frac{2(OS)\,Q_{LUP}}{2(OS) - (1 + \alpha^2)} = \frac{2(17.4)\,38.3}{2(17.4) - (1 + 0.85^2)}$$
   $$= 40.3$$
   Similarly, from Eq. (19)
   $$Q_{UUS} = \frac{2(IS)\,Q_{LUS}}{2(IS) - (1 + \alpha^2)} = \frac{2(9.09)\,38.3}{2(9.09) - (1 + 0.85^2)}$$
   $$= 42.3$$
8. Let $C_{ts} = 100$ pF; $X_{cs} = \dfrac{1}{\omega C_{ts}}$
   $$= \frac{1}{2\pi(10.7 \times 10^6)(100 \times 10^{-12})} = 149\ \Omega$$
   From Eq. (23)
   $$\frac{N4}{N3} = \left[\frac{Q_{UUS}Q_{LUS}X_{cs}}{(Q_{UUS} - Q_{LUS})\,r_{iep}}\right]^{1/2} = \left[\frac{42.3\,(38.3)\,149}{(42.3 - 38.3)\,600}\right]^{1/2}$$
   $$= 10.0$$
9. Let $C_{tp} = 47$ pF; $X_{cp} = \dfrac{1}{\omega C_{tp}} = \dfrac{1}{2\pi(10.7 \times 10^6)\ 47 \times 10^{-12}}$
   $$= 318\ \Omega$$
   Using Eq. (22)
   $$\frac{N1}{N2} = \left[\frac{Q_{UUP}Q_{LUP}X_{cp}}{(Q_{UUP} - Q_{LUP})\,r_{oep}}\right]^{1/2} = \left[\frac{40.3\,(38.3)\,318}{(40.3 - 38.3)\,65 \times 10^3}\right]^{1/2}$$
   $= 1.94$ (approximately center tapped)
   From Eq. (24),

$$C_n = \left(\frac{1}{\frac{N1}{N2} - 1}\right) C_{cb\ avg} = \left(\frac{1}{1.94 - 1}\right) 1.25$$

$= 1.33$ pF

Using the nearest standard 10% capacitor value,

$C_n = 1.2$ pF

10. AGC Primary

Assume $R_{agc} = 25$ kΩ

As before, $Q_{LUP} = 38.3$ (Loaded with 25 kΩ in parallel with 65 kΩ)

$Q_{UUP} = 40.3$ (loaded only with 25 kΩ)

Maintaining the same tuning capacitance and turns ratio as calculated for the other primaries,

$$C_{tp} = 47 \text{ pF and } \frac{N1}{N2} = 1.94$$

From Eq. (25) find

$$Q'_{UUP} = \frac{Q_{UUP}\, R_{agc}(N1/N2)^2}{R_{agc}\,(N1/N2)^2 - Q_{UUP}X_{cp}}$$

$$= \frac{40.3\ (25 \times 10^3)\ 1.94^2}{(25 \times 10^3)\ 1.94^2 - 40.3\ (318)}$$

$= 46.6$ (coil only—no loading resistance)

The complete IF amplifier then is shown in Fig. 7 with the coil information summarized below. All transistors in Fig. 7 are biased to an emitter current of approximately 2 mA.

	T 1	T 2		T 3		T 4		T 5
	sec	pri	sec	pri	sec	pri	sec	pri
C_t(pF)	100	47	100	47	100	47	100	47
Q_u	42.3	40.3	42.3	40.3*	42.3	40.3	42.3	40.3
	**	***	**	****	**	***	**	***
Q_L	38.3	38.3	38.3	38.3	38.3	38.3	38.3	38.3
N1/N2	—	1.94	—	1.94	—	1.94	—	1.94
N4/N3	10.0	—	10.0	—	10.0	—	10.0	—

*$R_{agc} = 25$ kΩ across N2 (with no loading $Q_U = 46.6$)

**$R_L = 600$ Ω across N3

***$R_g = 65$ kΩ across N2

****$R_g = 65$ kΩ in parallel with 25 kΩ across N2

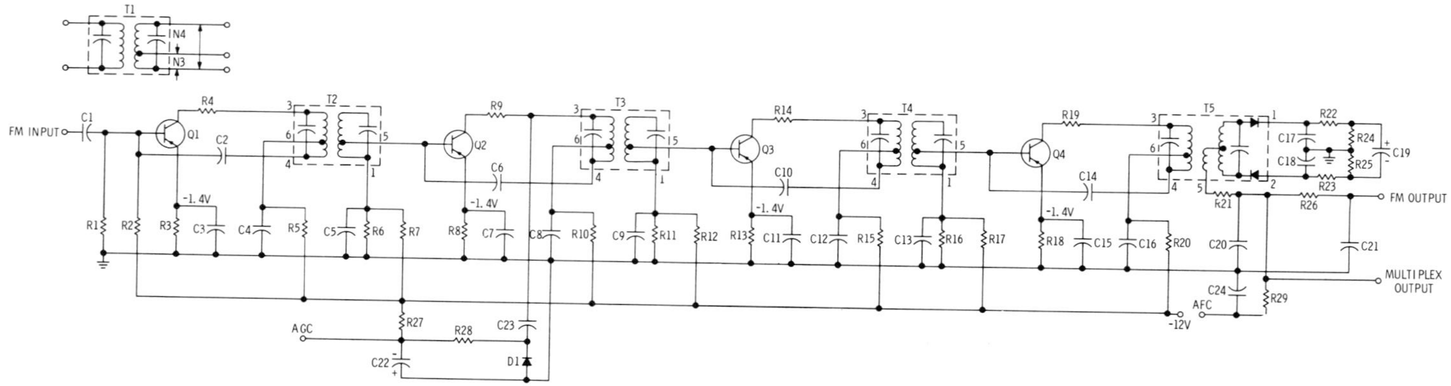

ALL RESISTOR VALUES IN OHMS

R1	10 000
R2	56 000
R3, R8, R13, R18	680
R4, R9, R14	220
R5, R10, R15, R20	330
R6, R11, R16	2 400
R7, R12, R17	12 000
R19	470
R21	62
R22	1 500
R23	1 000
R24, R25	8 200
R26	7 500
R27	33 000
R28	4 700
R29	200 000
C1	0.02 μF
C2, C6, C10, C14	1.8 pF
C3, C4, C5, C7, C8, C9, C11, C12, C13, C15, C16, C21	0.01 μF
C17, C18, C20	330 pF
C19	2 μF, 10 V, Electr.
C22	5 μF, 10 V, Electr.
C23	6.8 pF
C24	0.05 μF
D1	TI-6
Q1, Q2, Q3, Q4	TIXM204

Fig. 7. Four-stage Neutralized 10-7-MHz IF Amplifier Schematic

Three-Stage Unneutralized 10.7-MHz IF Amplifier

Transistor: NPN Epitaxial Planar Silicon SK5050A

Given:

1. SK5050A parameters at f = 10 MHz, V_{ce} = 10 V, I_e, = 2 mA.
   a. $r_{iep} = 1250\ \Omega$
   b. $r_{oep} = 80{,}000\ \Omega$
   c. $y_{fe} = 70$ mS
   *d. $C_{cb(max)} = 0.65$ pF
2. Center frequency $f_o = 10.7$ MHz
3. Coupling factor $\alpha = 0.85$
4. Required 6-dB overall bandwidth = 260 kHz
5. Number of stages = 3

Procedure:

1. The neutralizing capacitor is omitted, therefore $C_{fdbk} = C_{cb(max)}$
2. Using Eq. (13c)

$$\text{MUG} = 0.19 \left|\frac{y_{fe}}{\omega C_{fdbk}}\right| \alpha^2$$

$$= 0.19 \left[\frac{70 \times 10^{-3}}{2\pi\ (10.7 \times 10^6)\ 0.65 \times 10^{-12}}\right] 0.85^2$$

$$= 220 = 23.4 \text{ dB}$$

Total Power Gain = 3 (23.4) = 70.2 dB

3. From Table 6 or Eq. (3) with N = 3, $\alpha = 0.85$, $\rho = 2$(6dB):
   X = 1.0351
4. Using Eq. (5)

$$Q_L = \frac{Xf_o}{2\Delta f} = \frac{1.0351\ (10.7 \times 10^6)}{260 \times 10^3} = 42.6$$

and

$$Q_L = \sqrt{Q_{LUP} Q_{LUS}}$$

Let $Q_{LUP} = Q_{LUS}$

Hence $Q_{LUP} = Q_{LUS} = 42.6$

5. From Eq. (9c)

$$S = 1.31\ \omega\ C_{fdbk}\ y_{fe}\ r_{iep}\ r_{oep}$$

$$S = 1.31\ (2\pi\ 10.7 \times 10^6)\ (0.65 \times 10^{-12})\ 70 \times 10^{-3}\ (1250)\ 80 \times 10^3$$

$$= 400 = 26.0 \text{ dB}$$

6. As defined in Eq. (17)
   (OS) (IS) = S
   10 log (OS) + 10 log (IS) = 10 log S
   Let OS = 10.3 dB = 10.8 numeric
   IS = 15.7 dB = 37.0 numeric
7. Using Eq. (18)

$$Q_{UUP} = \frac{2(OS)\ Q_{LUP}}{2(OS) - (1 + \alpha^2)} = \frac{2\ (10.8)\ 42.6}{2\ (10.8) - (1 + 0.85^2)}$$

$$= 46.4$$

*C_{cb} is measured at f = 1 MHz, V_{cb} = 8 V, I_e = 0 mA.

Similarly, from Eq. (19)

$$Q_{UUS} = \frac{2(IS)\,Q_{LUS}}{2(IS) - (1 + \alpha^2)} = \frac{2\,(37.0)\,42.6}{2\,(37.0) - (1 + 0.85^2)}$$
$$= 43.6$$

8. Let $C_{ts} = 100$ pF; $X_{cs} = \frac{1}{\omega C_{ts}} = \frac{1}{2\pi(10.7 \times 10^6)\,100 \times 10^{-12}}$
$$= 149\ \Omega$$

From Eq. (23)

$$\frac{N4}{N3} = \left[\frac{Q_{UUS}Q_{LUS}X_{cs}}{(Q_{UUS} - Q_{LUS})\,r_{iep}}\right]^{1/2} = \left[\frac{43.6\,(42.6)\,149}{(43.6 - 42.6)\,1250}\right]^{1/2}$$
$$= 14.9$$

9. Let N1/N2 = 1.0
Using Eq. (26)

$$C_{tp} = \frac{Q_{UUP}Q_{LUP}}{2\pi f_o(Q_{UUP} - Q_{LUP})\,r_{oep}} = \frac{46.4\,(42.6)}{2\pi\ 10.7 \times 10^6\,(46.4 - 42.6)\,80 \times 10^3}$$

$$= 100\text{ pF}$$

10. AGC Primary
Assume $G_{agc} = 25$ kΩ
As before, $Q_{LUP} = 42.6$ (loaded with 25 kΩ in parallel with 80 kΩ)
$Q_{UUP} = 46.4$ (loaded with only 25 kΩ)
Maintaining the same tuning capacitance and turns ratio as calculated for the other primaries,
$C_{tp} = 100$ pF and N1/N2 = 1.0 (no tap)
From Eq. (25) find

$$Q'_{UUP} = \frac{Q_{UUP}R_{agc}\,(N1/N2)^2}{R_{agc}\,(N1/N2)^2 - Q_{UUP}X_{cp}}$$
$$= \frac{46.4\,(25 \times 10^3)\,1.0^2}{(25 \times 10^3)\,1.0^2 - 46.4\,(149)}$$
$$= 64.0 \text{ (coil only—no loading resistance)}$$

The complete IF amplifier then is shown in Fig. 8 with the coil information summarized below. All transistors in Fig. 8 are biased to an emitter current of approximately 2 mA.

	T 1	T 2		T 3		T 4
	sec	pri	sec	pri	sec	pri
C_t(pF)	100	100	100	100	100	100
Q_{UU}	43.6	46.4	43.6	* 46.4	43.6	46.4
Q_{LU}	** 42.6	*** 42.6	** 42.6	**** 42.6	** 42.6	*** 42.6
N4/N3	—	1.0	—	1.0	—	1.0
N1/N2	14.9	—	14.9	—	14.9	—

*$R_{agc} = 25\ k\Omega$ across N2 (with no loading $Q_{UU} = 64.0$)
**$R_L = 1250\ \Omega$ across N3
***$R_g = 80\ k\Omega$ across N2
****$R_g = 80\ k\Omega$ in parallel with $25\ k\Omega$ across N2

EQUATION DERIVATIONS

Equation (7). Minimum Capacitance Required to Cause Oscillation

With all impedances referred to the transistor's base and collector, an equivalent circuit is shown in Fig. 9. The circuit (Fig. 9a) may be simplified (Fig. 9b) by letting

$$R_{in} = \frac{(Q_{US}X_{cs})\,(r_{iep})}{Q_{US}X_{cs} + r_{iep}}$$

$$R_{out} = \frac{(Q_{UP}X_{cp})\,(r_{oep})}{Q_{UP}X_{cp} + r_{oep}}$$

The conditions for oscillation are satisfied when the closed-loop phase shift is 360° and the closed-loop gain is equal to unity. Neglecting the imaginary part of y_{fe}, the phase shift between the input voltage V_{be} and the current generator $y_{fe}V_{be}$ is 180°. Since the reactance of ΔC_{fb} is much larger than R_{in}, the feedback current leads the output voltage by 90°. To satisfy the phase shift requirement, the voltage across R_{out} must lead the current $y_{fe}V_{be}$ by 45° while the input voltage V_{be} must lead the feedback current by 45°. Oscillation, then, occurs when the reactance of the resonant circuits equals the resistance in parallel with them. At the frequency of oscillation, the forward and reverse voltage gains are

$$A_f = \frac{y_{fe}\,R_{out}}{\sqrt{2}}$$

$$A_r = \frac{\omega \Delta C_{fb}\,R_{in}}{\sqrt{2}}$$

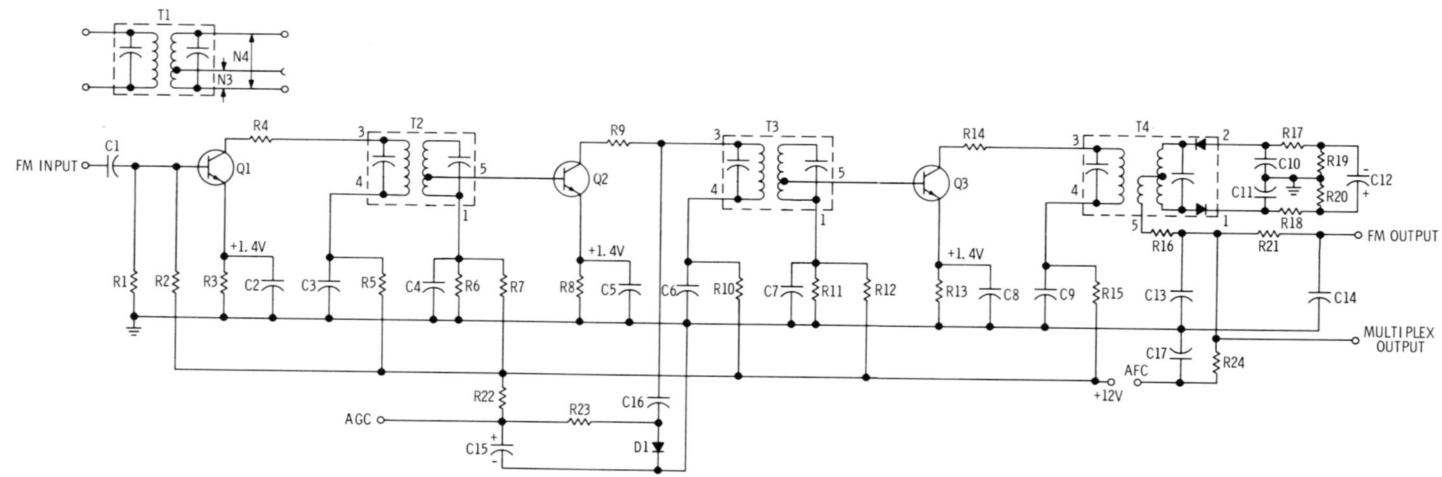

ALL RESISTOR VALUES IN OHMS

R1	9 100
R2	47 000
R3, R8, R13	680
R4, R9	220
R5, R10, R15	330
R6, R11	2 700
R7, R12	15 000
R14	470
R16	62
R17	1 000
R18	1 500
R19, R20	8 200
R21	7 500
R22	33 000
R23	4 700
R24	200 000
C1	0.02 μF
C2, C3, C4, C5, C6, C7, C8, C9, C14	0.01 μF
C10, C11, C13	330 pF
C12	2 μF, 10 V, electr.
C15	5 μF, 10 V, electr.
C16	6.8 pF
C17	0.05 μF
D1	TI-6
Q1, Q2, Q3	SK5050A

Fig. 8. Three-stage Unneutralized 10.7-MHz IF Amplifier Schematic

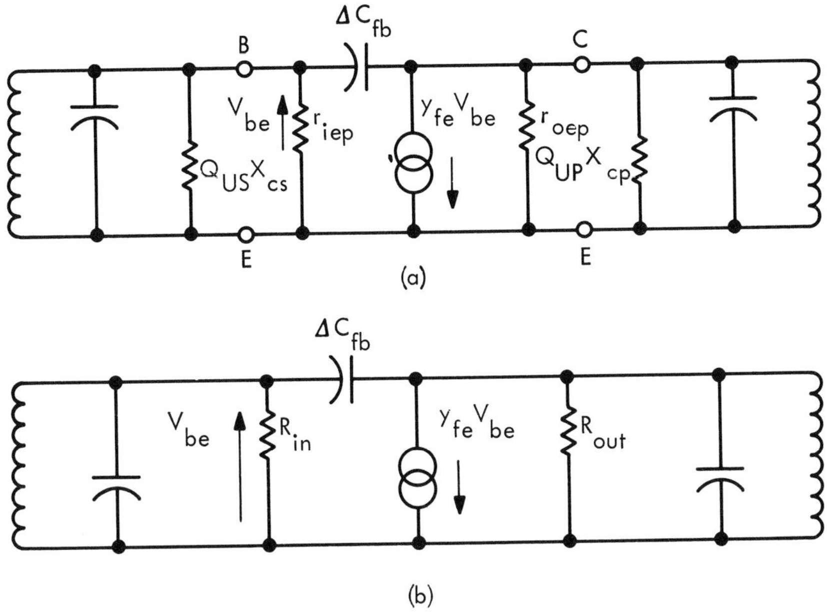

Fig. 9. Equivalent IF Amplifier Stage

and since

$$A_f A_r = 1$$

$$\frac{y_{fe}\ R_{out}}{\sqrt{2}} \left(\frac{\omega \Delta C_{fb}\ R_{in}}{\sqrt{2}} \right) = 1$$

Solving for ΔC_{fb}

$$\Delta C_{fb} = \frac{2}{\omega y_{fe}\ R_{in}\ R_{out}} \qquad \text{(a)}$$

For a complex conjugate match

$$Q_{US} X_{cs} = r_{iep} = 2\ R_{in}$$

$$Q_{UP} X_{cp} = r_{oep} = 2\ R_{out}$$

So that

$$\Delta C_{fb} = \frac{8}{\omega y_{fe}\ r_{iep}\ r_{oep}} \qquad \text{(b)}$$

Equation (18). Unloaded Coil Q

The unloaded coil Q can be related to the output stability factor, the loaded coil Q, and the coupling factor by recognizing first that

$$OS = \frac{r_{oep}\ (N1/N2)^2}{2R'}$$

where

$$R' = \text{total resistance shunting the transformer primary}$$

Now

$$R' = Q_{LCP}X_{cp}$$

where

$$Q_{LCP} = \text{loaded coupled Q of the primary}$$

$$X_{cp} = \frac{1}{2\pi f_o C_{tp}}$$

and since

$$Q_{LCP} = \frac{Q_{LUP}}{1 + \alpha^2}$$

then

$$OS = \frac{r_{oep}\left(\frac{N1}{N2}\right)^2}{2\left(\frac{Q_{LUP}X_{cp}}{1 + \alpha^2}\right)}$$

Or by rearranging terms,

$$\frac{Q_{LUP}X_{cp}}{r_{oep}\left(\frac{N1}{N2}\right)^2} = \frac{1 + \alpha^2}{2(OS)} \qquad \text{(c)}$$

Also,

$$Q_{LUP}X_{cp} = \frac{(Q_{UUP}X_{cp})\, r_{oep}\left(\frac{N1}{N2}\right)^2}{Q_{UUP}X_{cp} + r_{oep}\left(\frac{N1}{N2}\right)^2}$$

or

$$Q_{UUP} = \frac{Q_{LUP}\, r_{oep}\left(\frac{N1}{N2}\right)^2}{r_{oep}\left(\frac{N1}{N2}\right)^2 - Q_{LUP}X_{cp}}$$

$$= \frac{Q_{LUP}}{1 - \dfrac{Q_{LUP}\, X_{cp}}{r_{oep}\left(\frac{N1}{N2}\right)^2}} \qquad \text{(d)}$$

Therefore, substituting Eq. (c) into Eq. (d),

$$Q_{UUP} = \frac{Q_{LUP}}{1 - \left(\frac{1 + \alpha^2}{2(OS)}\right)} = \frac{2(OS)\, Q_{LUP}}{2(OS) - (1 + \alpha^2)} \tag{e}$$

Similarly,

$$Q_{UUS} = \frac{2(IS)\, Q_{LUS}}{2(IS) - (1 + \alpha^2)} \tag{f}$$

SYMBOLS

α	Coupling factor (Ratio of actual coefficient of coupling to the critical coefficient of coupling, k/k_c)
BW	Bandwidth
C_{cb}	Feedback capacitance from collector to base
C_{fdbk}	Feedback capacitance which limits the maximum usable gain
ΔC_{fb}	Minimum feedback capacitance under a conjugate matched condition to cause oscillation
C_n	Neutralizing capacitor
C_{tp}	Primary tuning capacitor
C_{ts}	Secondary tuning capacitor
f_m	Highest modulating frequency
Δf_{max}	Maximum signal deviation
f_o	Center or resonant frequency $\left(\frac{\omega}{2\pi}\right)$
f_T	Frequency at which $\lvert h_{fe}\rvert = 1$
IL	Insertion loss (Ratio of maximum available power to power actually delivered to the load)
IS	Input stability factor
k	Actual coefficient of coupling
k_c	Critical coefficient of coupling
K_n	Stability constant whose value depends upon the number of IF stages
K_s	Skew factor
MAG	Transistor maximum available power gain
MUG	Maximum usable power gain per stage
N	Number of identical transformers
N1	Total number of primary turns
N2	Number of primary turns from the transistor collector to a-c ground
N3	Number of secondary turns from the transistor base to a-c ground
N4	Total number of secondary turns
ω	$2\pi f_o$
OS	Output stability factor
Q_L	Effective loaded uncoupled Q of the coil
Q_{LCP}	Loaded coupled Q of the primary
Q_{LUP}	Loaded uncoupled Q of the primary
Q_{LUS}	Loaded uncoupled Q of the secondary
Q_{UUP}	Unloaded uncoupled Q of the primary
Q'_{UUP}	Unloaded uncoupled Q of the primary on an AGC stage

Q_{UUS}	Unloaded uncoupled Q of the secondary
R'	Loaded coupled resistance shunting the transformer primary
R_{agc}	Equivalent loading resistance of AGC network
R_g	Primary loading resistor simulating transistor r_{oep}
r_{iep}	Parallel-equivalent small-signal common-emitter input resistance with the output short circuited
R_L	Secondary loading resistor simulating transistor r_{iep}
r_{oep}	Parallel-equivalent small-signal common-emitter output resistance with the input short circuited
ρ	Attenuation at Δf Hz off resonance
S	Stability factor which must be maintained to ensure stable operation
θ	Phase shift between secondary current at resonance and the secondary current at Δf Hz off resonance
X	Bandwidth reduction factor
X_{cp}	Capacitive reactance of primary tuning capacitor at resonance
X_{cs}	Capacitive reactance of secondary tuning capacitor at resonance
y_{fe}	Small-signal common-emitter forward transfer admittance

REFERENCES

1. Maynard, J. E.: Universal Performance Curves for Tuned Transformers, *Electronics,* pp. 15-18, Feb. 1937.
2. Thompson, B. J.: Oscillation in Tuned RF Amplifiers, *Proc. IRE,* Vol. 19, pp. 421-437, March 1931.
3. Mergner, F. L.: A Survey of Performance Requirements and Design Techniques for Highest Quality FM Multiplex Reception, *Audio Engineering Society,* Preprint No. 345.
4. DeVries, Adrian J.: Design of Stereophonic Receiver for a Stereo System in the FM Band Using an AM Subcarrier, *IRE Transactions On Broadcast and Television Receivers,* July 1961.
5. Terman, F. E.: "Radio Engineering," pp. 349-351, McGraw-Hill Book Co., Inc., N. Y., 1947.
6. Langford-Smith, F.: "Radiotron Designers' Handbook," 4th Edition, RCA, Harrison, N. J. 1953.
7. Holmes, David D. and T. O. Stanley: Stability Considerations in Transistor IF Amplifiers, *Transistors I,* pp. 403-421, RCA Laboratories, Princeton, N. J., 1956.
8. Johnson, George: Power Gain and Stability in Linear Active Two Ports, pp. 31-66, *Texas Instruments RF Seminar Papers.*
9. Wolfendale, E.: "The Junction Transistor and Its Applications," pp. 180-186, The Macmillan Company, 1958.

10

AM/FM IF Amplifier Circuit Applications

INTRODUCTION

Three broadcast-band AM/FM IF strips with AM diode detectors and FM ratio detectors are described in this chapter. Also included is a five-transistor multiplex adapter designed for use with all IF strips described both in this chapter and in Chapter 11.

The IF strip designs should satisfy most requirements for IF gain and bandwidth from the low-cost table-model line-operated receivers to the component quality systems. All components requiring design are accurately specified and are commercially available from indicated sources. All strips have been constructed and tested to determine performance characteristics. The observed characteristics agree closely with the theoretical designs.

The transistors used in all applications are specifically designed and characterized for use in broadcast-band radio receivers. Transistors should be characterized with sufficient specifications to guarantee the usable gain; that is, the stable power gain achieved and used in the circuit.

Power gain at 10 MHz on these devices can be properly specified by guaranteeing small-signal common-emitter forward current transfer ratio $|h_{fe}|$, feedback capacitance, and collector-base time constant $r_b'C_c$. These parameters should be specified at or near the current and voltage conditions at which the device is to be operated in the final circuit. The maximum value of collector-base feedback capacitance should be specified since this represents maximum undesirable feedback. Power gain at high frequencies can be adequately specified by guaranteeing maximum $r_b'C_c$, minimum $|h_{fe}|$, and maximum feedback capacitance. The maximum $r_b'C_c$ restricts the minimum value of output impedance. The stability of the device is then dictated by the feedback capacitance specifications. These controls placed on transistor impedances help to guarantee minimum variation in gain and bandwidth in a properly designed amplifier stage. All the devices used in these applications exhibit these parameter controls and most of them guarantee both minimum and maximum values on the same parameter. This adds even tighter control on circuit performance. If the minimum $|h_{fe}|$ and maximum $r_b'C_c$ are unknown, the minimum value of transadmittance $|y_{fe}|$ should be specified at the operating current, voltage, and frequency. $|y_{fe}|$ is directly proportional to $|h_{fe}|$ and inversely proportional to the input impedance.

Neutralization is employed in one of the individual amplifier designs. This provides maximum economy while still maintaining sufficient power gain. The other two amplifiers are unneutralized. Due to low feedback capacitance, the devices used in the unneutralized applications inherently exhibit sufficient usable power gain without being neutralized. Individual IF strip bandwidth data is taken without the benefit of further bandwidth shrinkage and reduction which would result with the addition of T_1 (the first FM IF transformer). This transformer is included in the FM tuner. Addition of this transformer should yield overall 6 dB IF bandwidth of 260 kHz which is recommended for multiplex reception. Suggested design parameters for the secondary of T_1 are given in the parts lists. The recommended design should be followed by FM tuner manufacturers to ensure proper stability and over-all gain in the completed AM/FM or FM receiver.

Parameters for the TI403 PNP epitaxial diffused-base mesa germanium transistors used in the AM/FM designs in this chapter are as follows. All data were taken at $f = 10$ MHz, $|V_{CE}| = 6$ V and $|I_E| = 2$ mA except C_{cb} which was taken at $|V_{CB}| = 6$ V, $|I_E| = 0$ mA and $f = 1$ MHz. An equivalent circuit with typical design transistor parameters is shown in Chapter 9, Figure 5.

Device	r_{iep}(kΩ)	r_{oep}(kΩ)	y_{fe}(mS)	C_{cb}(pF)
TI 403	2.2	75	75	0.65

One additional feature of this chapter is the complete design of a printed circuit board for the four-stage FM, two-stage AM IF amplifier. A full-size photograph of this layout is shown to allow the radio engineer to use the work which has already been done.

FOUR-STAGE FM, TWO-STAGE AM UNNEUTRALIZED IF AMPLIFIER

This composite AM/FM IF strip uses the high-gain low-feedback capacitance TI403 series transistors. The amplifiers are unneutralized. This IF strip is designed to fulfill the needs for high-performance characteristics found in component quality receivers. Both maximum and minimum specifications are placed on some of the TI403 parameters to ensure production uniformity in receiver characteristics. The circuit is TI number 4038.

Typical Performance (4038)

FM PERFORMANCE

Data Taken with Input to Base of Transistor Q1
IF Frequency = 10.7 MHz

1. Overall 6 dB Bandwidth (Measured on Base of Q4) . . . . 290 kHz
2. Overall 40 dB Bandwidth (Measured on Base of Q4) . . . 770 kHz
3. Average Power Gain Per Stage . . . . 24 dB
4. FM 3 dB Limiting Level . . . . 25 μV input
5. AM Rejection at 3 dB Limiting Level . . . . >30 dB
6. Maximum Audio Recovery (Full Limiting),
   ± 22.5 kHz Deviation . . . . 280 mV
   ± 75 kHz Deviation . . . . 800 mV
7. Peak-to-Peak Separation of Ratio Detector . . . . . . >700 kHz

AM PERFORMANCE

Data Taken Using Measurements Corp. Signal Gen. Model 65-B with a 0.05-μF Capacitor and an 82-kΩ Resistor in Series with Generator Output
IF Frequency = 455 kHz

1. Overall 6 dB Bandwidth . . . . 9.2 kHz
2. Overall 20 dB Bandwidth . . . . 17.6 kHz
3. Maximum Audio Output (30% Modulation) . . . . 250 mV
4. AM Sensitivity for 20 mV Output (30% Modulation) . 1.6 mV input at Pin 3 of T6

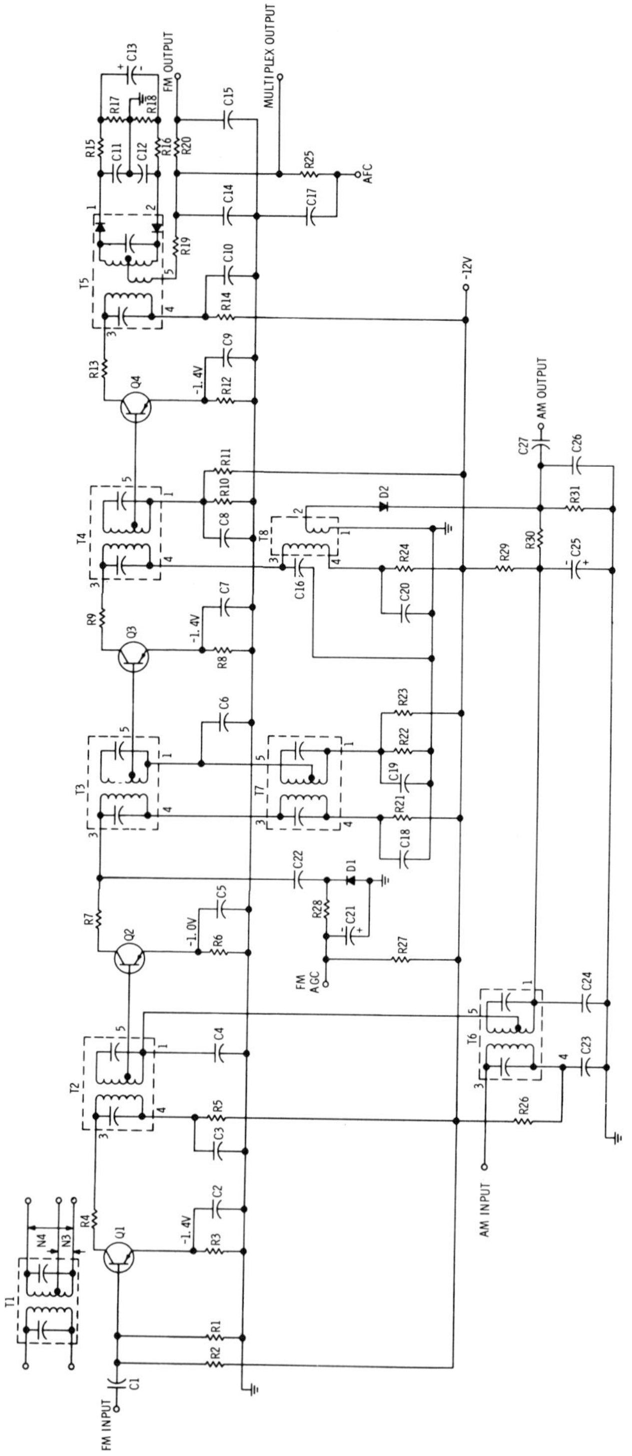

Schematic (4038)

Parts List (4038)

ALL RESISTOR VALUES IN OHMS

R1	10 000
R2	56 000
R3, R8, R12	680
R4, R7, R9	220
R5, R14, R21, R24, R26	330
R6	560
R10, R22	2 400
R11, R23	15 000
R13	470
R15	1 500
R16	1 000
R17, R18	8 200
R19	62
R20	7 500
R25	200 000
R27	33 000
R28	4 700
R29	68 000
R30	6 800
R31	2 700
C1	0.02 μF
C2, C3, C8, C9, C10, C15	0.01 μF
C4, C6	0.001 μF
C5, C7, C17, C18, C19, C20, C23, C24, C26	0.05 μF
C11, C12, C14	330 pF
C16	0.0015 μF
C13	2 μF, 10 V, electr.
C21, C25	5 μF, 10 V, electr.
C22	6.8 pF
C27	5 μF, 15 V, electr.
D1	TI-6
D2	1N295
Q1, Q2, Q3, Q4	T1403
J1, J2	Jumper
T2, T4	TRW #18746 or equiv.
T3	TRW #18747 or equiv.
T5	TRW #20061 or equiv.
T6	TRW #18302-R2 or equiv.
T7	TRW #18303-R2 or equiv.
T8	TRW #18304-R1 or equiv.

RECOMMENDED T1 SECONDARY

C_T	56 pF
Q_u	42
Q_L	41.5

($R_L = 2.2$ kΩ across N3)

$\frac{N4}{N3}$	18

Photograph (4038)

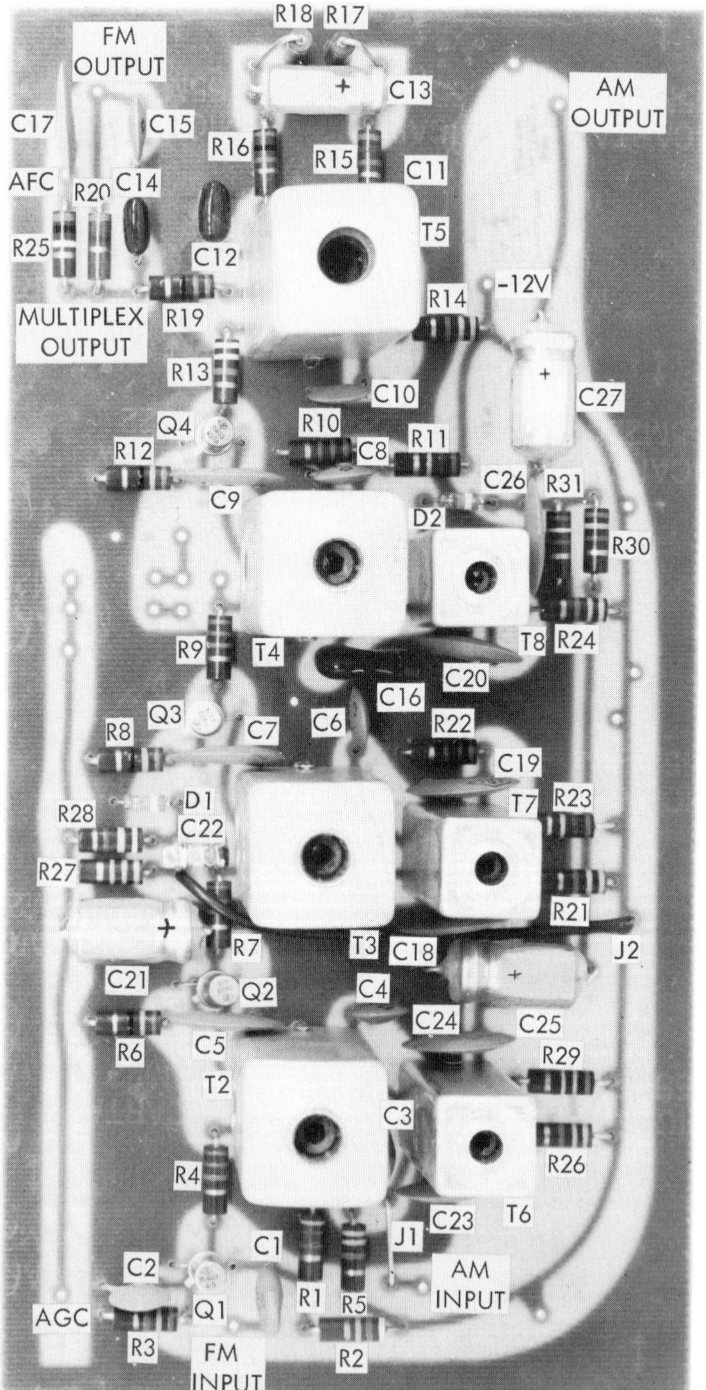

Printed Circuit (4038)

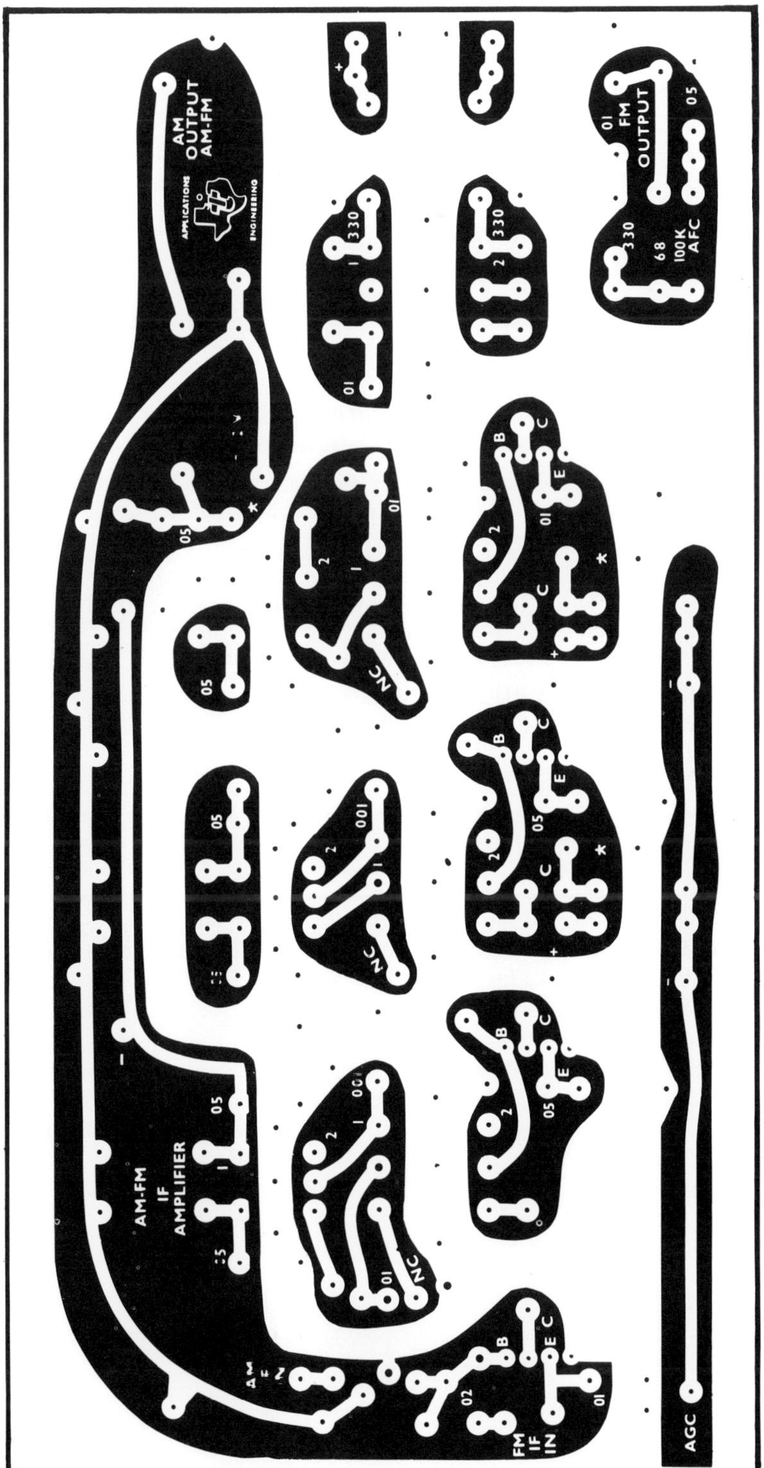

THREE-STAGE FM, TWO-STAGE AM UNNEUTRALIZED IF AMPLIFIER

The high-gain and low-feedback capacitance characteristics of the TI403 series devices are used in this composite AM/FM design to provide a reliable three-stage amplifier. High gain is attained without the need for neutralization. Sufficient gain and selectivity are present to satisfy the requirements of most stereo console systems. The circuit board for TI circuit 4038 can be redesigned and modified to accommodate this three-stage IF strip. This would require that the FM input be on the base of the second stage on the circuit board with the AM transformers connected according to the schematic for TI circuit number 4037.

Typical Performance (4037)

FM PERFORMANCE

Data Taken with Input on Base of Transistor Q1
IF Frequency = 10.7 MHz

1. Overall 6 dB Bandwidth (Measured on Base of Q3) . . . . 300 kHz
2. Overall 40 dB Bandwidth (Measured on Base of Q3) . . 1100 kHz
3. Average Power Gain Per Stage . . . . 24 dB
4. FM 3 dB Limiting Level . . . . 600 μV input
5. AM Rejection at 3 dB Limiting Level . . . . >30 dB
6. Maximum Audio Recovery (Full Limiting),
   ± 22.5 kHz Deviation . . . . 280 mV
   ± 75 kHz Deviation . . . . 800 mV
7. Peak-to-Peak Separation of Ratio Detector . . . . . . >700 kHz

AM PERFORMANCE

Data Taken Using Measurements Corp. Signal Gen. Model 65-B with a 0.05-μF Capacitor and an 82-kΩ Resistor in Series with Generator Output
IF Frequency = 455 kHz

1. Overall 6 dB Bandwidth . . . . 9.2 kHz
2. Overall 20 dB Bandwidth . . . . 17.6 kHz
3. Maximum Audio Output (30% Modulation) . . . . 250 mV
4. AM Sensitivity for 20 mV Output (30% Modulation) . 1.6 mV input at Pin 3 of T5

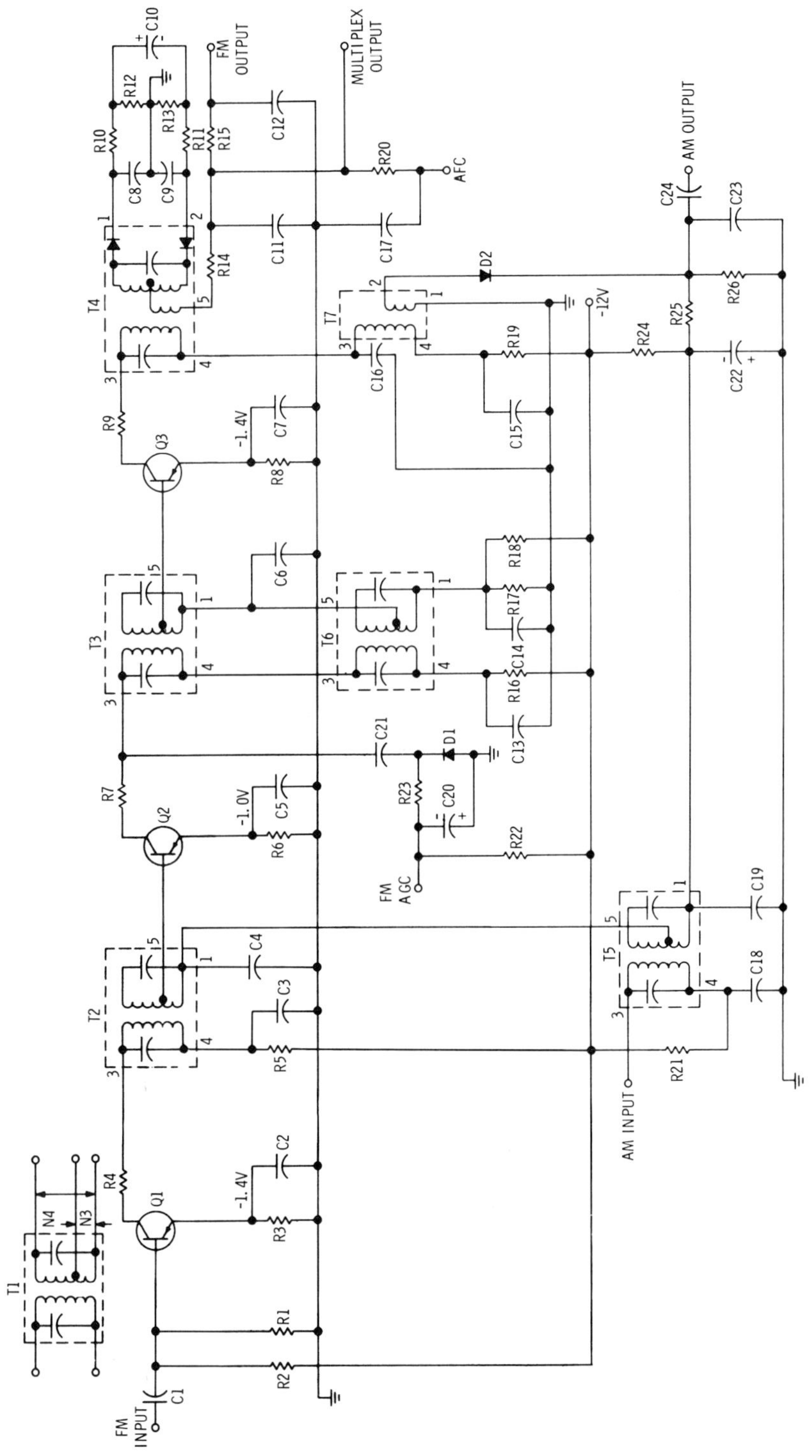

Schematic (4037)

Parts List (4037)

ALL RESISTOR VALUES IN OHMS

R1	10 000
R2	56 000
R3, R8	680
R4, R7	220
R5, R16, R19, R21	330
R6	560
R9	470
R10	1 500
R11	1 000
R12, R13	8 200
R14	62
R15	7 500
R17	2 400
R18	15 000
R20	200 000
R22	33 000
R23	4 700
R24	68 000
R25	6 800
R26	2 700
C1	0.02 μF
C2, C3, C12	0.01 μF
C4, C6	0.001 μF
C5, C7, C13, C14, C15, C17, C18, C19, C23	0.05 μF
C8, C9, C11	330 pF
C10	2 μF, 10 V, electr.
C16	0.0015 μF
C20, C22	5 μF, 10 V, electr.
C21	6.8 pF
C24	5 μF, 15 V, electr.
D1	TI-6
D2	1N295
Q1, Q2, Q3	TI403
T2	TRW #18252-R1 or equiv.
T3	TRW #18748-R1 or equiv.
T4	TRW #20061 or equiv.
T5	TRW #18302-R2 or equiv.
T6	TRW #18303-R2 or equiv.
T7	TRW #18304-R1 or equiv.

RECOMMENDED T1 SECONDARY

C_T	56 pF
Q_u	43
Q_L	42

($R_L = 2.2\ k\Omega$ across N3)

$\frac{N4}{N3}$	14.4

TWO-STAGE FM, TWO-STAGE AM NEUTRALIZED IF AMPLIFIER

This composite AM/FM IF strip uses a pair of neutralized TI403 transistors. Neutralization is required to provide sufficient gain without the use of more amplifiers. This strip is designed primarily for use in low-cost line-operated AM/FM receivers and portables. Printed circuit board layout for this IF strip would be similar to that used for TI circuit 4038. The FM input should be fed to the base of Q3 (reference circuit board for TI circuit 4038). Neutralization should be added and AM transformers connected in composite according to the schematic for TI circuit number 4048.

Typical Performance (4048)

FM PERFORMANCE

Data Taken with Input on Base of Transistor Q1
IF Frequency = 10.7 MHz

1. Overall 6 dB Bandwidth (Measured on Base of Q2) . . . . 345 kHz
2. Overall 40 dB Bandwidth (Measured on Base of Q2) . . 2800 kHz
3. Average Power Gain Per Stage . . . . 30 dB
4. FM 3 dB Limiting Level . . . . 2000 μV input
5. AM Rejection at 3 dB Limiting Level . . . . >30 dB
6. Maximum Audio Recovery (Full Limiting),
   ± 22.5 kHz Deviation . . . . 280 mV
   ± 75 kHz Deviation . . . . 800 mV
7. Peak-to-Peak Separation of Ratio Detector . . . . . . >600 kHz

AM PERFORMANCE

Data Taken Using Measurements Corp. Signal Gen. Model 65-B with a 0.05-μF Capacitor and an 82-kΩ Resistor in Series with Generator Output

1. Overall 6 dB Bandwidth . . . . 9.2 kHz
2. Overall 20 dB Bandwidth . . . . 17.6 kHz
3. Maximum Audio Output (30% Modulation) . . . . 250 mV
4. AM Sensitivity for 20 mV Output (30% Modulation) . 1.6 mV input at Pin 3 of T4

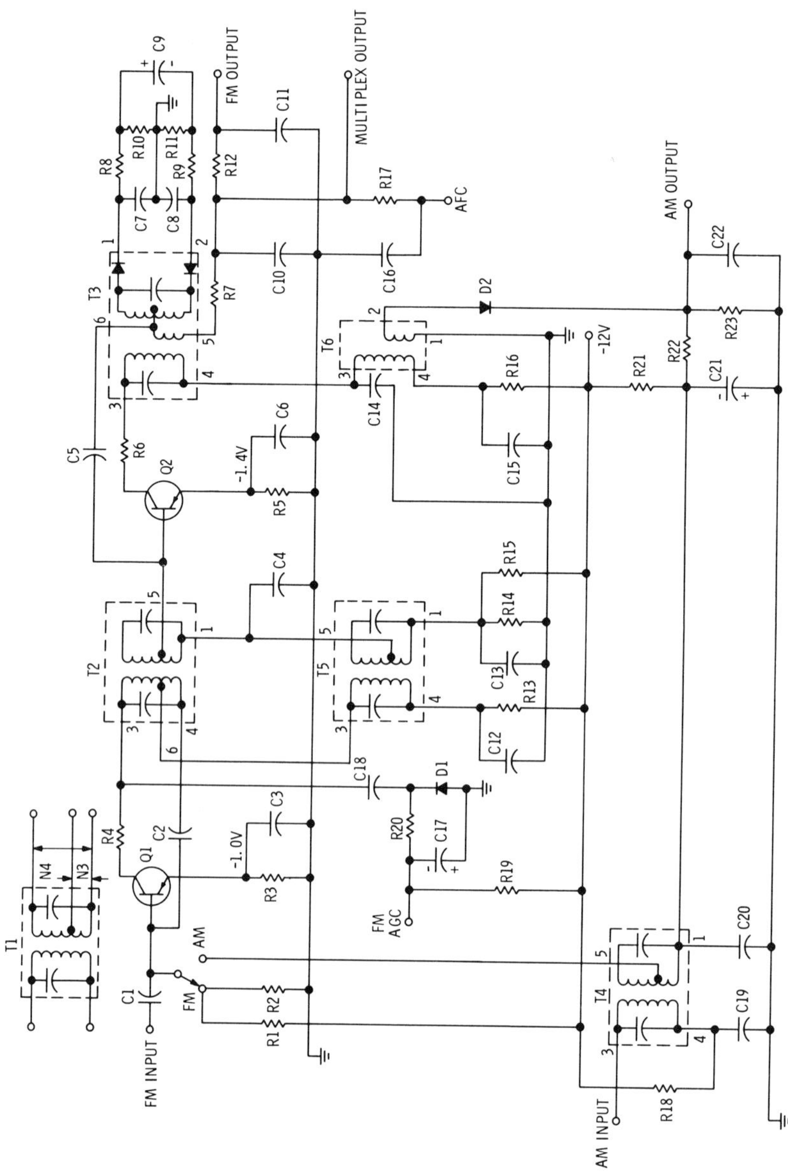

Schematic (4048)

Parts List (4048)

ALL RESISTOR VALUES IN OHMS

R1	82 000
R2	10 000
R3	560
R4, R6	220
R5	680
R7	62
R8	1 500
R9	1 000
R10, R11	8 200
R12	7 500
R13, R16, R18	330
R14	2 400
R15	15 000
R17	200 000
R19	33 000
R20	4 700
R21	68 000
R22	6 800
R23	2 700
C1	0.02 μF
C2	1.5 pF
C3, C6, C12, C13, C15, C16, C19, C20, C22	0.05 μF
C4	0.001 μF
C5	2.2 pF
C7, C8, C10	330 pF
C9	2 μF, 10 V, electr.
C11	0.01 μF
C14	0.0015 μF
C17, C21	5 μF, 10 V, electr.
C18	6.8 pF
C23	5 μF, 15 V, electr.
D1	TI-6
D2	1N295
Q1, Q2	TI403
T2	TRW #19749-R2 or equiv.
T3	TRW #19059-R1 or equiv.
T4	TRW #18302-R2 or equiv.
T5	TRW #18303-R2 or equiv.
T6	TRW #18304-R1 or equiv.

RECOMMENDED T1 SECONDARY

C_T	82 pF
Q_u	59.1
Q_L	55.9

($R_L = 2.2$ kΩ across N3)

$\frac{N4}{N3}$	9.2

MULTIPLEX ADAPTER

This multiplex adapter uses five TI germanium small-signal transistors and four TI silicon diodes. Separation versus input signal and frequency are design requirements and the overall achievement in this adapter. Automatic switching from monaural to multiplex is a feature made possible with the use of transistor Q2. High input impedance makes this adapter versatile when driven from both discriminators and ratio detectors. The stereo indicator is an added feature which most stereo listeners demand. The circuit is TI number 4026-R1.

Typical Performance (4026-R1)

38-kHz FILTER USED WHEN MEASURING SEPARATION

I. SEPARATION VERSUS SIGNAL LEVEL

MPX SIGNAL INPUT	SEPARATION IN dB
0.6 V peak to peak	35
1.0 V peak to peak	38
2.0 V peak to peak	39
2.5 V peak to peak	36

II. SEPARATION OPTIMIZED 2 V VERSUS SIGNAL

MPX SIGNAL INPUT	SEPARATION IN dB
0.6 V peak to peak	18
1.0 V peak to peak	28
2.0 V peak to peak	39
2.5 V peak to peak	28

III. SEPARATION OPTIMIZED 2 V PEAK-TO-PEAK VERSUS FREQUENCY

MPX SIGNAL INPUT	SEPARATION IN dB
100 Hz	30
1000 Hz	39
5000 Hz	30
10000 Hz	21
15000 Hz	10

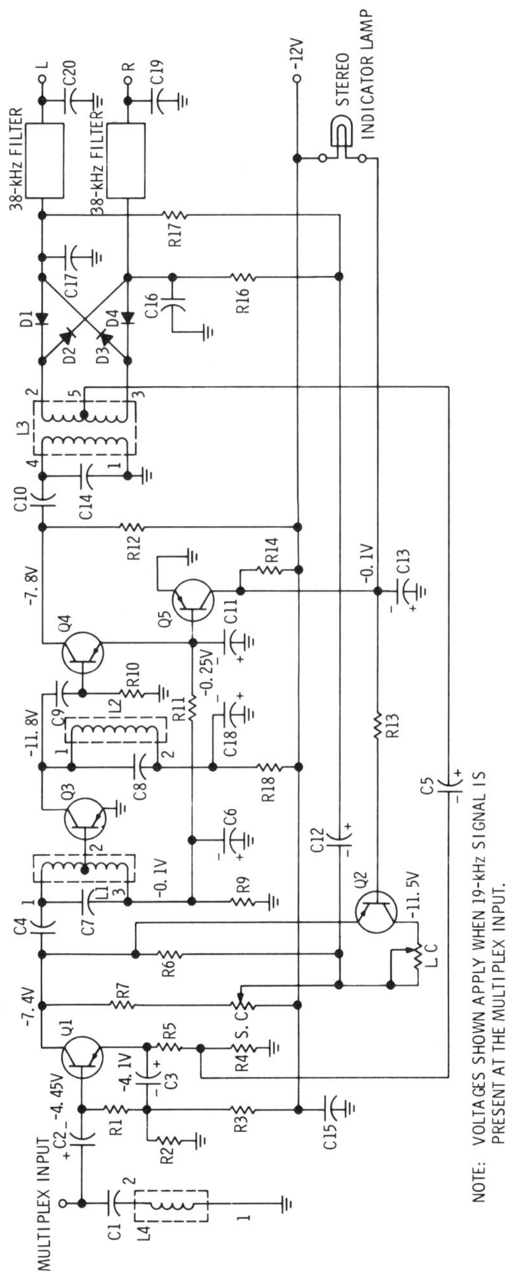

Schematic (4026-R1)

Parts List (4026-R1)

ALL RESISTOR VALUES IN OHMS

R1, R7, R10	15 000, ½ W, 10%
R2	10 000, ½ W, 10%
R3	12 000, ½ W, 10%
R4	680, ½ W, 10%
R5, R18	220, ½ W, 10%
R6	1 000, ½ W, 10%
R8, R15	Omitted
R9	8 200, ½ W, 10%
R11	18 000, ½ W, 10%
R12	6 800
R13	100 000
R14*, R16, R17	27 000*
C1	130 pF Dura Mica.
C2, C5, C12	20 μF, 15V, Electr.
C3	50 μF, 6V, Electr.
C4, C9	680 pF Dura Mica.
C6, C11	5 μF, 10V, Electr.
C7, C8	2000 pF Dura Mica.
C10	82 pF Dura Mica.
C13, C18	5 μF, 15V, Electr.
C14	1300 pF Dura Mica.
C15	50 μF, 15V, Electr.
C16, C17, C19, C20	0.001 μF, Disc Ceramic
D1, D2, D3, D4	TI-6
Q1	GC 1098
Q2, Q3, Q5	GC 1099
Q4	GC 1100
L1	TRW #18256 or Equiv. 19 kHz input
L2	TRW #18257 or Equiv. 19 kHz output
L3	TRW #18258 or Equiv. 38 kHz Doubler
L4	TRW #18259 or Equiv. 67 kHz Trap
NFL, NFR	CRL #5A35 or Equiv. 38 kHz Notch Filter
L.C.	CTS-X201(5kΩ) or Equiv. Level Control
S.C.	CTS-X201(5kΩ) or Equiv. Separation Control

*R14—27kΩ resistor across Stereo Indicator Lamp. Not shown on P.C. Board Layout. Stereo Indicator Lamp—GE1869 or equiv.

Photograph (4026-R1)

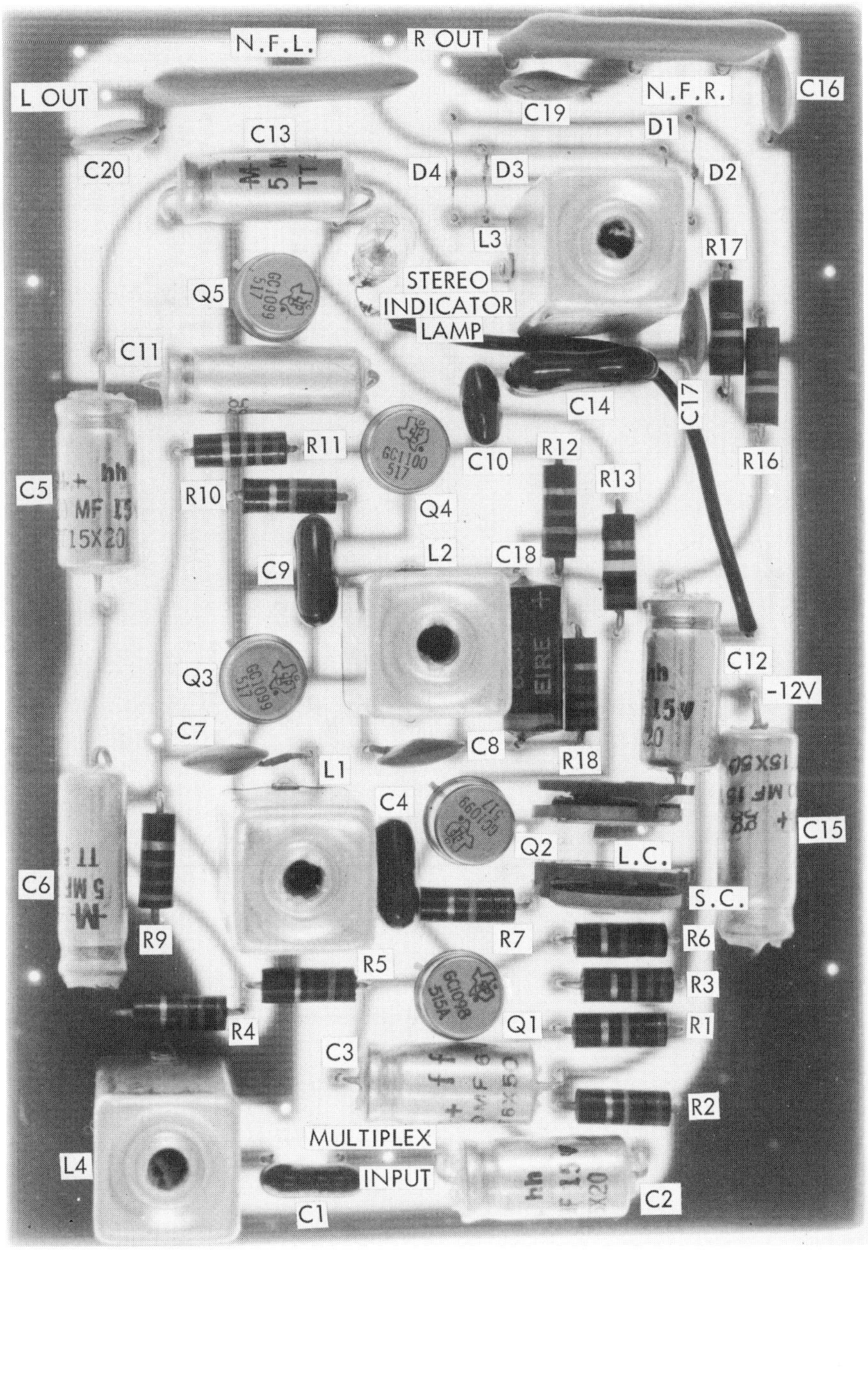

Printed Circuit (4026-R1)

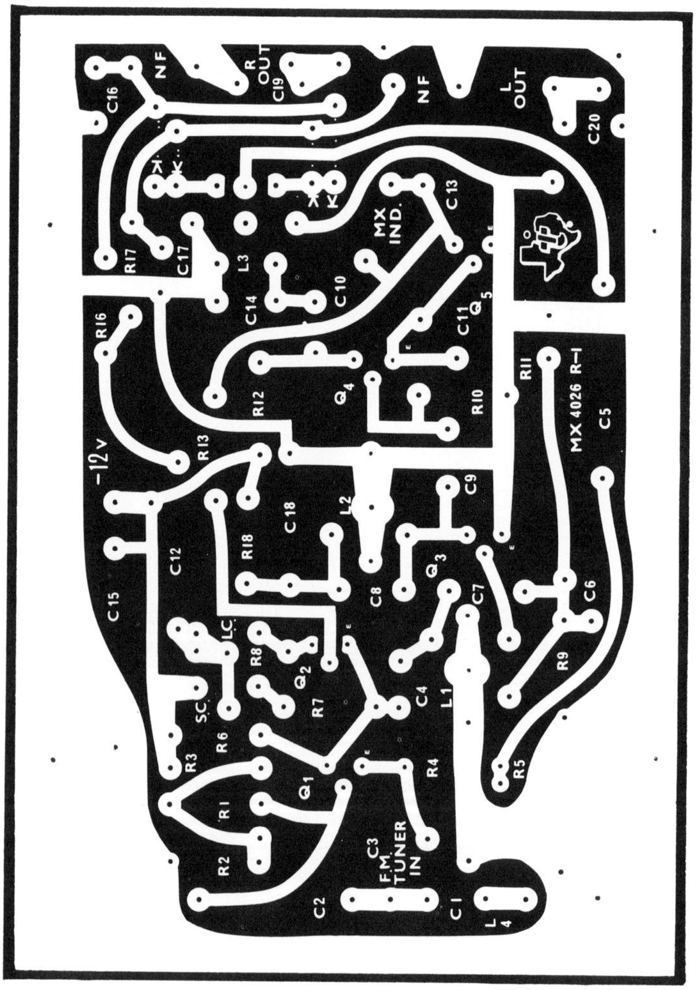

11

FM IF Amplifier Circuit Applications

INTRODUCTION

Five FM IF strips with ratio detectors are presented here. These designs meet present-day requirements for IF gain and bandwidth from low-cost receivers to component quality systems.

A discussion of transistor characteristics given in the introduction to Chapter 10 applies equally to this chapter. Parameters for the transistors used here are as follows. All data were taken at $f = 10$ MHz, $|V_{CE}| = 6$ V and $|I_E| = 2$ mA except C_{cb} which was taken at $|V_{CB}| = 6$ V, $|I_E| = 0$ mA and $f = 1$ MHz. An equivalent circuit with typical design transistor parameters is shown in Chapter 9, Fig. 5.

Device	$r_{iep}(k\Omega)$	$r_{oep}(k\Omega)$	$y_{fe}(mS)$	$C_{cb}(pF)$
2N3826	1.2	7	60	2.6
TI 408	1.6	45	70	1.1
TIXM04	2.2	40	60	0.9
TIXM204	0.6	65	70	1.3

A feature employed also in this chapter is the complete design of printed circuit boards. Full-size photographs of these layouts are shown to allow the radio engineer to use the layout work which has already been acccomplished.

FOUR-STAGE NEUTRALIZED AMPLIFIER USING SILICON TRANSISTORS 2N3826

This 10.7-MHz strip (TI circuit number 4050) uses four of the 2N3826 NPN planar silicon transistors and is satisfactory for use in component quality receivers. Neutralization is used to achieve high gain with low-cost devices. This IF strip can be used as composite AM/FM by using the same composite technique used in TI circuit 4038 (see Chapter 10).

Typical Performance (4050)

Data Taken with Input on Base of Transistor Q1

1. Overall 6 dB Bandwidth (Measured on Base of Q4) . . . . 290 kHz
2. Overall 40 dB Bandwidth (Measured on Base of Q4) . . . 770 kHz
3. Average Power Gain Per Stage . . . . 22 dB
4. FM 3 dB Limiting Level . . . . 100 μV input
5. AM Rejection at 3 dB Limiting Level . . . . >30 dB
6. Maximum Audio Recovery (Full Limiting),
   ± 22.5 kHz Deviation . . . . 250 mV
   ± 75 kHz Deviation . . . . 700 mV
7. Peak-to-Peak Separation of Ratio Detector . . . . . . >400 kHz

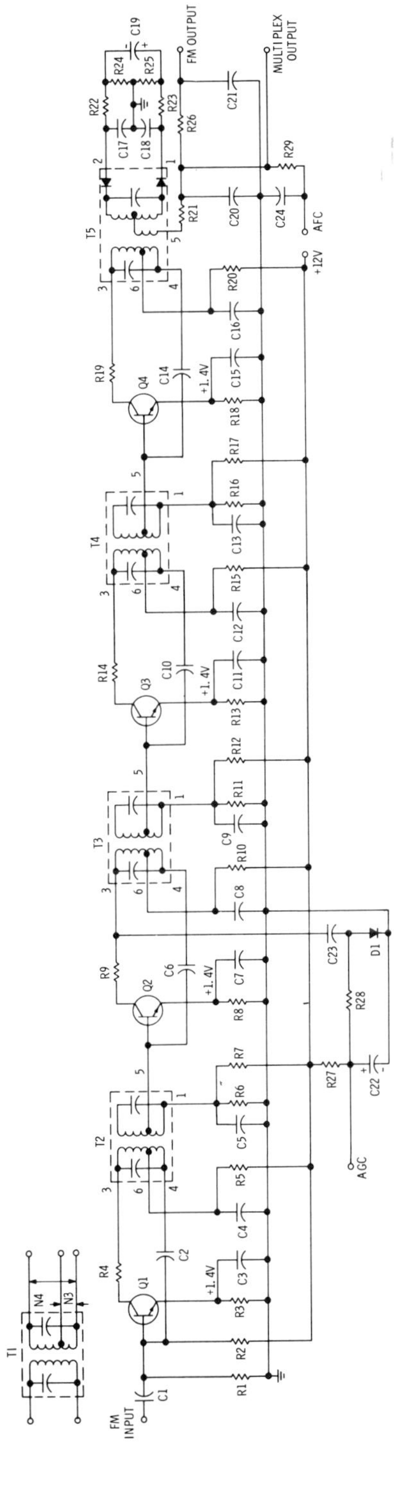

Schematic (4050)

Parts List (4050)

ALL RESISTOR VALUES IN OHMS

R1	9 100
R2	47 000
R3, R8, R13, R18	680
R4, R9, R14	220
R5, R10, R15, R20	330
R6, R11, R16	2 700
R7, R12, R17	15 000
R19	470
R21	62
R22	1 000
R23	1 500
R24, R25	8 200
R26	7 500
R27	33 000
R28	4 700
R29	200 000
C1	0.02 μF
C2, C6, C10	1.8 pF
C3, C4, C5, C7, C8, C9, C11, C12, C13, C15, C16, C21	0.01 μF
C14	3.3 pF
C17, C18, C20	330 pF
C19	2 μF, 10 V, electr.
C22	5 μF, 10 V, electr.
C23	6.8 pF
C24	0.05 μF
D1	TI-6
Q1, Q2, Q3, Q4	2N3826
J1	Jumper
T2	TRW #19827-R1 or equiv.
T3	TRW #19828-R1 or equiv.
T4	TRW #19829-R1 or equiv.
T5	TRW #19753-R1 or equiv.

RECOMMENDED TI SECONDARY

C_T	39 pF
Q_u	40.7
Q_L	38.3 ($R_L = 1.2$ kΩ across N3)
$\frac{N4}{N3}$	11.3

Photograph (4050)

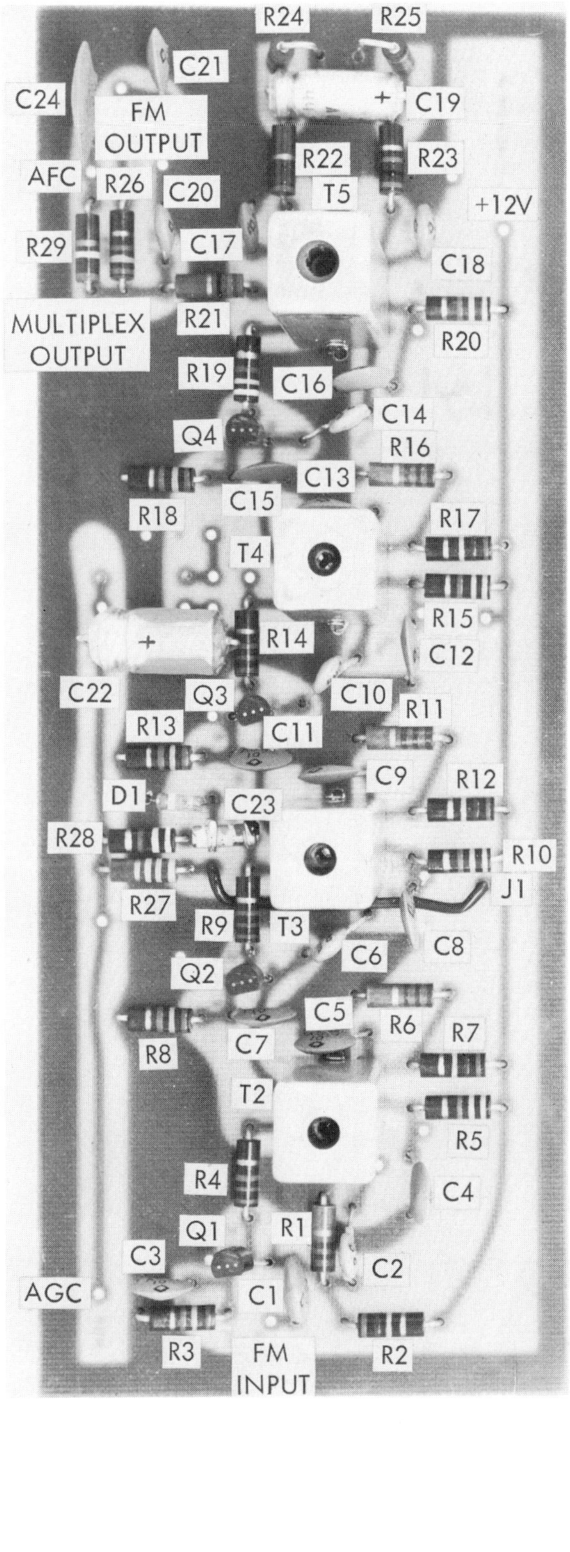

Printed Circuit (4050)

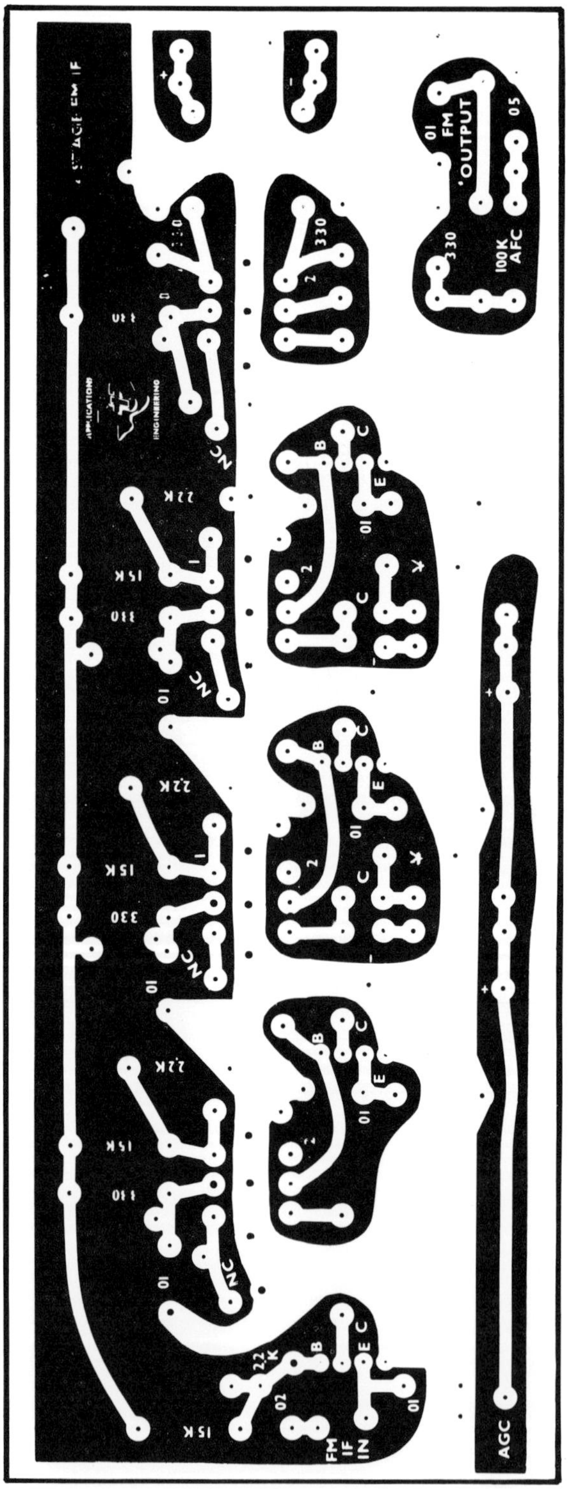

THREE-STAGE NEUTRALIZED AMPLIFIER USING SILICON TRANSISTORS 2N3826

This 10.7-MHz strip, using three 2N3826 NPN planar silicon transistors, is ideally suited for most low-cost medium-performance stereo receiver applications. AM/FM composite design can be applied to this circuit by using the composite techniques used in TI circuit 4037 (see Chapter 10). Neutralization has been used to obtain a high performance-to-cost ratio. The circuit is TI number 4049.

Typical Performance (4049)

Data Taken with Input on Base of Transistor Q1

1. Overall 6 dB Bandwidth (Measured on Base of Q3) . . . 300 kHz
2. Overall 40 dB Bandwidth (Measured on Base of Q3) . . 1100 kHz
3. Average Power Gain Per Stage . . . . 22 dB
4. FM 3 dB Limiting Level . . . . 1200 μV input
5. AM Rejection at 3 dB Limiting Level . . . . $>$30 dB
6. Maximum Audio Recovery (Full Limiting)
   - $\pm$ 22.5 kHz Deviation . . . . 250 mV
   - $\pm$ 75 kHz Deviation . . . . 700 mV
7. Peak-to-Peak Separation of Ratio Detector . . . . . . $>$400 kHz

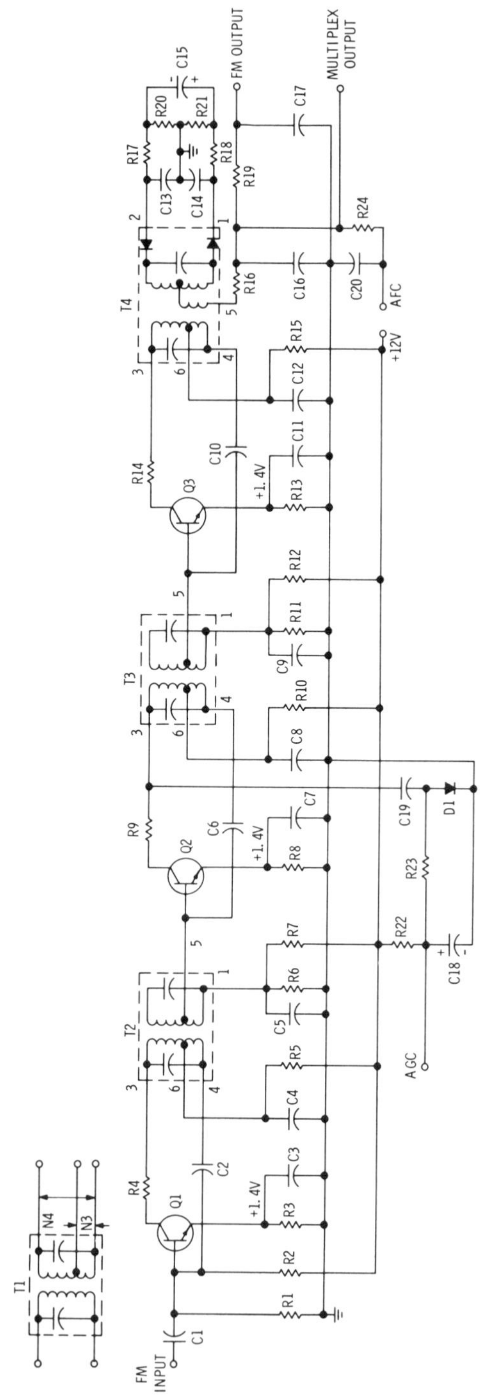

Schematic (4049)

Parts List (4049)

ALL RESISTOR VALUES IN OHMS

R1	9 100
R2	47 000
R3, R8, R13	680
R4, R9, R14	220
R5, R10, R15	330
R6, R11	2 700
R7, R12	15 000
R16	62
R17	1 000
R18	1 500
R19	7 500
R20, R21	8 200
R22	33 000
R23	4 700
R24	200 000
C1	0.02 μF
C2, C6	1.5 pF
C3, C4, C5, C7, C8, C9, C11, C12, C17	0.01 μF
C10	4.7 pF
C13, C14, C16	330 pF
C15	2 μF, 10 V, electr.
C18	5 μF, 10 V, electr.
C19	6.8 pF
C20	0.05 μF
J1	Jumper
D1	TI-6
Q1, Q2, Q3	2N3826
T2	TRW #19751-R1 or equiv.
T3	TRW #19752-R2 or equiv.
T4	TRW #19753-R1 or equiv.

RECOMMENDED T1 SECONDARY

C_T	39 pF
Q_u	44.1
Q_L	42.6 ($R_L = 1.2\ k\Omega$ across N3)
$\frac{N4}{N3}$	10.9

Photograph (4049)

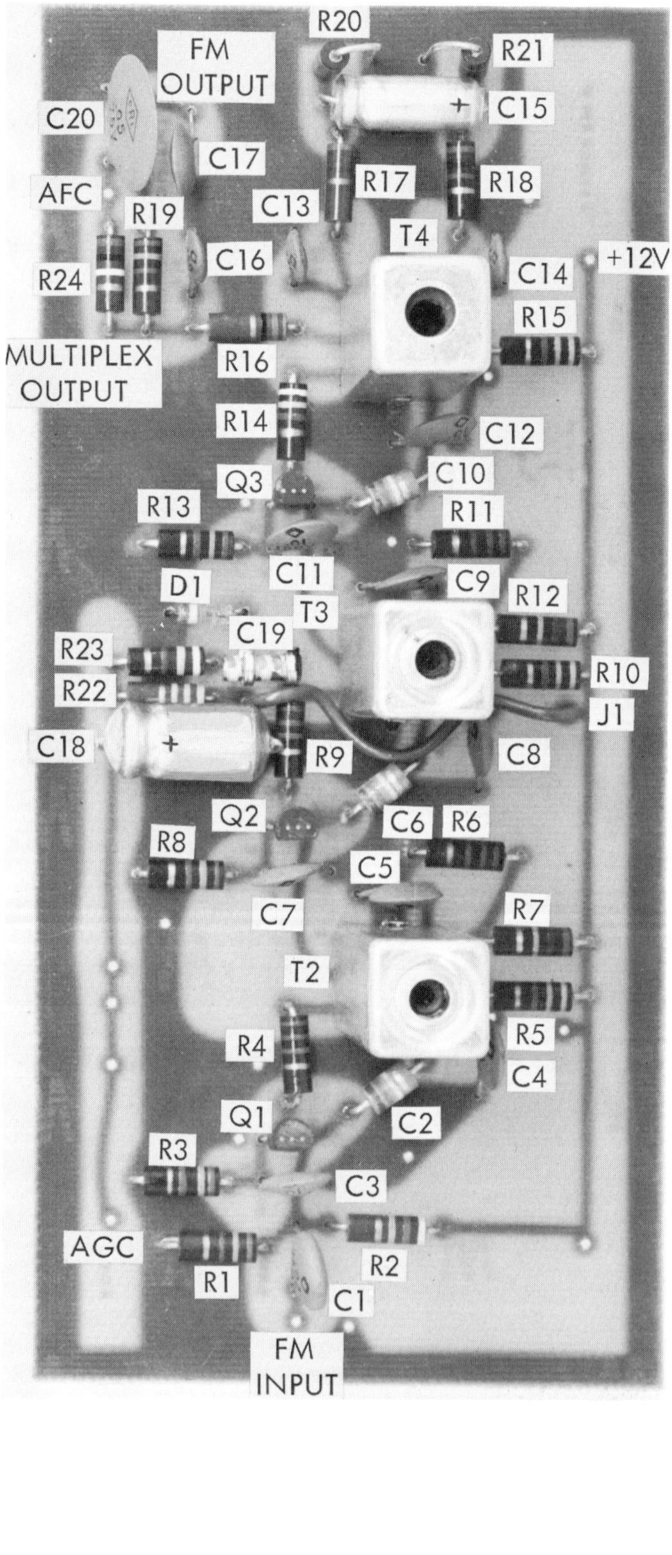

Printed Circuit (4049)

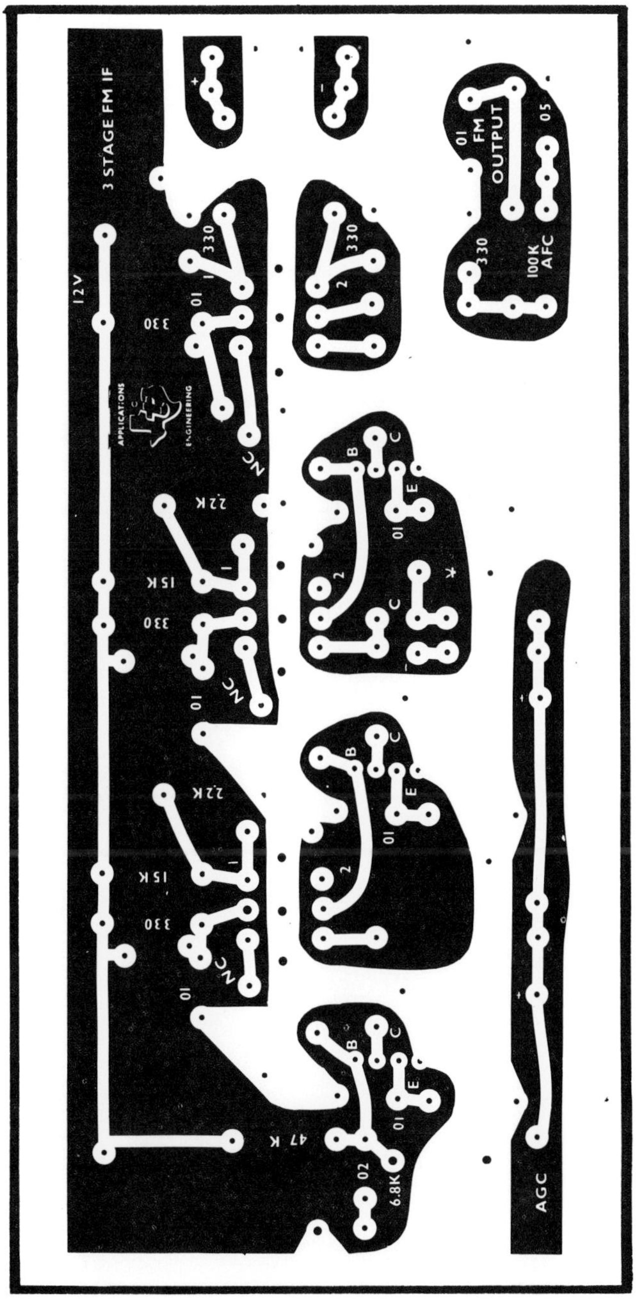

THREE-STAGE NEUTRALIZED AMPLIFIER USING SILICON TRANSISTORS TI408

This 10.7-MHz strip (TI circuit number 4051) uses three TI408 NPN planar silicon transistors in the neutralized condition. Neutralization is necessary to achieve high usable-gain-per-stage characteristics. Satisfactory stereo receiver characteristics are a feature in this IF strip. The printed circuit board layout used with TI circuit 4049 is recommended for use with this circuit.

Typical Performance (4051)

Data Taken with Input on Base of Transistor Q1

1. Overall 6 dB Bandwidth (Measured on Base of Q3) . . . . 300 kHz
2. Overall 40 dB Bandwidth (Measured on Base of Q3) . . 1100 kHz
3. Average Power Gain Per Stage . . . . 26 dB
4. FM 3 dB Limiting Level . . . . 400 μV input
5. AM Rejection at 3 dB Limiting Level . . . . $>$30 dB
6. Maximum Audio Recovery (Full Limiting)
   - $\pm$ 22.5 kHz Deviation . . . . 280 mV
   - $\pm$ 75 kHz Deviation . . . . 800 mV
7. Peak-to-Peak Separation of Ratio Detector . . . . . . $>$600 kHz

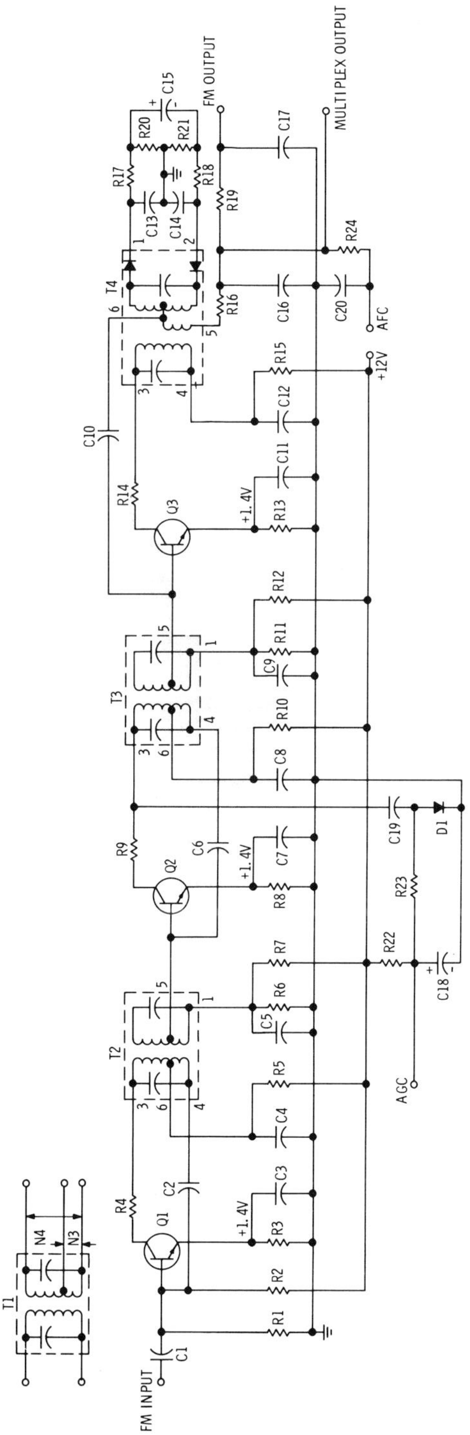

Schematic (4051)

Parts List (4051)

ALL RESISTOR VALUES IN OHMS

R1	9 100
R2	47 000
R3, R8, R13	680
R4, R9, R14	220
R5, R10, R15	330
R6, R11	2 700
R7, R12	15 000
R16	62
R17	1 500
R18	1 000
R19	7 500
R20, R21	8 200
R22	33 000
R23	4 700
R24	200 000
C1	0.02 μF
C2, C6, C10	2.2 pF
C3, C4, C5, C7, C8, C9, C11, C12, C17	0.01 μF
C13, C14, C16	330 pF
C15	2 μF, 10 V, electr.
C18	5 μF, 10 V, electr.
C19	6.8 pF
C20	0.05 μF
D1	TI-6
Q1, Q2, Q3	TI408
T2	TRW #19897-R1 or equiv.
T3	TRW #19898-R1 or equiv.
T4	TRW #19059-R1 or equiv.

RECOMMENDED T1 SECONDARY

C_T	47 pF
Q_u	44.5
Q_L	42.6 ($R_L = 1.6$ kΩ across N3)
$\frac{N4}{N3}$	14.1

FOUR-STAGE UNNEUTRALIZED AMPLIFIER USING GERMANIUM TRANSISTORS TIXM04

This four-stage 10.7-MHz amplifier (TI circuit 4047) uses the new PNP epitaxial planar germanium transistors in their unneutralized application. The application of these devices provides sufficient gain and a high degree of selectivity which are basic requirements for multiplex reception. The printed circuit board used with TI circuit 4050 may be used with the following modifications:

(1) Remove neutralization capacitors C2, C6, C10, C14.
(2) Change pin 4 with pin 6 on T2, T3, T4, T5.
(3) Change voltage polarity as shown on schematic 4047.
(4) Reverse D1 and C18 polarity.

Typical Performance (4047)

Data Taken with Input on Base of Transistor Q1

1. Overall 6 dB Bandwidth (Measured on Base of Q4) . . . 290 kHz
2. Overall 40 dB Bandwidth (Measured on Base of Q4) . . 770 kHz
3. Average Power Gain Per Stage . . . . 21 dB
4. FM 3 dB Limiting Level . . . . 80 μV input
5. AM Rejection at 3 dB Limiting Level . . . . $>$30 dB
6. Maximum Audio Recovery (Full Limiting)
   $\pm$ 22.5 kHz Deviation . . . . 280 mV
   $\pm$ 75 kHz Deviation . . . . 800 mV
7. Peak-to-Peak Separation of Ratio Detector . . . . . . $>$600 kHz

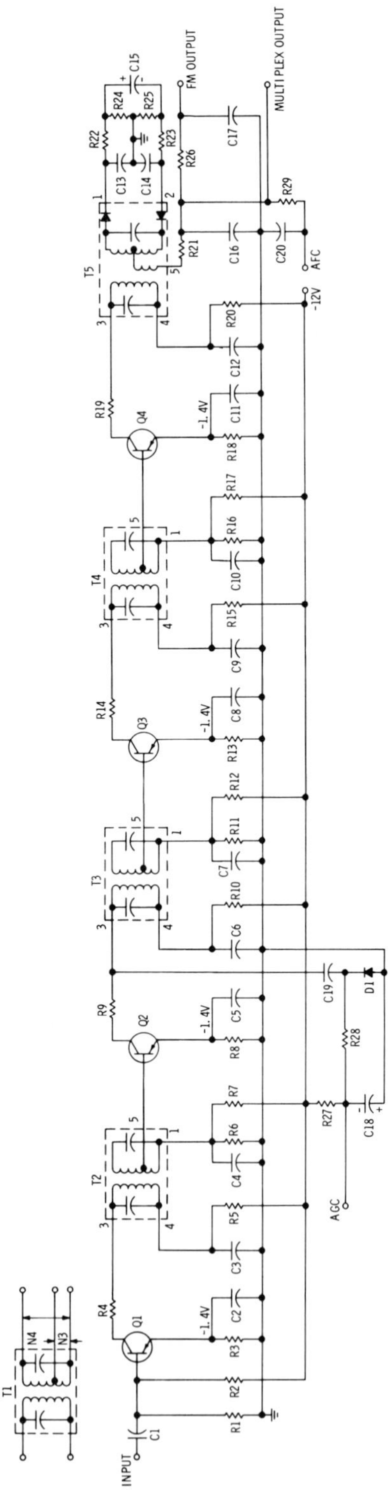

Schematic (4047)

Parts List (4047)

ALL RESISTOR VALUES IN OHMS

R1	10 000
R2	56 000
R3, R8, R13, R18	680
R4, R9, R14	220
R5, R10, R15, R20	330
R6, R11, R16	2 400
R7, R12, R17	15 000
R19	470
R21	62
R22	1 500
R23	1 000
R24, R25	8 200
R26	7 500
R27	33 000
R28	4 700
R29	200 000
C1	0.02 μF
C2, C3, C4, C5, C6, C7, C8, C9, C10, C11, C12, C17	0.01 μF
C13, C14, C16	330 pF
C15	2 μF, 10 V, electr.
C18	5 μF, 10 V, electr.
C19	6.8 pF
C20	0.05 μF
D1	TI-6
Q1, Q2, Q3, Q4	TIXM04
T2	TRW #19745-R1 or equiv.
T3	TRW #19746-R1 or equiv.
T4	TRW #19747-R1 or equiv.
T5	TRW #19059-R1 or equiv.

RECOMMENDED T1 SECONDARY

C_T	82 pF
Q_u	38.6
Q_L	38.3 ($R_L = 2.2$ kΩ across N3)
$\frac{N4}{N3}$	18.8

THREE-STAGE NEUTRALIZED AMPLIFIER USING GERMANIUM TRANSISTORS TIXM204

This three-stage 10.7-MHz strip (TI circuit 4044) is designed for use in stereo console receivers. The TIXM204 PNP epitaxial diffused-base mesa devices are neutralized to achieve the highest gain-to-cost ratio attainable. The printed circuit board used with TI circuit 4051 may be used with this circuit if the following modifications are made:

(1) Change voltage polarity.
(2) Reverse polarity of D1 and C18.

Typical Performance (4044)

Data Taken with Input on Base of Transistor Q1

1. Overall 6 dB Bandwidth (Measured on Base of Q3) . . . . 300 kHz
2. Overall 40 dB Bandwidth (Measured on Base of Q3) . . 1100 kHz
3. Average Power Gain Per Stage . . . . 26 dB
4. FM 3 dB Limiting Level . . . . 400 μV input
5. AM Rejection at 3 dB Limiting Level . . . . >30 dB
6. Maximum Audio Recovery (Full Limiting)
   ± 22.5 kHz Deviation . . . . 280 mV
   ± 75 kHz Deviation . . . . 800 mV
7. Peak-to-Peak Separation of Ratio Detector . . . . >600 kHz

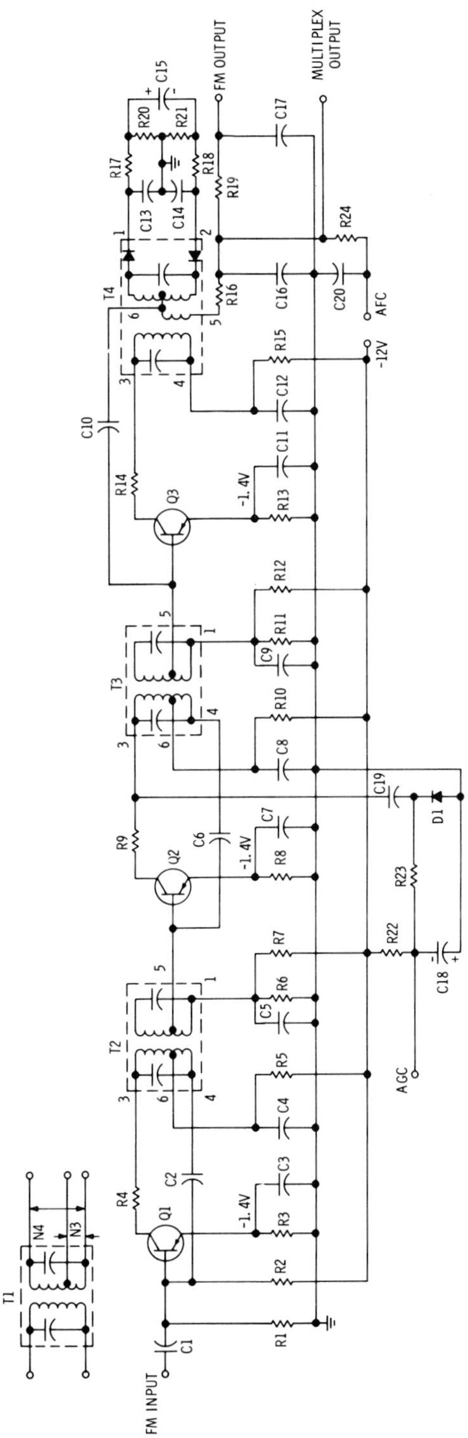

Schematic (4044)

Parts List (4044)

ALL RESISTOR VALUES IN OHMS

R1	10 000
R2	56 000
R3, R8, R13	680
R4, R9, R14	220
R5, R10, R15	330
R6, R11	2 400
R7, R12	15 000
R16	62
R17	1 500
R18	1 000
R19	7 500
R20, R21	8 200
R22	33 000
R23	4 700
R24	200 000
C1	0.02 μF
C2, C6	1.8 pF
C3, C4, C5, C7, C8, C9, C11, C12, C17	0.01 μF
C10	3.3 pF
C13, C14, C16	330 pF
C15	2 μF, 10 V, electr.
C18	5 μF, 10 V, electr.
C19	6.8 pF
C20	0.05 μF
D1	TI-6
Q1, Q2, Q3	TIXM204
T2	TRW #19699-R1 or Equiv.
T3	TRW #19700-R1 or Equiv.
T4	TRW #19059-R1 or Equiv.

RECOMMENDED T1 SECONDARY

C_T	47 pF
Q_u	48.8
Q_L	42.6 ($R_L = 600\ \Omega$ across N3)
$\frac{N4}{N3}$	15.1

PART III

Television Design

12

UHF TV Tuners

The FCC requirement, that all sets manufactured for sale in interstate commerce in the United States be equipped for all-channel reception, resulted in accelerated efforts to redesign UHF tuners for economy and performance. Tuner circuits, consisting of an oscillator tuned with a variable capacitor, a double-tuned preselector circuit, and a diode mixer, remain essentially unchanged since the early 1950's. One most important change, from the standpoint of both economy and performance, has been the use of a transistor in the oscillator circuit. The economical considerations are:

1. Cost of the transistor versus the cost of the tube,
2. Elimination of the tube socket, and
3. Elimination of the heater circuit with its bypass capacitors and troublesome chokes.

The performance considerations are:

1. Less thermal drift with the elimination of the heaters, and
2. Less radiation because of much smaller power generation.

Practically all UHF tuners now being manufactured have transistor oscillators. The typical U.S. tuner has a noise figure of 9 to 13 dB with a conversion loss of 6 to 10 dB. To make up the loss, which occurs in the diode mixer, the IF output of the UHF tuner is fed to channel one of the VHF tuner where it is amplified before being applied to the IF strip.

Figure 1 is a diagram of a typical UHF tuner. The chassis is divided into three compartments consisting of the oscillator tuned circuit and two preselector tuned circuits slightly overcoupled to provide a bandwidth of approximately 20 MHz. The preselector circuits are quarter-wave transmission lines

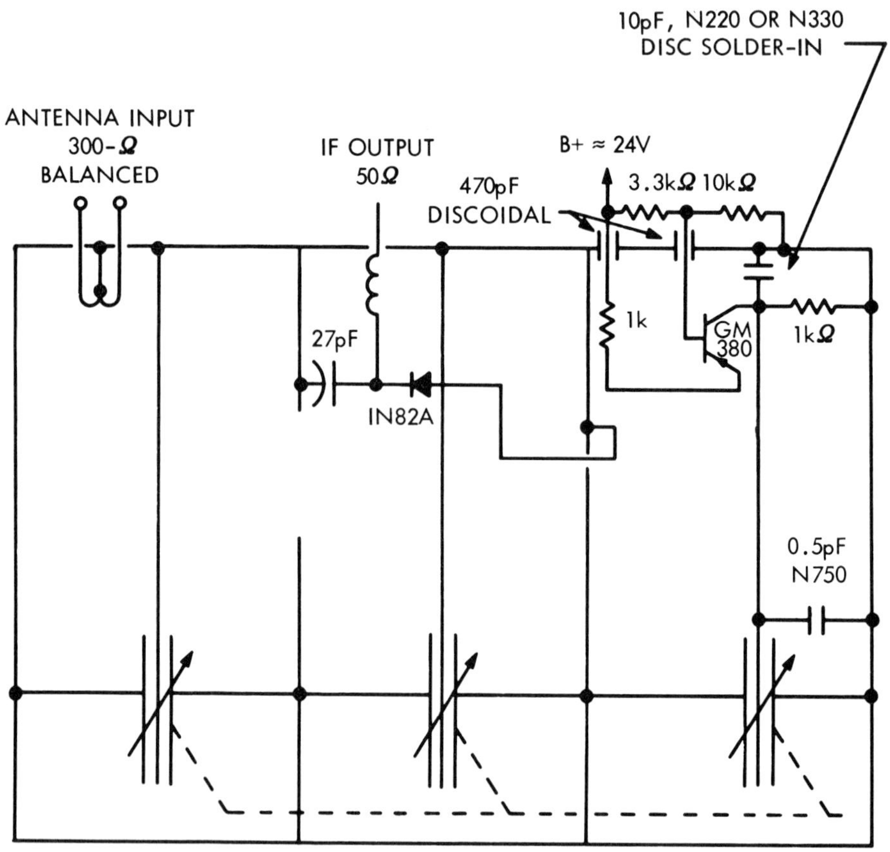

Fig. 1. UHF TV Tuner

tuned with variable air capacitors which are shaped to provide a linear relationship between frequency and angle of shaft rotation. The preselectors tune over the UHF band, 470 to 890 MHz.

The oscillator circuit may be considered in two ways:

1. It is a half-wave transmission line terminated on both ends by capacitances, one of which is the variable tuning element; the other is the parallel combination of the transistor C_{ob}, the "swamping capacitor," and the low-frequency trimmer if used.
2. It is a simple inductance tuned with the two series capacitances. It has a virtual ground since the common point of the two series capacitances is grounded.

The voltage node (or virtual ground) moves along the line (or inductance) as the tuning capacitance is varied. Since operation of the oscillator tuned circuit cannot conveniently be described mathematically by either lumped-

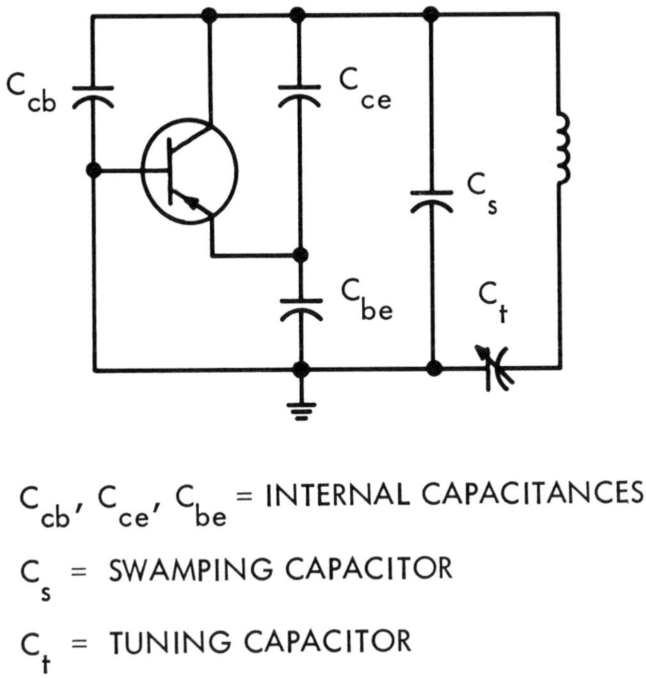

Fig. 2. Simplified A-C Circuit of Variable-capacitance Oscillator

constant or distributed-constant theory, its design is almost entirely empirical. The criteria of oscillator performance are:

1. Proper oscillator injection into the mixer diode at all frequencies over the band, and
2. Frequency stability over a specified range of temperature and power supply variations.

Many transistors have been developed for this application. One of the first germanium devices was the GM380 which is presently used in about one-third of U.S. tuner production. The output reactance change, with changes in temperature and supply voltage, is somewhat greater than in more recently developed germanium devices; however, at least three major tuner manufacturers have successfully compensated for this drift. The TIS18 silicon transistor has inherent drift characteristics about four times better than GM380; that is, without compensation, the frequency drift of the TIS18 is ¼ that of the uncompensated GM380. Economical germanium devices have been designed which equal the performance of the TIS18.

Figure 2 is a simplified schematic which represents a great majority of the tuners being presently manufactured. Capacitor C_s represents the circuit technique that gives the transistor oscillator its superior drift characteristics. The value of C_s ranges from 7 to 18 pF, depending upon the tuner manufacturer, and effectively "swamps out" the transistor output reactance changes. C_s also provides thermal compensation for the changes in the tuned line. It usually has a temperature characteristic of N220 to N470 depending upon the mechanical configuration of the line.

The greatest problem in the design of UHF tuners is the avoidance of parasitic resonances within the oscillator tuning range, 515 to 935 MHz. A resonance causing the diode injection current to fall below a certain value, depending upon the diode used, will cause a very significant increase in noise figure at that frequency. It also disturbs the IF impedance of the diode which results in IF bandpass skewing and distortion of color information.

At the time of this writing a Schottky-barrier diode has just been sampled to the industry to replace the point-contact diode which has been used as a mixer. Objections to the point-contact which the new diode appears to have overcome are:

1. The inherent fragility which has resulted in unreliability, and
2. The non-uniformity of performance.

An additional bonus is the very low noise figure of the Schottky-barrier: 5.5 to 7.0 dB compared to 6.5 to 12 dB with the point-contact.

A refinement of the UHF tuner for use in some color sets is the addition of an AFC circuit. A discriminator or quadrature detector in the set provides a voltage proportional to the frequency deviation from a preselected value. This voltage is applied to a varactor diode which parallels the oscillator tuned circuit. The capacity of the varactor varies with the applied AFC voltage to bring the frequency back toward the preset value. The diode most used in this application is Texas Instruments A660.

13

VHF TV Tuners

TUNER FUNCTIONS

The principal function of a tuner is to convert an RF signal to an IF signal. This must be accomplished while preserving the signal-to-noise ratio, supplying power gain, and maintaining selectivity. Also, the tuner must provide the antenna line with a proper termination, prevent local-oscillator radiation, assure frequency stability and have controllable gain (AGC). A block diagram of a typical VHF TV tuner is shown in Fig. 1.

The ultimate limitation on sensitivity is the noise figure.[1] Of course, sufficient power gain is necessary to provide the video drive required by the cathode ray tube, but a usable picture also demands a high signal-to-noise ratio. A receiver noise figure of 6 dB meets this requirement.

TV receiver manufacturers usually desire a 30-dB power gain in the tuner. This somewhat arbitrary value should be considered a minimum since greater stable gain is advantageous.

All image and IF rejection is obtained by circuits in front of the mixer or first detector. The IF rejection is enhanced by placing a resonant circuit or filter near the tuner input. A similar trap for commercial FM broadcasts may be included since this service uses the 88- to 108-MHz band which is adjacent to channel 6.

Cross-modulation may be reduced by providing selectivity in front of the RF amplifier. However, this technique is limited by noise figure degradation.

With few exceptions, receiving antennas require a 300-ohm balanced termination. Ordinarily, a broadband VHF balun is used to transform this balanced

300 ohms into an unbalanced 75-ohm load. The Federal Communications Commission has decreed that the maximum allowable local-oscillator radiation measured at 100 feet from the TV receiver shall not exceed 50 microvolts per meter on channels 2 through 6, or 150 microvolts per meter on channels 7 through 13. Judicious use of shielding and filtering techniques can control radiation.

RF AMPLIFIER

The most common RF amplifier configuration is shown in Fig. 2. L_1, C_1, C_2, the output capacitance of the input filter, and the input capacitance of the transistor form a single-tuned resonant circuit that transforms the antenna impedance into a value that yields the lowest noise figure. Loss of power gain due to impedance mismatch is small because the mismatch factor is small. The loaded Q of this circuit must be quite low to keep the insertion loss small. (Loss is a direct function of the ratio of loaded Q to unloaded Q.) This is very necessary because the tuner noise will be proportionally degraded by the loss in all the circuits in front of the transistor.

This minimum loss consideration dictates such a small loaded Q that bandwidths of about 50 MHz are encountered. However, from the standpoint of cross-modulation, the bandwidth should be as small as possible, i.e., 6 MHz, but this would degrade the noise figure about 6.2 dB (assuming the unloaded Q of L_1 is 70). Clearly this is unacceptable for all but some few specialized applications. A common compromise (channel 13) is a bandwidth of 20 MHz with the attendant 1-dB noise figure degradation.

Another method used to improve cross-modulation is to unbypass a portion of the emitter resistance. Thanos[2] achieved a 6- to 12-dB improvement in cross-modulation with an attendant 2-dB increase in noise figure by adding a 56-ohm unbypassed emitter resistor.

Friis[3] has shown that, if the noise figures of the RF and mixer stages are similar and the numeric power gain of the RF stage is at least 10, the tuner noise figure is just the noise figure of the RF amplifier.

Voltage standing wave ratio (VSWR) is another consideration of the input circuitry. Ideally, the VSWR should be near 1:1; however, 3:1 is acceptable. Higher VSWR's will produce "ghosts" if the antenna lead-in line exceeds 40 feet, or, if shorter, make the picture appear as though the high-frequency video response is poor.[4]

Figure 3 shows the desired response curve of the double-tuned overcoupled transformer that drives the mixer. Almost all of the system's image

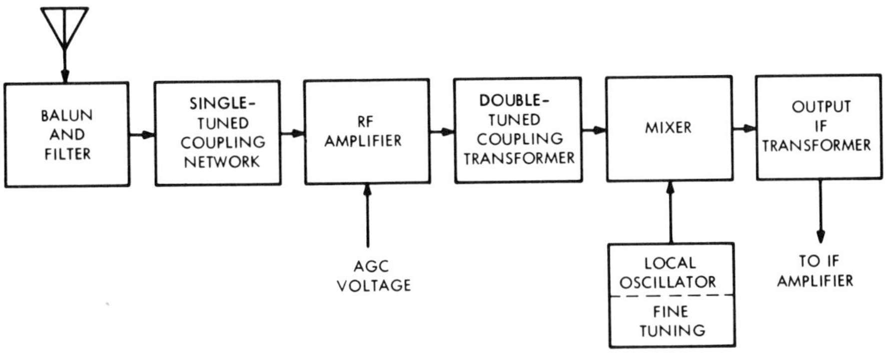

Fig. 1. VHF TV Tuner Block Diagram

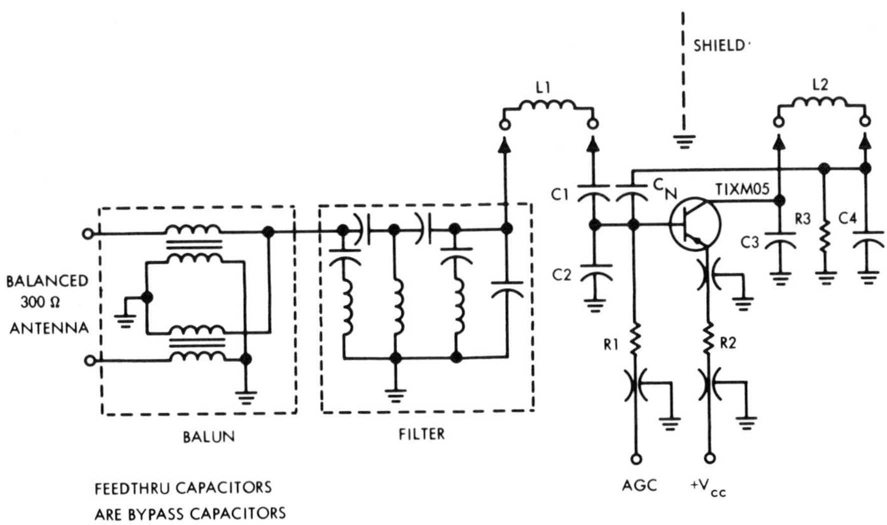

Fig. 2. RF Amplifier

rejection is obtained here. L_2 and L_3 (on Fig. 5) are selected to have the same unloaded uncoupled Q. Also, for equal amplitude response peaks, the loaded uncoupled Q must be equal. C_4 is much larger than C_3 (see neutralization discussion below); therefore, the primary tank consists essentially of L_2, C_3, transistor output impedance, and strays. The secondary tank consists of L_3, C_5, C_6, transistor input impedance, and strays. C_5 and C_6 are selected to obtain proper loaded uncoupled Q.

Neutralization[5] may be accomplished by the addition of C_N and C_4. C_4 is chosen so that the series resonant frequency of C_4 and L_2 is much less than that

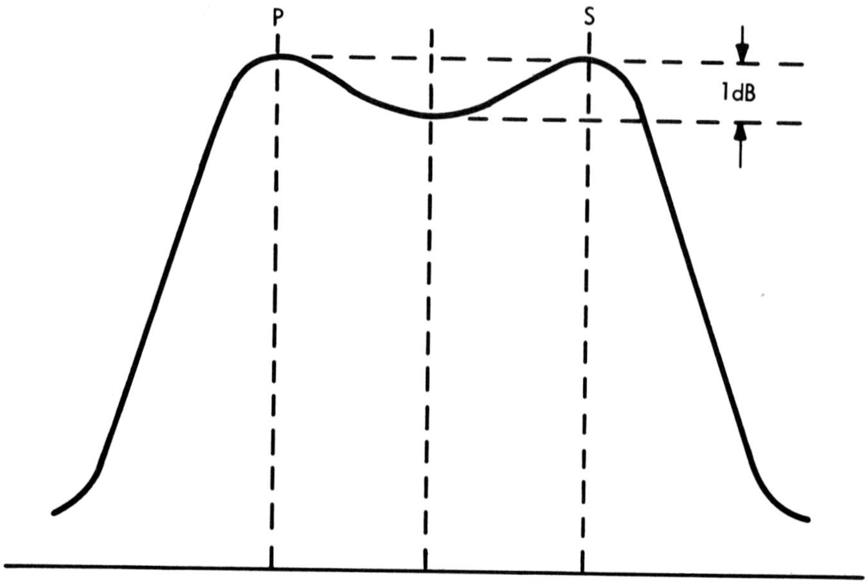

Fig. 3. Mixer Driver Transformer Response Curve

of L_2 in parallel with C_{TOTAL}, the total effective parallel capacitance across L_2. The voltage across C_4 will lag the collector voltage by about 180 degrees.

Transistors designed for forward automatic gain control (AGC) achieve this ability from their characteristics of decreasing h_{fe} as I_C increases (Fig. 4). Impedance-transformation gain and h_{fe} gain both contribute at maximum-gain bias conditions. However, as I_C increases, the transistor input and output impedance and h_{fe} are greatly reduced yielding a large reduction in gain. At least 40 dB of gain reduction is usually required.

R_1 is usually about 2000 ohms, a value large enough to prevent shunting a significant portion of the signal yet small enough to be compatible with the AGC amplifier.

MIXER

A transistor mixer needs high gain at the IF frequency and an efficient emitter-base diode at the RF signal frequency. The total conversion gain will be about 2 to 6 dB less than the IF frequency gain. Figure 5 shows a popular configuration for a neutralized mixer. This neutralizing scheme is the same as for the RF amplifier stage. The d-c collector current is initially selected for maximum gain as in an ordinary IF amplifier, then varied empirically for the

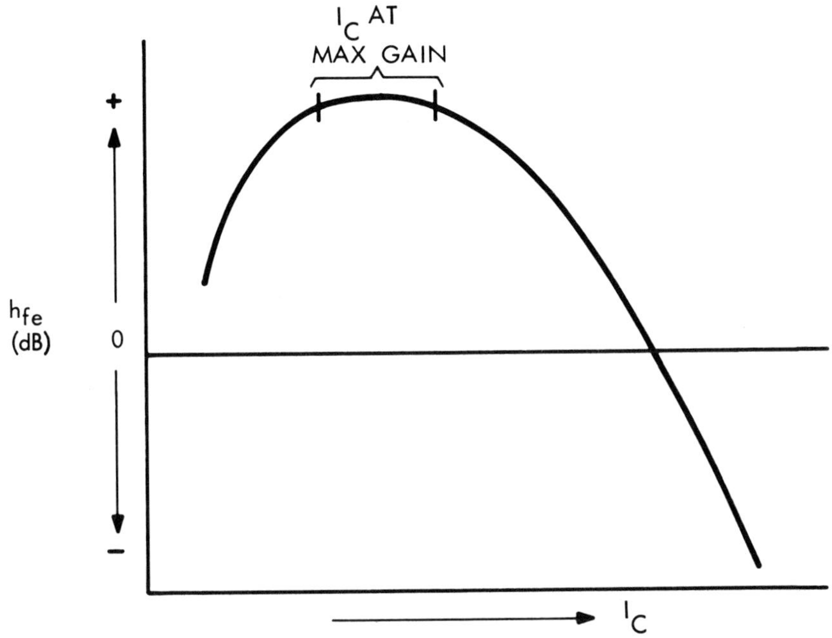

Fig. 4. Transistor h_{fe} versus I_c

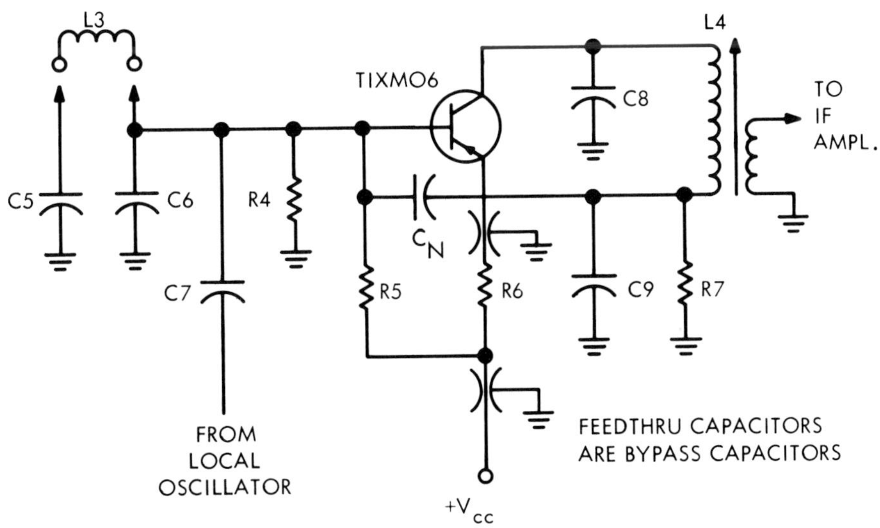

Fig. 5. Mixer

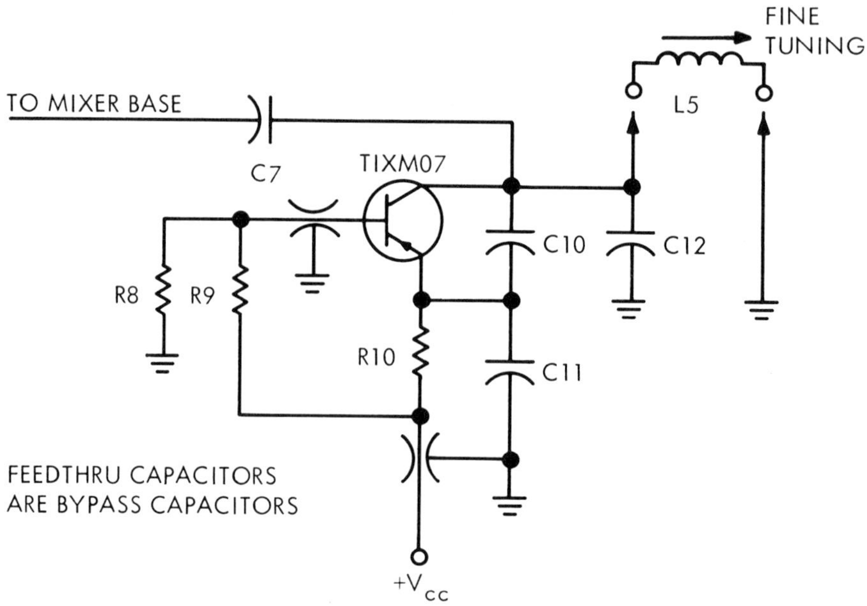

Fig. 6. Local Oscillator

best conversion gain because the d-c operating point is a compromise between IF frequency gain and diode efficiency.

The diode conversion loss helps to ensure stable operation, allowing the collector load to be nearly matched for maximum power gain. The collector load for the mixer is usually a double-tuned overcoupled transformer whose primary is parallel-tuned while the secondary is a series-tuned inductor and capacitor located in the IF amplifier. The coupling link "taps" the collector impedance at 50 ohms so that a 50-ohm coaxial cable may be used to couple the tuner to a remotely located IF amplifier. The tuner output response curve should have a 3-dB bandwidth, without sound traps, of about 5.5 MHz.

OSCILLATOR

The common-base transistor amplifier is usually regenerative at VHF frequencies making it very suitable for an oscillator. Figure 6 shows a common-base oscillator where C_{10} and C_{11} provide positive feedback energy and L_5, C_{10}, C_{11}, C_{12}, and the transistor output impedance form the frequency determining parallel resonant circuit. C_7 is selected to supply the required level of injection voltage to the mixer. Mixer injection is about 0.1 to 0.2 volts and is best determined empirically for optimum conversion gain.

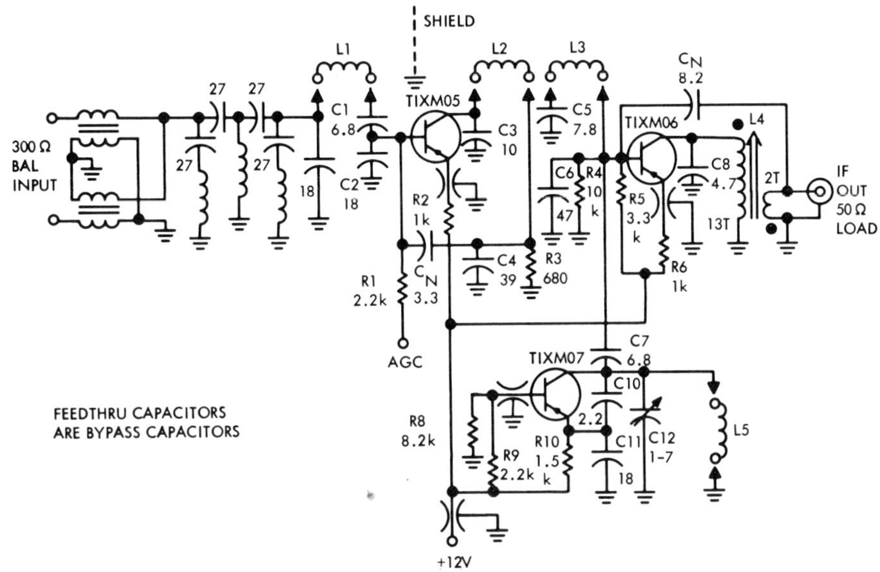

Fig. 7. Channel 13 Prototype Schematic using Transistor Series TIXM05

A fine tuning range of ± 1.5 MHz minimum is obtainable by varying L_5 or C_{12}. Varying L_5 by a movable slug provides desirable memory tuning.

Typical frequency stability requirements are $+100$ kHz and -300 kHz maximum frequency change for an environmental temperature change of $+30°C$. C_{12} can be a temperature-compensating capacitor whose temperature coefficient is selected empirically to control frequency drift with temperature change.

DESIGN EXAMPLE

The most difficult case, channel 13 (210-216 MHz), is presented in this design example. With the exception of IF rejection, performance can be expected to improve on lower channels. A channel 13 schematic using the transistor series TIXM05 is shown in Fig. 7.

Mixer. The IF center frequency f_0 is 43.5 MHz with a required 3-dB bandwidth BW of about 5.5 MHz. The Q loaded and coupled Q_{LC} is

$$Q_{LC} = \frac{f_0}{BW} = \frac{43.5}{5.5} = 7.9 \tag{1}$$

The maximum available gain MAG at f_0 is

$$\text{MAG (dB)} = 20 \log\left(\frac{f_{(max)}}{f_0}\right) \tag{2}$$

where $f_{(max)}$ is the maximum frequency of oscillation and is

$$f_{(max)} = \sqrt{\frac{f_T}{8\pi\ r_b'\ C_c}} \tag{3}$$

where f_T is the frequency where h_{fe} equals zero dB.

The TIXM06 is a PNP epitaxial planar germanium transistor designed for television application as a VHF mixer. The minimum value of f_T and the maximum value of $r_b'C_c$ are given on the data sheet as 380 MHz and 10 picoseconds. Therefore,

$$f_{(max)} = \sqrt{\frac{380 \times 10^6}{8\pi\ 10 \times 10^{-12}}} = 1231 \text{ MHz}$$

and

$$\text{MAG(dB)} = 20 \log \left(\frac{1231}{43.5}\right) = 29.06 \text{ dB}$$

The conversion gain $PG_{(CONV)}$ of a neutralized mixer is MAG, at f_o, minus all circuit losses. One such loss is the transistor emitter-base diode conversion loss DCL. A useful empirical formula relating DCL to RF signal frequency for TV mixers is,

$$\text{DCL(dB)} = 20 \log \left(1.054 + \frac{f_{(RF)}}{320 \times 10^6}\right) \tag{4}$$

For channel 13, $f_{(RF)} = 213.5$ MHz. Therefore,

$$\text{DCL(dB)} = 20 \log \left(1.054 + \frac{213.5 \times 10^6}{320 \times 10^6}\right) = 4.46 \text{ dB}$$

The other important source of losses is the transformer insertion loss IL:

$$\text{IL} = \text{mismatch Loss ML} + \text{Coil Loss CL} \tag{5}$$

where

$$\text{ML(dB)} = 10 \log \left[\frac{(1 + a)^2}{4a}\right] \tag{6}$$

and the mismatch factor "a" is greater than one.
Also,

$$\text{CL(dB)} = 20 \log \left(\frac{Q_{UU}}{Q_{UU} - Q_{LC}}\right) \tag{7}$$

where Q_{UU} is the unloaded uncoupled Q of the transformer primary and Q_{LC} is the loaded coupled Q of the mixer collector circuit.

Stability usually requires that a certain amount of loss be designed into the stage. This loss can be calculated by temporarily assuming that the transistor's input admittance equals the source admittance and that all the required loss will be mismatched loss in the collector circuit.

Stern's stability expression is

$$(g_{11} + G_g)\,(g_{22} + G_L) = \frac{k}{2}\,(L + M) \tag{8}$$

Where stability is achieved when k is equal to or greater than one and where

$g_{11} = \text{Re}\,(y_{11})$
$g_{22} = \text{Re}\,(y_{22})$
$G_g = \text{Re}$ (source admittance)
$G_L = \text{Re}$ (load admittance)
$L = |y_{12}\,y_{21}|$
$M = \text{Re}\,(y_{12}\,y_{21})$

Solving Eq. (8) for G_L yields

$$G_L = \frac{k(L + M)}{2(g_{11} + G_g)} - g_{22} \tag{9}$$

By choosing k equal to one, Eq. (9) becomes an expression for the minimum value of load conductance that is necessary to introduce the loss required to assure stability.

At 45 MHz with V_{CE} equal to 10 volts and I_c equal to 2 mA, the TIXM08 has

$g_{11} = 1.2 \times 10^{-3}$ S
$g_{22} = 0.025 \times 10^{-3}$ S
$y_{12} = (0.15 + j0) \times 10^{-3}$ S
$y_{21} = (45 + j10) \times 10^{-3}$ S

Therefore, $L = 6.57 \times 10^{-6}$
$M = -1.43 \times 10^{-6}$

and since G_g equals g_{11},

then $G_g = 1.2 \times 10^{-3}$ S

Solving Eq. (9) for G_L,

$$G_L = \frac{(6.57 - 1.43) \times 10^{-6}}{2(1.2 + 1.2) \times 10^{-3}} - 0.025 \times 10^{-3}$$
$$= 1.045 \times 10^{-3}\ \text{S}$$

The required loss for stability is

$$\text{Loss} = 10 \log \left[\frac{(1 + a)^2}{4a}\right]$$

where $$a = \frac{G_L}{g_{22}}$$

$$= \frac{1.045 \times 10^{-3}}{0.025 \times 10^{-3}} = 41.7$$

Therefore, $$\text{Loss} = 10 \log \left[\frac{1 + (41.7)^2}{4 \times 41.7}\right]$$

$$\approx 10 \text{ dB}$$

Thus 10 dB of power gain must be "thrown away" to assure stability. Now the actual reflected load resistance is calculated.

The 10 dB loss may be divided thus:

$$\text{ML} = 4.0 \text{ dB, CL} = 1.5 \text{ dB, with a DCL of } 4.46 \text{ dB.}$$

Because $$\text{ML(dB)} = 4 \text{ dB} = 10 \log \left[\frac{(1 + a)^2}{4a}\right]$$

then $$a = 8$$

Also $$a = \frac{r_{oep}}{R_L'}$$

where R_L' is the load reflected to the primary and r_{oep} is the transistor's parallel equivalent resistance (40 kΩ).

Therefore, $$R_L' = \frac{r_{oep}}{8} = \frac{40 \text{ k}\Omega}{8} = 5 \text{ k}\Omega$$

Since $$\text{CL(dB)} = 1.5 \text{ dB} = 20 \log \left(\frac{Q_{UU}}{Q_{UU} - Q_{LC}}\right)$$

then $$Q_{UU} = 50$$

The total conversion gain $PG_{(CONV)}$ for the stage is

$$PG_{(CONV)} = \text{MAG} - (\text{ML} + \text{CL} + \text{DCL}) \qquad (11)$$
$$= 29.06 - (4.0 + 1.5 + 4.46) = 19.1 \text{ dB}$$

The equivalent circuit of the mixer collector circuit (refer to Fig. 7) is

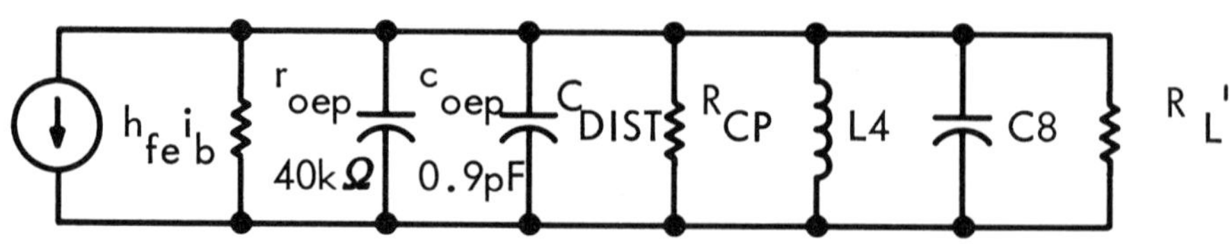

The equivalent parallel resistance R_{CP} of the transformer primary is found by assuming that the primary open-circuit inductance L_4 is equal to 2 microhenries. Therefore,

$$R_{CP} = Q_{UU}X_{L4} = Q_{UU}\omega L_4 \tag{12}$$
$$= 50\ (2.73 \times 10^8)\ (2 \times 10^{-6}) = 27\ 300\Omega$$

The total equivalent parallel resistance R_T is

$$\frac{1}{R_T} = \frac{1}{r_{oep}} + \frac{1}{R_{CP}} + \frac{1}{R_L{}'} \tag{13}$$
$$= \frac{1}{40\ k\Omega} + \frac{1}{27.3\ k\Omega} + \frac{1}{5\ k\Omega}$$
$$R_T = 3820\ \Omega$$

Since $Q_{LC} = \omega\ C_T R_T$, the total effective parallel capacitance C_T is

$$C_T = \frac{Q_{LC}}{\omega\ R_T} = \frac{7.9}{(2.73 \times 10^8)\ 3820} = 7.57\ \text{pF} \tag{14}$$

The distributed capacitance C_{DIST} is about 1.5 pF; therefore, the collector capacitor C_8 is

$$C_8 = C_T - (c_{oep} + C_{DIST}) = 7.57 - (0.9 + 1.5)$$
$$= 5.17\ \text{pF}$$

The nearest standard value of 4.7 pF is selected.

Now L_4 may be found.

$$L_4 = \left[C_T\ \omega^2\right]^{-1} = \left[7.57 \times 10^{-12}\ (2.73 \times 10^8)^2\right]^{-1}$$
$$= 1.77\ \mu\text{H}$$

which is sufficiently close to the assumed value that no recomputation of R_{CP} is required.

The mixer output transformer may now be wound to satisfy the requirements of $L_4 = 1.77\ \mu$H, $Q_{UU} = 50$ and $R_L{}' = 5000\ \Omega$. The transformer in this example was close wound with No. 31 enameled wire on a 3/16-inch O.D. form with a ferrite core, has 13 primary turns and a 2-turn coupling link close wound to the "cold" end of the primary. This transformer may be empirically designed with the aid of a Boonton Radio Co. Type 250-A RX Meter and Type 260-A Q Meter.

The TIXM05 series data sheet shows that h_{fe} reaches a maximum when I_C is in the range of 1.5 to 2.5 mA and V_{CE} is about 10 volts. Therefore, the quiescent bias point is chosen as I_C equal to 2 mA and V_{CE} equal to 10 volts. The supply voltage is the typical plus 12 volts. The emitter resistor R_6 is

$$R_6 = \frac{V_E}{I_E} = \frac{V_{CC} - V_{CE}}{I_E} = \frac{12\ V - 10\ V}{2\ mA} = 1000\ \Omega$$

A common-emitter circuit has a useful simplified expression for current stability. The stability factor S is

$$S = 1 + \frac{R_5\ (R_4)}{(R_5 + R_4)R_6} \tag{15}$$

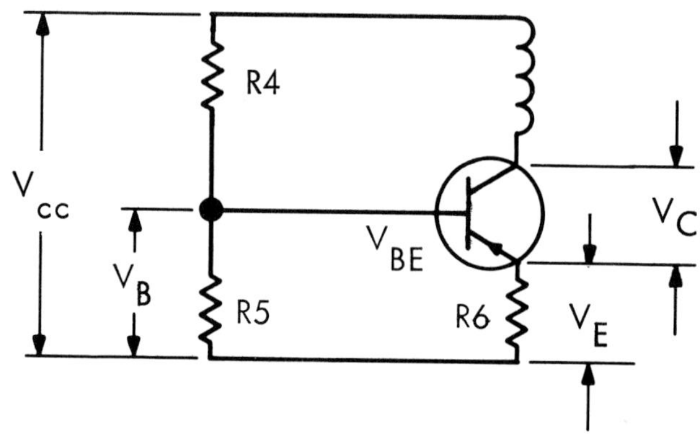

Solving for R_5,

$$R_5 = \frac{(S-1)\ R_6\ (R_4)}{R_4 - (S-1)R_6}$$

Also,

$$\frac{V_{CC}}{R_5 + R_4} = \frac{V_B}{R_5} \tag{16}$$

Therefore,

$$R_4 = \left(\frac{V_{CC} - V_B}{V_B}\right) R_5 \tag{17}$$

Combining Eq. (16) and (17) and solving for R_5 yields

$$R_5 = R_6\ (S-1) \left(1 + \frac{V_B}{V_{CC} - V_B}\right) \tag{18}$$

Also,

$$V_B = V_E + V_{BE} = I_E R_6 + V_{BE} \tag{19}$$

A stability factor S of 3 will enable germanium to operate at +60°C ambient. V_{BE} for the TIXM05 series is about 0.3 volts. Therefore,

$$V_B = 2\text{ mA } (1000\ \Omega) + 0.3\text{ V} = 2.3\text{ V}$$

and

$$R_5 = 1000\ (3-1)\left(1 + \frac{2.3}{12-2.3}\right) = 2474\ \Omega$$

Therefore,

$$R_4 = \left(\frac{12-2.3}{2.3}\right) 2474 = 10\ 400\Omega$$

These values of R_5 and R_4 are the initial choice and are adjusted empirically for maximum conversion gain in the mixer.

Maximum power dissipation should now be checked against the data sheet limit.

$$P_{DISP(max)} = I_E V_{CE} = 2\text{ mA } (10\text{ V}) = 20\text{ mW} \qquad (20)$$

Power derating factor (to +100°C) = 1 mW/°C

Rated device dissipation (+25°C) = 75 mW

Therefore, maximum allowable ambient temperature $T_{(max)}$ is

$$T_{(max)} = \frac{75\text{ mW} - P_{DISP(max)}}{1\text{ mW/°C}} + 25°\text{C} \qquad (21)$$

$$= \frac{75-20}{1} + 25 = 80°\text{C}$$

The neutralizing scheme used here is different from the one previously discussed to show another circuit configuration. The value of the neutralizing capacitor C_N can be easily found by applying a 43.5-MHz sweep generator to the transformer output winding and observing the base signal with an oscilloscope. The 8.2-pF neutralizing capacitor is selected for minimum base signal at 43.5 MHz.

RF Amplifier. The RF amplifier can be designed by first considering the equivalent collector circuit when it is loaded but uncoupled.

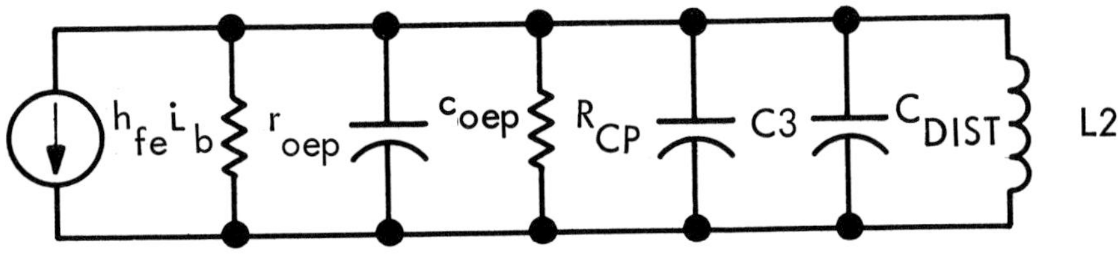

For channel 13, $\omega = 13.4 \times 10^8$. The TIXMO5 has these typical values at 213.5 MHz:

$r_{oep} = 10\ k\Omega$

$c_{oep} = 0.9\ pF$

$r_{iep} = 160\ \Omega$

$c_{iep} = 7\ pF$

Let L_2 (Fig. 7) = 0.065 μH and $Q_{UU} = 70$.
Therefore,

$$R_{CP} = Q_{UU}\ \omega\ L_2 = 70(13.4 \times 10^8)\ (0.065 \times 10^{-6}) = 6\ k\Omega$$

and

$$R_T = \frac{r_{oep}\ R_{CP}}{r_{oep} + R_{CP}} = \frac{10(6)}{10+6} = 3.75\ k\Omega$$

The bandwidth of the primary alone is about 5 MHz as is the secondary which, when coupled, will yield a peak-to-peak bandwidth of 6 MHz.

$$Q_{LU} = \frac{f_o}{BW} = \frac{213.5}{5.0} = 42.7 \tag{22}$$

The total effective parallel capacitance C_T is

$$C_T = \frac{Q_{LU}}{\omega R_T} = \frac{42.7}{(13.4 \times 10^8)\ 3750} = 8.6\ pF$$

Therefore,

$$L_2 = \left[C_T\omega^2\right]^{-1} = \left[8.6 \times 10^{-12}\ (13.4 \times 10^8)^2\right]^{-1}$$

$$= 0.065\ \mu H$$

Thus the assumed value of L_2 is good. Next C_4 is resonated with L_2 at a frequency much lower than 213.5 MHz; i.e., 100 MHz.

$$C_4 = \left[L_2\omega^2\right]^{-1} = \left[0.065 \times 10^{-6}\ (6.28 \times 10^8)^2\right]^{-1}$$

$$= 39.1\ pF$$

Select standard value, $C_4 = 39$ pF.

$$C_3 = \frac{C_4C_T}{C_4 - C_T} - (c_{oep} + C_{DIST}) = \frac{39(8.6)}{39-8.6} - (0.9 + 0.5)$$

$$\approx 10\ pF$$

Select standard value, $C_3 = 10$ pF.

From the transistor data sheet, $C_{cb} = 1.0$ pF max.

$$C_N = \frac{C_{cb}C_4}{C_3 + C_{oep} + C_{DIST}} = \frac{1.0(39)}{10 + 0.9 + 0.5} = 3.42 \text{ pF}$$

Select standard value, $C_N = 3.3$ pF.

The RF transformer secondary is now considered when it is loaded but uncoupled.

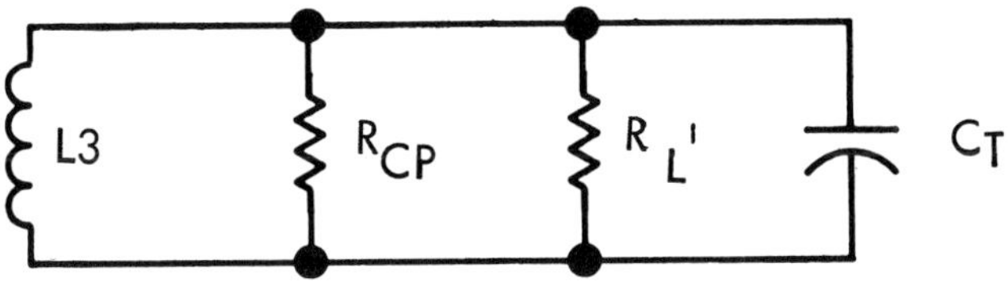

Let L_3 (Fig. 7) $= 0.08\ \mu$H and $Q_{UU} = 70$.

Thus

$$R_{CP} = Q_{UU}\omega L_3 = 70\ (13.4 \times 10^8)\ (0.08 \times 10^{-6}) = 7.6\ \text{k}\Omega$$

Since $L_3 \approx 0.08\ \mu$H

$$C_T = \left[L_3\omega^2\right]^{-1} = \left[0.08 \times 10^{-6}\ (13.4 \times 10^8)^2\right]^{-1}$$

$$\approx 7.0 \text{ pF}$$

For both peaks of the RF response curve (Fig. 3) to have equal amplitude, the total equivalent parallel resistances R_T of the primary and secondary must be equal. R_T of the primary was found to be 3.75 kΩ. R_T of the secondary is

$$R_T = \frac{R_{CP}\, R_L'}{R_{CP} + R_L'} = 3.75\ \text{k}\Omega \qquad (23)$$

Solving for R_L',

$$R_L' = \frac{R_{CP}\, R_T}{R_{CP} - R_T} = \frac{7.6(3.75)}{7.6 - 3.75} = 7.4\ \text{k}\Omega$$

Capacitors C_5 and $C_{6(TOTAL)}$ form a divider that transforms impedances according to the ratios

$$\frac{C_{6(TOTAL)}}{C_5} = \sqrt{\frac{R_L'}{r_{iep}}}$$

$$C_{6(TOTAL)} = \sqrt{\frac{7400}{160}}\, C_5 = 6.8\ C_5$$

C_5 reduced by $C_{6(TOTAL)}$ in series will equal C_T.

$$C_5 = \frac{C_{6(TOTAL)}\, C_T}{C_{6(TOTAL)} - C_T}$$

But $C_{6(TOTAL)} = 6.8\ C_5$

Therefore,

$$C_5 = \frac{6.8C_5\ C_T}{6.8C_5 - C_T} = 1.15\ C_T$$

where

$C_T \approx 7$ pF and

$C_5 \approx 1.15(7.0) \approx 8$ pF

and

$C_{6(TOTAL)} \approx 6.8(8.0) \approx 55$ pF

Referring to Fig. 5,

$$C_{6(TOTAL)} = C_6 + c_{iep} + C_{DIST} + C_{osc}$$

where

c_{iep} = Mixer input parallel capacitance at 213.5 MHz = 7 pF

C_{DIST} = Distributed capacitance $\approx$ 1 pF

C_{osc} = Capacitance load of oscillator $\approx$ 2 pF

C_6 = Fixed capacitor

$$C_6 = C_{6(TOTAL)} - (c_{iep} + C_{DIST} + C_{osc})$$
$$= 55 - (7 + 1 + 2) = 45 \text{ pF}$$

Select standard value, $C_6 = 47$ pF.

The bandwidth of the secondary is chosen to be 5.8 MHz which will ultimately aid in selecting standard component values.

$$Q_{LU} = \frac{f_o}{BW} = \frac{213.5}{5.8} = 36.4$$

and

$$C_T = \frac{Q_{LU}}{\omega\ R_T} = \frac{36.4}{(13.4 \times 10^8)3750} = 7.2 \text{ pF}$$

Therefore,

$$L_3 = \left[C_T\omega^2\right]^{-1} = \left[7.2 \times 10^{-12}\ (13.4 \times 10^8)^2\right]^{-1}$$

$$= 0.078\ \mu\text{H}$$

Thus the assumed value of L_3 is sufficiently close that R_{CP} need not be recomputed.

Finally, $$C_5 = \frac{C_{6(TOTAL)}C_T}{C_{6(TOTAL)} - C_T} = \frac{55\ (7.2)}{55 - 7.2} = 8.1 \text{ pF}$$

Select standard values 5.6 and 2.2 in parallel for C_5.

The RF transformer loss TL is approximately

$$TL(dB) = 20 \log \left(\frac{Q_{UU}}{Q_{UU} - Q_{LC}}\right)$$

where

$$Q_{LC} = \frac{f_o}{BW_{(peak\text{-}to\text{-}peak)}} = \frac{213.5}{6.0} = 35.6$$

Therefore,

$$TL(dB) = 20 \log \left(\frac{70}{70 - 35.6}\right) \approx 6 \text{ dB}$$

The TIXM05 data sheet gives minimum f_T = 450 MHz and maximum $r_b'C_c = 7.5$ picoseconds.

Therefore,

$$f_{(max)} = \sqrt{\frac{450 \times 10^6}{8\pi\ 7.5 \times 10^{-12}}} = 1547 \text{ MHz}$$

and

$$MAG(dB) = 20 \log \left(\frac{1547}{213.5}\right) = 17.2 \text{ dB.}$$

The total power gain PG for the RF amplifier is,

$$PG(dB) = MAG - TL = 17.2 - 6 = 11.2 \text{ dB.} \qquad (24)$$

The total tuner gain is 11.2 plus 19.1 dB or 30.3 dB which agrees closely with the design example prototype's gain of 33 dB.

In designing the input matching circuit, capacitor C_1 (Fig. 7) is initially selected to be 6.8 pF in order that L_1 may be a practical size for channel 13. Capacitor C_2 is selected to be 18 pF or about three times larger than c_{iep} at 213.5 MHz.

The input circuit,

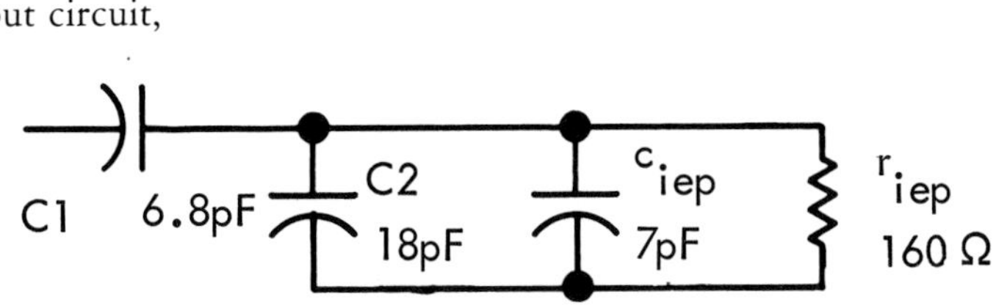

is reduced to an equivalent single R and C in series:

$$Q = \omega C_p R_p = (13.4 \times 10^8)(18 + 7) \times 10^{-12}\,(160) = 5.36$$

$$R_s = \frac{R_p}{1 + Q^2} = \frac{160}{1 + 5.36^2} = 5.01\ \Omega$$

$$C_s = C_p \left(\frac{1 + Q^2}{Q^2}\right) = 25 \left(\frac{1 + 5.36^2}{5.36^2}\right) = 25.8\ \text{pF}$$

Therefore, the series equivalent circuit is

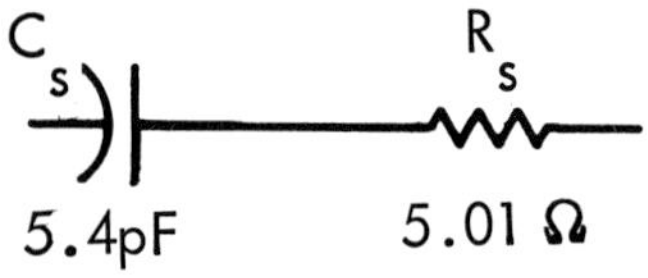

which is now reduced to the parallel equivalent circuit:

$$Q = \frac{1}{\omega C_s R_s} = \frac{1}{13.4 \times 10^8\ (5.4 \times 10^{-12})\ 5.01} = 27.5$$

$$R_p = R_s\ (1 + Q^2) = 5.01\ (1 + 27.5^2) = 3790\ \Omega$$

$$C_p = C_s \left(\frac{Q^2}{1 + Q^2}\right) = 5.4 \left(\frac{27.5^2}{1 + 27.5^2}\right) = 5.4\ \text{pF}$$

Therefore the parallel equivalent input circuit is

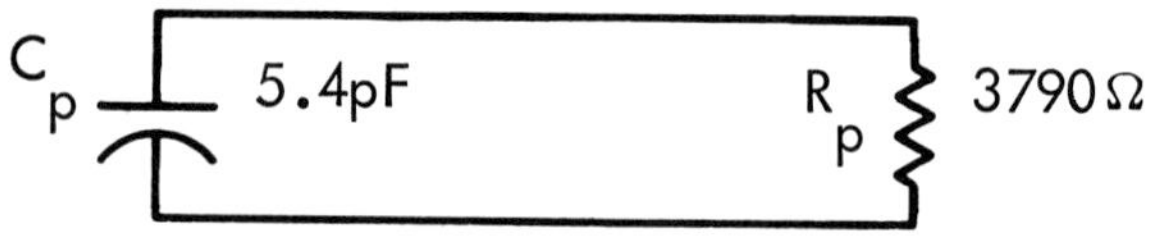

The output impedance of the balun and filter is approximately

and converting to a series equivalent circuit,

$$Q = \omega C_p R_p = 13.4 \times 10^8\ (25 \times 10^{-12})\ 50 = 1.68$$

$$R_s = \frac{R_p}{1 + Q^2} = \frac{50}{1 + 1.68^2} = 13.1\ \Omega$$

$$C_s = C_p \left(\frac{1 + Q^2}{Q^2}\right) = 25 \times 10^{-12} \left(\frac{1 + 1.68^2}{1.68^2}\right) = 33.9\ \text{pF}$$

Thus the series equivalent circuit is,

which is now combined with the equivalent input circuit:

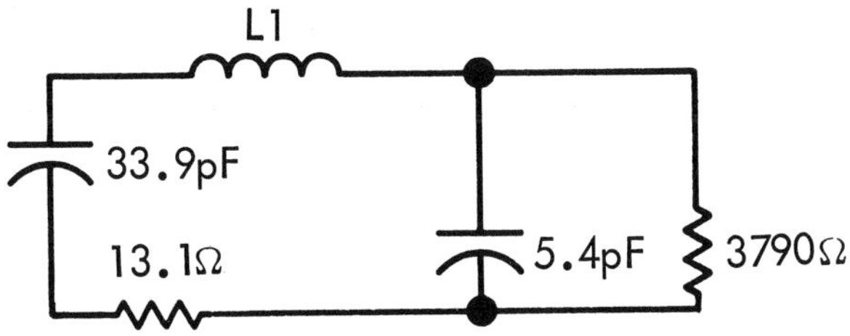

The total effective capacitance C_T shunting L_1 is approximately

$$C_T \approx \frac{33.9\ (5.4)}{33.9 + 5.4} \approx 4.7\ \text{pF}$$

Therefore,

$$L_1 \approx \left[C_T\omega^2\right]^{-1} \approx \left[4.7 \times 10^{-12}\ (13.4 \times 10^8)^2\right]^{-1}$$

$$\approx 0.12\ \mu\text{H}$$

The circuit

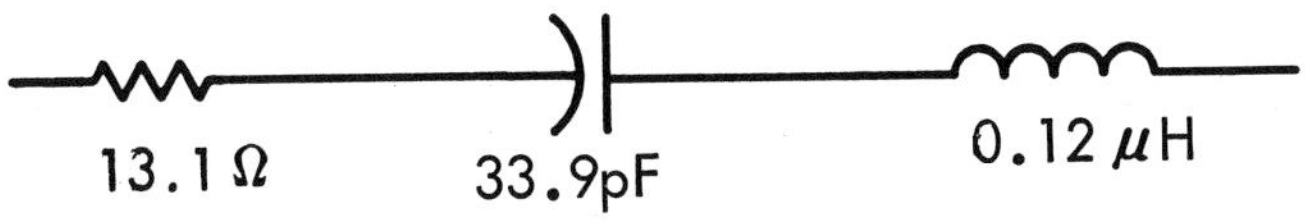

is equivalent to

at 213.5 MHz, of which the parallel equivalent is

$$Q = \frac{\omega L_S}{R_S} = \frac{13.4 \times 10^8\ (0.103 \times 10^{-6})}{13.1} = 10.2$$

$$R_p = R_S\ (1 \times Q^2) = 13.1\ (1 + 10.2^2) = 1410\ \Omega$$

$$L_p = L_s\left(\frac{1+Q^2}{Q^2}\right) = 0.103 \times 10^{-6}\left(\frac{1+10.2^2}{10.2^2}\right) = 0.104\ \mu\text{H}$$

Finally, the complete equivalent circuit is

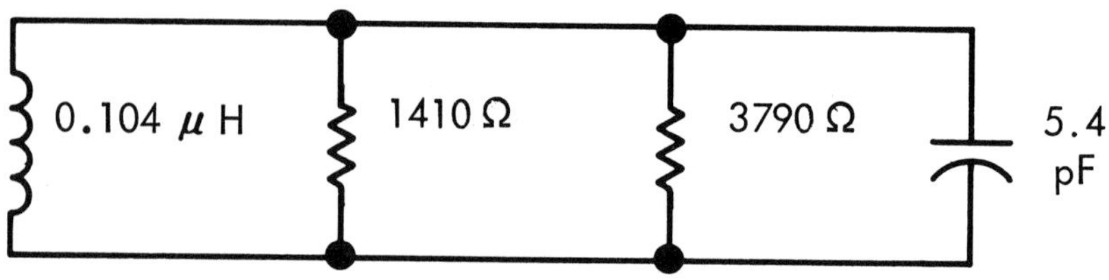

Now the circuit losses are computed. The total effective parallel resistance R_T is

$$R_T = \frac{R_A\, R_B}{R_A + R_B}$$

where $R_A = R_p$ transformed from the antenna and $R_B = R_p$ transformed from the transistor. Then,

$$R_T = \frac{1400\ (3790)}{1410 + 3790} = 1030\ \Omega$$

Therefore,

$$Q_L = \omega C_T R_T = 13.4 \times 10^8\ (4.7 \times 10^{-12})\ 1030$$
$$= 6.45$$

Assuming the coil unloaded Q is 70, the coil loss CL is

$$\text{CL(dB)} = 20 \log\left(\frac{Q_U}{Q_U - Q_L}\right) = 20 \log\left(\frac{70}{70-6.45}\right) = 0.84\ \text{dB}$$

Input bandwidth BW is

$$\text{BW} = \frac{f_o}{Q_L} = \frac{213.5}{6.45} = 33\ \text{MHz}$$

Mismatch loss ML is

$$\text{ML(dB)} = 10 \log\left[\frac{(1+a)^2}{4a}\right]$$

where

$$a = \frac{R_B}{R_A} = \frac{3790}{1410} = 2.69$$

$$\text{ML(dB)} = 10 \log \left[\frac{(1 + 2.69)^2}{4(2.69)}\right] = 1.0 \text{ dB}$$

Total input circuit losses = CL + ML = 1.84 dB.

Although the input matching circuit was designed to have low insertion loss and small VSWR, the value of L_1 should be adjusted for the lowest noise figure. The prototype (Fig. 7) has a channel 13 noise figure of 5.5 dB which agrees quite well with the calculated value of 5.34 dB, the sum of the transistor noise figure (3.5 dB) and the input circuit losses (1.84 dB).

Balun. The balun and filter circuit can have many configurations. It may include FM and IF traps but must have very low loss in the pass band to prevent noise figure degradation.

The power dissipation of the RF stage is computed by considering that the bias conditions for maximum gain for the TIXM05 are V_{CE} about 8.5 volts and I_C about 2 mA.

$$R_3 + R_2 = \frac{V_{CC} - V_{CE}}{I_E} = \frac{12 - 8.5}{2 \times 10^{-3}} = 1.75 \text{ k}\Omega$$

Maximum power dissipation $P_{DISP(max)}$ when $R_3 + R_2 = 1.75$ kΩ is

$$P_{DISP(max)} = \frac{(V_{CC}/2)^2}{R_3 + R_2} = \frac{(12/2)^2}{1750} = 20 \text{ mW} \qquad (25)$$

Since $P_{DISP(max)}$ for the RF stage is the same as the mixer, and the transistors have the same thermal specifications, the maximum ambient temperature is likewise +80°C.

Resistor R_3 should be an order of magnitude larger than the reactance of capacitor C_4 at the center frequency of channel 2.

$$X_{C4} = \frac{1}{\omega C_4} = \frac{1}{2\pi\,(57 \times 10^6)\,39 \times 10^{-12}} = 72\ \Omega$$

Therefore, let R_3 be 680 ohms and R_2 be 1000 ohms.

Before leaving the RF amplifier, it should be noted that the amount of gain reduction available is primarily a function of transistor characteristics, so the circuit designer is somewhat limited. However, to achieve all possible gain reduction, care should be taken to provide input/output shielding. The prototype tuner has a 41-dB gain reduction.

Oscillator. The TIXM07 with a 2-mA emitter current generates many times the few hundred microwatts of injection power needed by the mixer. This increases the protection against oscillator "frequency pulling" by allowing very light loading. Capacitor C_7 sets the optimum level of injection voltage which is the minimum voltage that yields maximum conversion gain and minimizes distortion and oscillator radiation. Capacitors C_{10} and C_{11} provide the external feedback that assures dependable starting at the lowest frequency, 101 MHz for channel 2.

The total effective parallel capacitance C_T across L_5 is about 5 pF, therefore L_5 is 0.077 μH. The emitter resistor R_{10} must be large enough to prevent loss of the feedback signal and, with R_8 and R_9, effect a small stability factor. Choosing R_{10} equal to 1500 ohms and a stability factor S of 2, the values of R_8 (8.3 kΩ) and R_9 (2.2 kΩ) are computed in the same manner as the mixer bias resistors.

Packaging. This single-channel tuner (Fig. 7) was built as an "engineering breadboard" with strict attention given to component placement, lead dress, shielding and decoupling. These items must be considered here as in any VHF design. Packaging a multichannel tuner in practical hardware is indeed an art. Such things as the inductance and capacitance of switch contacts modify the values of small reactive components. Unfortunately, the required compromises tend to degrade some aspects of performance. Experience with a wide variety of tuner hardware, both domestic and foreign, has shown that performance decreases with the tuner's physical size. Some manufacturers are interested in printed-circuit-board tuners for the implicit ease of fabrication and cost reduction. While such tuners have been produced, it seems pertinent to point out that the adaptation of complex VHF circuitry, with feedback amplifiers and strict coupling requirements, to printed-circuit boards is at best a formidable task, often plagued with the necessity of trial and redesign.

REFERENCES

1. North, D. O.: The Absolute Sensitivity of Radio Receivers, *RCA Rev.*, Vol. 6, p. 332, January, 1942.
2. Thanos, H.: Crossmodulation in Transistorized TV Tuners, *IEEE Trans. on Broadcast and TV Receivers,* vol. BRT-9, p. 41, November, 1963.
3. Friis, H. T.: Receiver Noise Figure,*Proc. IRE,* vol. 32, p. 419, July, 1944.
4. Deutsch, S.: "Theory and Design of Television Receivers," p. 110, McGraw-Hill Book Company Inc., New York, 1951.
5. Cooke, H. F.: Designing TV Tuners with Mesa Transistors, *Electronics,* p. 64, April, 1960.

14

Video IF Amplifier

GAIN

The major requirement of the video IF amplifier is to provide an amount of power gain commensurate with satisfactory receiver performance in fringe or low-signal areas. One hundred microvolts across 50 ohms (200 picowatts) at the IF input for 1.0 volt dc across a detector load of 3.3 kilohms (303 microwatts) represents an amplifier power gain PG of

$$PG(\text{dB}) = 10 \log \left(\frac{303 \times 10^{-6}}{200 \times 10^{-12}}\right) = 10 \log (1.51 \times 10^{6}) = 62 \text{ dB} \qquad (1)$$

The chosen input voltage for the IF results from the assumption that a typical VHF tuner will, for the worst case, provide 30 dB power gain. Thirty decibels below 200 picowatts is 0.2 picowatts which, for a 300-ohm tuner input impedance, is approximately an 8.0-microvolt signal level. This represents adequate receiver sensitivity.

The foregoing calculation does not account for circuit losses due to traps and coupling networks. Logically, the IF will be used to make up for these losses; therefore, approximately 10 dB should be added to the IF power gain requirement.

POWER OUTPUT

In addition to providing sufficient drive to the video detector, the last IF stage is required to handle, linearly, large signal and noise impulses. Failure to meet this requirement may result in AGC lock-out or improper action of noise immunity circuits.

From practical considerations a 10-volt peak-to-peak swing across the detector load is an accepted criterion. Because of this, the power output of the detector for a load of 3.3 kilohms is

$$P_o = \frac{10^2}{3.3 \times 10^3} = 30.2 \text{ mW} \tag{2}$$

Assuming a detector efficiency of 60% ($\eta = 0.6$) the load presented to the drive transformer is

$$R_{IN} = \frac{3.3 \times 10^3}{2(0.6)} = 2.75 \text{ k}\Omega \tag{3}$$

and the peak RF excursion at the transformer secondary may be found as

$$RF_{(PK)} = \frac{\text{d-c out}}{\eta} = \frac{10}{0.6} = 16.7 \text{ V} \tag{4}$$

The secondary peak current is

$$I_{(PK)} = \frac{16.7}{2.75 \times 10^3} = 6.07 \text{ mA} \tag{5}$$

These values and the voltage, current and dissipation ratings of the last IF transistor allow a suitable selection of coupling transformer turns ratio.

A-C STABILITY

In designing a multistage amplifier the nonunilateral nature of transistors gives rise to problems of interaction between stages (tuning and terminations of one stage affecting another), bandpass skewing, and, in the worst case, oscillation. These problems may be largely overcome either by unilateralization or by mismatching the transistor terminal impedances.

Unilateralization on a repeatable basis is a virtual impossibility since one must contend with parametric variations in both transistor and circuit components and bias changes due to supply drift and the AGC requirement. However, a combination of partial unilateralization (neutralization) and mismatching will provide the desired results. Since the price one must pay for stability through mismatch is an attendant power loss, the problem becomes one of how much power must be lost to attain stability.

A. P. Stern[1] has defined a stability factor "k" in the relation

$$(g_{11} + G_g)\,(g_{22} + G_L) = \frac{k(L+M)}{2} \tag{6}$$

where

k is greater than one for stability,

g_{11} and g_{22} are real parts of transistor input and output admittance respectively,

G_g and G_L are real parts of source and load admittance respectively,

$L = |y_{12}y_{21}|$ product of transistor transadmittances, and

$M =$ real part of $(y_{12})(y_{21})$.

If it is initially assumed that $g_{22} = G_L$ the following expression may be derived:

$$G_g = \frac{k(L+M)}{4g_{22}} - g_{11} \tag{7}$$

For a given value of k and from device y-parameter data, G_g may be calculated. Power loss due to mismatch may be seen to be

$$\text{loss(dB)} = 10 \log \frac{G_g}{4g_{11}} \text{ for } \frac{G_g}{g_{11}} >> 1. \tag{8}$$

Once this number is ascertained, the actual mismatching may be done on either the input or output side or both.

In a tuned amplifier the total load seen by the generator is a combination of the actual load, usually the reflected impedance of the following stage, and the effective impedance of the tuning elements. Hence it is possible to introduce loss (and stability) by mismatch of actual reflected load to generator or through increased loading of tuning elements. For the condition of maintaining a given bandwidth, it has been shown[2/] that, regardless of loading due to coil, maximum power is delivered to the load for reflected load impedance equal to generator impedance.

DESIGN PROCEDURE

An interstage-design procedure may now be outlined:

1. Determine necessary loss for stability considerations.
2. Select transformer turns ratio to match source impedance to load impedance.
3. Determine unloaded Q from

$$Q_U = \frac{Q_L \sqrt{IL}}{\sqrt{IL} - 1} \tag{9}^*$$

4. Determine tuning capacitance (hence inductance) from

$$C = \frac{2g_{22}\, Q_U Q_L}{\omega(Q_U - Q_L)} \tag{10}$$ *

5. Determine damping resistance to be added from (this assumes effective coil resistance much greater than R_a)

$$R_a = Q_U X \tag{11}$$

DESIGN EXAMPLE

This interstage design example is based on the use of TIXM08 transistors shown in the IF amplifier circuits of Fig. 1 and Fig. 2.

At the operating point of 8 V and 1.5 mA the y-parameters are:

$y_{11} = 0.667 \times 10^{-3} + j2.83 \times 10^{-3}$ S

$y_{22} = 0.0143 \times 10^{-3} + j0.24 \times 10^{-3}$ S

$y_{12} = 0 - j0.16 \times 10^{-3}$ S

$y_{21} = (45 - j15) \times 10^{-3} = 47.5 \angle -18.4^\circ$ mS

$y_{12}y_{21} = (45 - j15) \times 10^{-3} (-j0.16) \times 10^{-3}$

$= (-2.4 - j7.2) \times 10^{-6}$

$L = 7.6 \times 10^{-6} \angle -108^\circ$

$M = -2.4 \times 10^{-6}$

$L + M = 5.2 \times 10^{-6}$

For stability

$$(g_{11} + G_g)(g_{22} + G_L) = \frac{k\,(L + M)}{2}$$

Assuming k = 1 and $g_{22} = G_L$

$$2g_{22}g_{11} + 2g_{22}\, G_g = 2.6 \times 10^{-6}$$

$$G_g = \frac{2.6 \times 10^{-6} - 2g_{22}g_{11}}{2g_{22}}$$

$$= \frac{2.6 \times 10^{-6} - 2(0.0143 \times 10^{-3})(0.667 \times 10^{3})}{2(0.0143 \times 10^{-3})}$$

$$= 90.5 \text{ mS}$$

*See Equation Derivation Section.

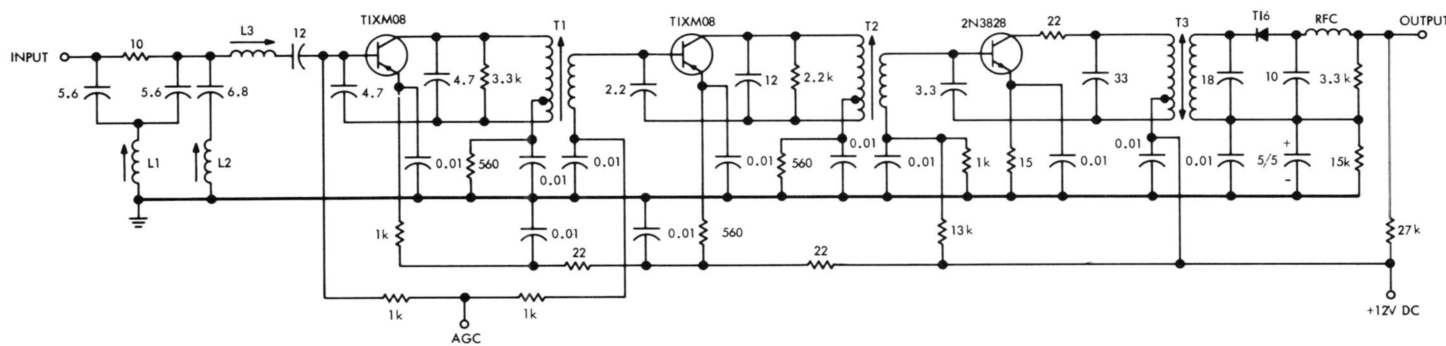

COIL AND TRANSFORMER DATA

L1 — 19 Turns No. 25
L2 — 20 Turns No. 25
L3 — 10 Turns No. 33

T1 Primary — 14 Turns No. 33
Secondary — 2 Turns No. 33
Nominal Inductance 1.5 microhenry

T2 Primary — 7 Turns No. 33
Secondary — 2 Turns No. 33
Nominal Inductance 0.6 microhenry
Transformers T1 and T2 Bifilar wound using EL-RAD form 30B801
base 2B100
can 11B100-1

T3 Primary — 6 Turns No. 25 w/tap at 2 Turns
Secondary — 4 Turns No. 25
Both windings on common 1/4" form with secondary adjustable.

PERFORMANCE DATA

3 dB BW = 3.5 MHz
Power Gain = 62 dB
Output Capability
8.0 volts p-p

AGC Voltage
Max. Gain occurs at 8.5 volts
Min. Gain occurs at 3.6 volts

Fig. 1. Three-stage 45-MHz IF Amplifier

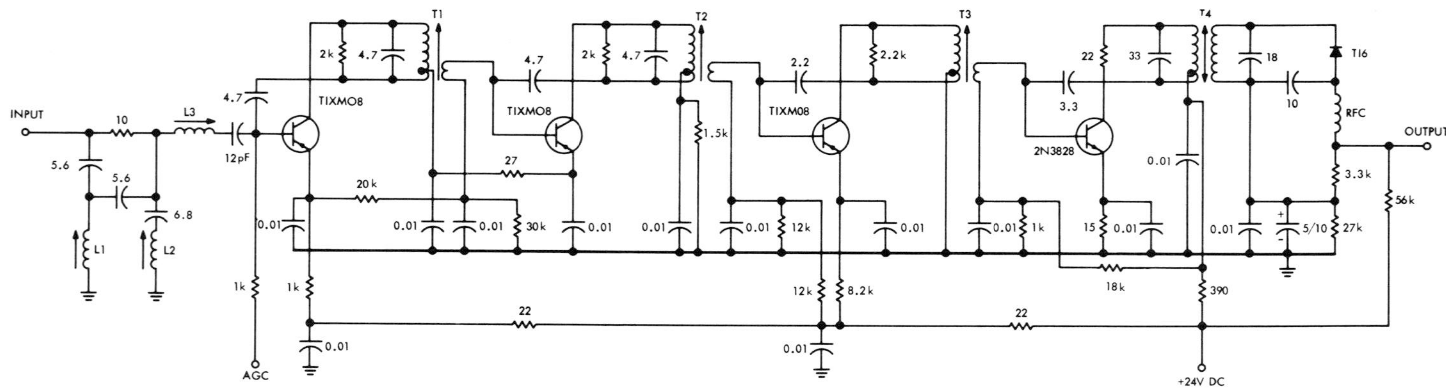

COIL AND TRANSFORMER DATA

L_1 19 Turns No. 25
L_2 20 Turns No. 25
L_3 10 Turns No. 33

T1 Primary — 14 Turns No. 33
Secondary — 2 Tunrs No. 33
Nominal Inductance 1.5 microhenry

T2 = T1
Transformers T1, T2, and T3 bifilar wound using EL-RAD form 30B801
Base 2B100—Can 11B100-1

T3 Primary — 7 Turns No. 33
Secondary — 2 Turns No. 33
Nominal Inductance 1.5 microhenry

T4 Primary — 6 Turns No. 25 w/Tap at 2 Turns
Secondary — 8 Turns No. 25
Both windings on common 1/4" form with secondary adjustable

PERFORMANCE DATA

3 dB BW = 4 MHz
Power Gain = 78 dB
AGC Voltage
Max. Gain occurs at 21 volts
Min. Gain occurs at 13 volts
Output Capability 10 volts p-p

Fig. 2. Four-stage 45-MHz IF Amplifier

Insertion loss IL is given by

$$\mathrm{IL} = 10 \log \left[\frac{(g_{11} + G_g)^2}{4g_{11}\, G_g}\right]$$

$$= 10 \log \left[\frac{(1 + G_g/g_{11})^2}{4\,(G_g/g_{11})}\right]$$

and for $G_g >> g_{11}$

$$\mathrm{IL} = 10 \log \left[\frac{G_g}{4g_{11}}\right]$$

$$= 10 \log \left[\frac{90.5}{4(0.667)}\right] \approx 15 \text{ dB}$$

This amount of gain must be thrown away to acquire stability.

From $\dfrac{y_{21}{}^2}{4g_{11}g_{22}}$ the TIXM08 available power gain is found to be 48 dB.

This less 15 dB amounts to 33 dB power gain per stage. Three stages of gain must be used because approximately 75 dB is the overall requirement. Therefore, any additional amount up to 8 dB per stage may be taken as loss. Amplifier stability is thereby enhanced because the greater the loss, the larger the stability factor k becomes.

On the basis of desired overall bandwidth and response skewing with AGC, the interstage bandwidth is chosen to be 6.0 MHz or a loaded Q of 7.5. Using 21.5 dB or a power ratio of 140 for insertion loss, the unloaded Q may be found:

$$Q_U = \frac{Q_L \sqrt{\mathrm{IL}}}{\sqrt{\mathrm{IL}} - 1} = \frac{7.5\ (11.85)}{10.85} = 8.2$$

Total tuning capacity is determined from

$$C = \frac{2g_{22}Q_UQ_L}{\omega\,(Q_U - Q_L)} = \frac{2(0.0143 \times 10^{-3})\ 8.2(7.5)}{2\pi(45 \times 10^6)(8.2 - 7.5)} = 8.87 \text{ pF}$$

and $L = 1.4\ \mu\mathrm{H}$.

A damping resistance to be added across the primary is given by

$$R_a = \frac{1/2g_{22}\ (Q_LX)}{1/2g_{22} - Q_LX} = 3.26 \text{ k}\Omega \qquad (13)$$

$$\text{Transformer turns ratio} = N_p/N_s = (g_{11}/g_{22})^{1/2} = 6.84. \qquad (14)$$

The last IF transistor exhibits terminal parameters different from those of the TIXM08. For this reason the coupling network into this stage is also different, but values are derived in the same manner.

A double-tuned transformer is used to couple the last IF to the video detector. It was shown earlier that the peak current and voltage on the secondary of this transformer will be 6.07 milliamperes and 16.7 volts. For a 12-volt supply the transistor collector-to-emitter operating voltage can not be much in excess of 10 volts which dictates a peak swing of 10 volts. From voltage considerations the transformer turns ratio is

$$\frac{N_p}{N_s} = \frac{E_p}{E_s} = \frac{10}{16.7} \text{ or } 1.67\ N_p = N_s \qquad (15)$$

Peak current on the primary side may then go to 1.67 × 6.07 = 10.1 mA, which shows that the quiescent transistor current must be approximately 10 mA. Transistor nonlinearities and transformer efficiency will cause an increase in power required into the last IF device.

EQUATION DERIVATIONS

Equation (9).

The interstage equivalent circuit is shown below:

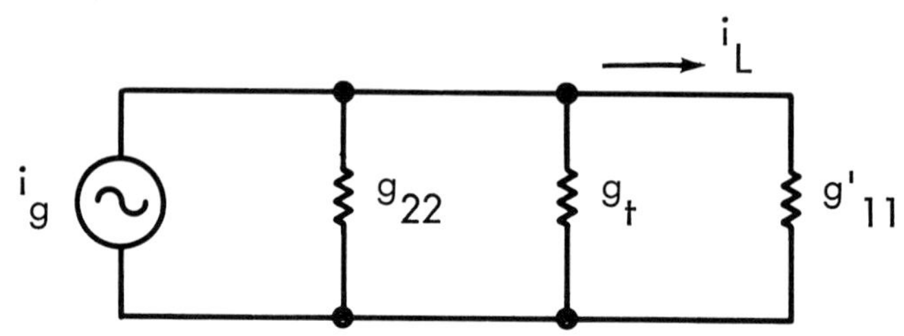

Insertion loss IL is given by:

$$IL = \frac{\text{maximum available power}}{\text{power to load}}$$

Maximum available power MAP is given by:

$$MAP = \frac{i_g^2}{4g_{22}}$$

and

$$\text{Power to load} = \frac{\left(i_g \dfrac{g_{11}'}{g}\right)^2}{g_{11}'}$$

Total conductance g is given by:

$$g = g_{22} + g_t + g_{11}'$$

So that

$$IL = \frac{i_g^2/4g_{22}}{i_g^2(g_{11}'/g^2)} = \frac{g^2}{4g_{11}'g_{22}}$$

For $g_{22} = g_{11}'$

$$g = 2g_{22} + g_t \text{ or } g_{22} = \frac{g - g_t}{2}$$

$$IL = \frac{g^2}{4\left(\dfrac{(g-g_t)}{2}\right)^2} = \frac{g^2}{(g-g_t)^2}$$

and

$$\sqrt{IL} = \frac{g}{g - g_t}$$

Because

$$Q = \frac{1}{gX} \text{ or } g = \frac{1}{QX}$$

$$\sqrt{IL} = \frac{1/Q_L X}{\dfrac{1}{Q_L X} - \dfrac{1}{Q_U X}} = \frac{Q_U}{Q_U - Q_L}$$

or

$$Q_U = \frac{Q_L \sqrt{IL}}{\sqrt{IL} - 1} \tag{9}$$

Equation (10).

$$Q_L = \frac{\omega C}{g_{22} + g_t + g_{11}'}$$

For

$$g_{22} = g_{11}'$$

$$Q_L = \frac{\omega C}{2g_{22} + g_t}$$

Because

$$Q_U = \frac{\omega C}{g_t} \text{ or } g_t = \frac{\omega C}{Q_U}$$

$$Q_L = \frac{\omega C}{2g_{22} + \omega C/Q_U}$$

or

$$C = \frac{2g_{22}\, Q_U\, Q_L}{\omega(Q_U - Q_L)} \tag{10}$$

REFERENCES

1. Stern, A. P.: Stability and Power Gain of Tuned Transistor Amplifiers, *Proc. IRE,* March, 1957.
2. Webster, Roger R.: How to Design IF Transistor Transformers, *Electronics.*

15

TV Automatic Gain Control

AGC figure-of-merit is described as the number of dB reduction in the input signal, below 100 000 μV, required to reduce the picture output voltage by 10 dB.[1] Both the sound and the picture AGC figure-of-merit are described in this manner. This chapter discusses the AGC of the tuner RF amplifier and video IF amplifiers and the resultant video output.

The job of the AGC amplifier is to control the gain of the tuner RF and the video IF amplifiers to maintain a nearly constant output level at the video detector while the input signal to the tuner varies. This may be done by using either forward or reverse AGC techniques; although, more recent devices work best when forward AGC is applied.[2] Figure 1 shows a typical h_{fe} versus I_C curve.

REQUIREMENTS

The AGC amplifier power requirements and total loop gain needed to provide sufficient control of the stages being AGC'd are of prime importance. The curves in Fig. 2 show the requirements of a typical tuner and IF strip.

In the circuit considered here, the AGC'd stages are PNP devices in a circuit operating from a positive supply. (See Fig. 3.) The bias voltage on the AGC'd devices will be $V_{CC}-V_{AGC}$, so that, as the AGC voltage decreases, the actual bias voltage is increasing, resulting in an increase in I_B.

The AGC power required by the AGC'd stages is described as follows:[2]

$$P_{AGC} = \frac{\Delta V_B \Delta I_E}{h_{FE} + 1} \tag{1}$$

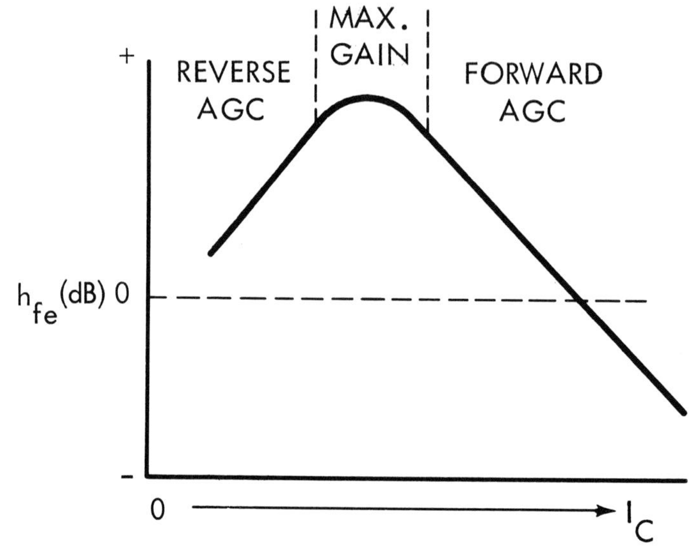

Fig. 1. Typical h_{fe} versus I_c Curve

where

$$\frac{\Delta I_E}{h_{FE} + 1} = \Delta I_B$$

ΔV_B and ΔI_B represent the changes in base voltage and base current over the full AGC range.

Note that the power required to maintain the AGC'd stages in their maximum-gain condition is delivered from fixed d-c bias circuitry. The AGC amplifiers are not in the control mode during this time. Power required for this low-signal, maximum-gain condition is not referred to as AGC power and is subtracted from the maximum power requirements to yield AGC power.

The V_{AGC} minimum and maximum limits are given by the tuner requirements. The tuner voltage calculations are as follows:

The required voltage limits are: $V_{AGC(min)} = 3.9$ V and $V_{AGC(max)} = 8.15$ V

$$V_{B(max)} = V_{CC} - V_{AGC(min)} = 12\text{ V} - 3.9\text{ V} = 8.1\text{ V} \qquad V_{B(min)} = V_{CC} - V_{AGC(max)} = 12\text{ V} - 8.15\text{ V} = 3.85\text{ V}$$

$$\Delta V_B = 4.25\text{ V}$$

Similarly with the current:

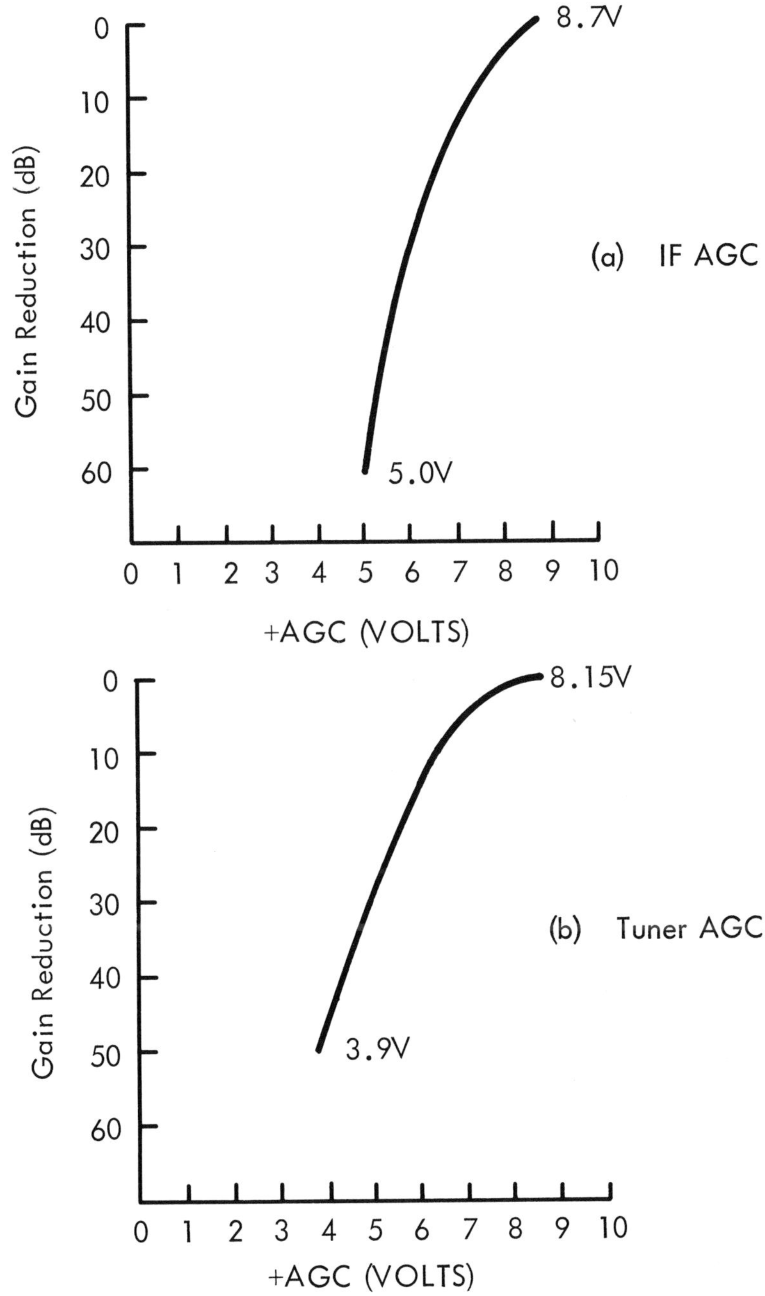

Fig. 2. AGC Requirements

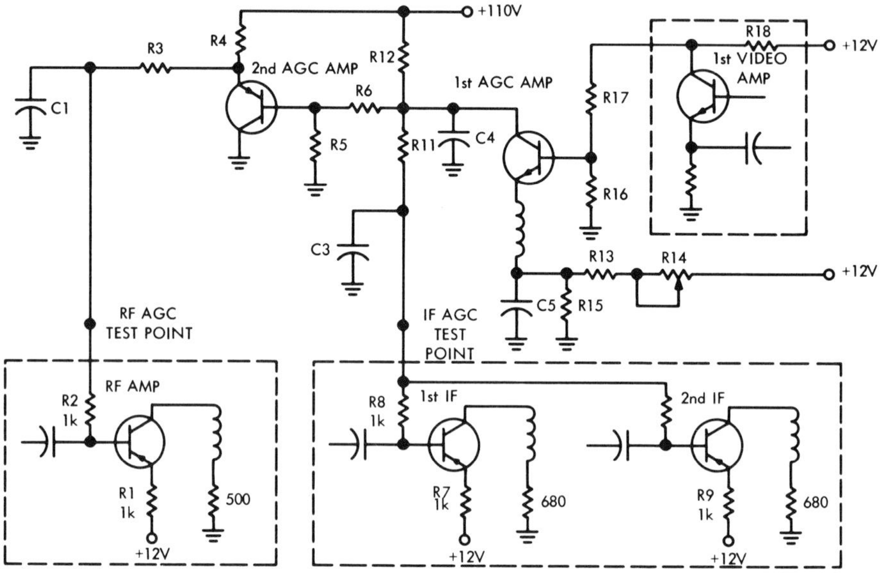

Fig. 3. Schematic of AGC'd Stages

$$I_{E(min)} = \frac{V_{B(min)} - V_{BE}}{R_E} = \frac{3.85\ V - 0.35\ V}{1\ k\Omega} = 3.5\ mA$$

$$I_{E(max)} = \frac{V_{B(max)} - V_{BE}}{R_E} = \frac{8.1\ V - 0.35\ V}{1\ k\Omega} = 7.75\ mA$$

$$\Delta I_E = 4.25\ mA$$

Using Eq. (1) where $h_{FE} = 50$

$$P_{AGC(tuner)} = \frac{\Delta V_B \Delta I_E}{h_{FE} + 1} = \frac{4.25\ V\ (4.25\ mA)}{50 + 1} = 354\ \mu W$$

Given the voltage and resultant total base current requirements in the two AGC'd IF stages, their P_{AGC} requirements may be determined.

$$V_{B(min)} = 3.3\ V \qquad V_{B(max)} = 7.0\ V$$

$$2I_{B(min)} = 117.6\ \mu A \qquad 2I_{B(max)} = 263\ \mu A$$

$$P_{AGC(IF)} = \Delta V_B \Delta I_B = 3.7\ V\ (145.4\ \mu A) = 538\ \mu W \tag{2}$$

The total AGC power is then

$$P_{AGC(total)} = P_{AGC(IF)} + P_{AGC(tuner)} = 538\ \mu W + 354\ \mu W = 892\ \mu W\ \text{(under maximum AGC conditions)} \tag{3}$$

The AGC circuit drive power is the input power at maximum AGC conditions.

The signal level on the base of the first AGC amplifier is 1.4 volts maximum under full AGC.

$$I_B = \frac{I_{C(max)}}{h_{FE}} \tag{4}$$

Where $I_{C(max)}$ is 0.5 mA and h_{FE} is 100

$$I_B = \frac{0.5\ \text{mA}}{100} = 5\ \mu\text{A}$$

$$P_{AGC(drive)} = I_B V_B \tag{5}$$

where I_B and V_B are the drive levels applied to the base of the first AGC amplifier.

$$P_{AGC(drive)} = 5\ \mu\text{A}(1.4\ \text{V}) = 7.0\ \mu\text{W}$$

The power gain PG requirements for the AGC circuit are therefore

$$PG = \frac{P_{AGC(total)}}{P_{AGC(drive)}} = \frac{892\ \mu\text{W}}{7.0\ \mu\text{W}} = 127.4 \tag{6}$$

A power gain of 127.4 is best obtained by the use of two AGC amplifiers providing sufficient gain and better isolation between RF and IF stages. The AGC circuit configuration in Fig. 3 is discussed in this design procedure.

The first AGC amplifier is a germanium NPN alloy type 2N1308 and is selected for its low V_{BE} requirements, low cost, and high minimum h_{FE} of 80 at 10 mA. The second AGC amplifier is a germanium PNP alloy type TIXA04 with a minimum h_{FE} of 100.

CIRCUIT EVALUATION

An AGC voltage swing from +8.15 volts with no signal applied, to +3.9 volts with maximum signal applied is required for the tuner. The direct currents for the base of the RF unit are furnished by the AGC circuit. In this design example, the RF device is a PNP type TIXM05. The circuit in Fig. 4 shows the d-c loading values.

With an AGC voltage of 8.15 volts, the base current may be found by this expression:

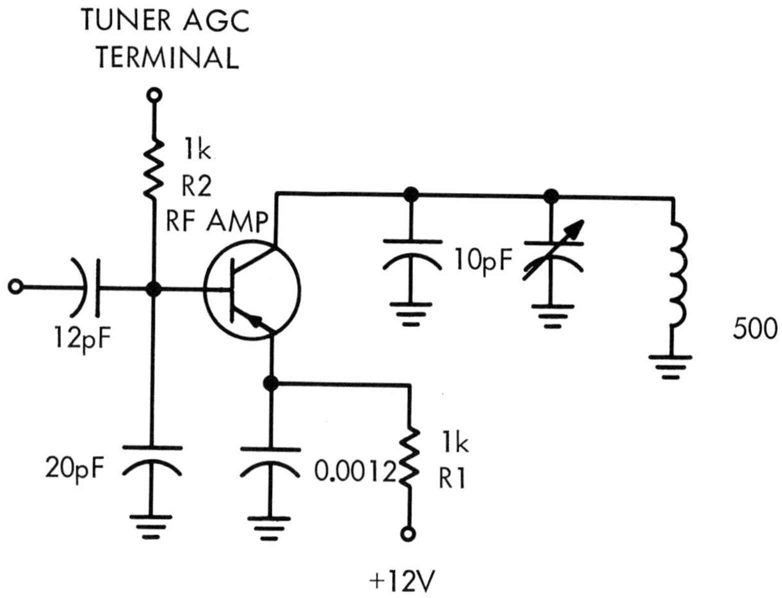

Fig. 4. D-C loading Values for RF Amplifier

$$I_B = \frac{V_{CC} - (V_{AGC} + V_{BE})}{h_{FE}R_1 + R_2} - I_{CBO} \tag{7}$$*

In this example, $V_{CC} = +12$ volts, $V_{AGC(max)} = +8.15$ volts (no signal condition), $V_{BE} = 0.3$ volts, $h_{FE} = 50$, $R_1 = 1\ k\Omega$, $R_2 = 1\ k\Omega$, and $I_{CBO} = 10\ \mu A$:

$$I_{B(min)} = \frac{12\ V - (8.15\ V + 0.3\ V)}{50(1\ k\Omega) + 1\ k\Omega} - 10\ \mu A$$

$$= 59.6\ \mu A$$

Under full AGC'd conditions, the AGC voltage is specified as 3.9 volts. The resulting I_E is therefore

$$I_E = \frac{V_{CC} - (V_{AGC(min)} + V_{BE})}{R_1} \tag{8}$$

$$= \frac{12\ V - (3.9\ V + 0.3\ V)}{1\ k\Omega}$$

$$= 7.8\ mA\ \text{(under full AGC)}$$

---

*See Equation Derivations at the end of this chapter.

The values in the equation for $I_{B(max)}$ are as follows: $V_{CC} = 12$ V, $V_{AGC} = 3.9$ volts (maximum signal condition), $V_{BE} = 0.3$ volts, $h_{FE} = 50$, $R_1 = 1$ kΩ, $R_2 = 1$ kΩ, and I_{CBO} is negligible.

From Fig. 4 it can be seen that if I_E is 7.8 mA, V_{CE} will be near zero.

$$I_{B(max)} = \frac{V_{CC} - (V_{AGC} + V_{BE})}{h_{FE}R_1 + R_2} \qquad (9)^*$$

$$= \frac{12\text{ V} - (3.9\text{ V} + 0.3\text{ V})}{50\ (1\text{ k}\Omega) + 1\text{ k}\Omega}$$

$$= 153\ \mu\text{A}$$

The calculated RF unit base current flows through R_3. The voltage swing of the emitter of the second AGC amplifier may now be calculated.

$$V_{E(min)} = V_{AGC(min)} - I_{B(max)}\ R_3 \qquad (10)$$

where V_E is the second AGC amplifier emitter voltage, I_B is the RF base current and R_3 is the resistor tied to the emitter of the second AGC amplifier as shown in Fig. 3.

$$V_{E(min)} = 3.9\text{ V} - (153\ \mu\text{A})(1\text{ k}\Omega)$$
$$= 3.75\text{ V}$$

R_3 was chosen as 1 kΩ. This value affords enough isolation for good decoupling but is not so large as to cause significant AGC power loss.

$$V_{E(max)} = V_{AGC(max)} - I_{B(min)}\ R_3 \qquad (11)$$
$$= 8.15\text{ V} - (60\ \mu\text{A})(1\text{ k}\Omega)$$
$$= 8.09\text{ V}$$

The resulting second AGC amplifier emitter swings from + 3.75 volts to + 8.09 volts or about 4.34 volts.

R_4 (see Fig. 3) is connected from the emitter to a + 110 volt supply. The value of this resistor must be large enough to swing the above voltage with the current drive available to it.

Although 10 mA is the current level at which the device h_{FE} is maximum, the drive power requirements at this level are too high. If the current is reduced by ⅓, the h_{FE} is still 90% of maximum and the drive required is reduced by ⅓. Thus, the approximate value of I_E chosen is 4 mA. The voltage drop across R_4 is a little more than 100 volts.

*See Equation Derivations at the end of this chapter.

$$\frac{100\ \text{V}}{4\ \text{mA}} = 25\ \text{k}\Omega$$

A standard value of 27 kΩ is chosen for this circuit.

The resulting value of I_B for the second AGC amplifier is found as follows:

$$I_{B(max)} = \frac{\dfrac{V_{CC} - V_{E(min)}}{R_4} + I_{AGC(max)}}{h_{FE} + 1} \tag{12}*$$

$$= \frac{\dfrac{110\ \text{V} - 3.75\ \text{V}}{27\ \text{k}\Omega} + 153\ \mu\text{A}}{100 + 1}$$

$$= 40.4\ \mu\text{A}$$

$$I_{B(min)} = \frac{\dfrac{V_{CC} - V_{E(max)}}{R_4} + I_{AGC(min)}}{h_{FE} + 1} \tag{13}*$$

$$= \frac{\dfrac{110\ \text{V} - 8.09\ \text{V}}{27\ \text{k}\Omega} + 59.6\ \mu\text{A}}{100 + 1}$$

$$= 37.9\ \mu\text{A}$$

V_B is equal to $V_E - 0.3$ volts; therefore, $V_{B(max)} = 7.79$ volts at $I_{B(min)}$ and $V_{B(min)} = 3.45$ volts at $I_{B(max)}$.

The first AGC amplifier must perform several functions: First, it acts as the AGC detector and amplifier. Second, it must furnish the AGC for the video IF stages, and third, it drives the second AGC amplifier.

The IF stages to be AGC'd, like the tuner, have a range of bias voltages which are required for proper AGC action. In this example the voltages are 8.7 and 5.0 volts.

It is necessary at this point to look at the video levels driving the first AGC amplifier and other circuit considerations. The AGC amplifiers are to hold the video output level to within 10 dB over as wide a range of input levels as possible. In this example the maximum peak-to-peak video output on the collector of the first video amplifier is given as 2.8 volts. The desired minimum level at −10 dB will be 0.887 volts where:

---

*See Equation Derivations at the end of this chapter.

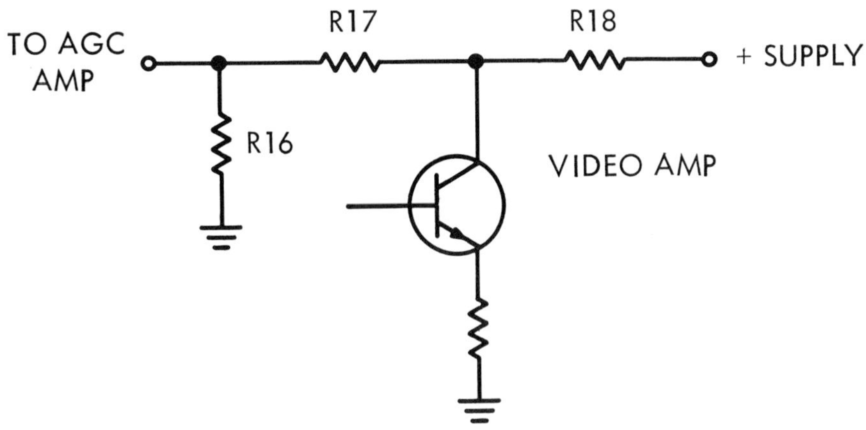

Fig. 5. Voltage Divider Network for Video Amplifier

$$\text{Video}_{(\text{min})} = \frac{\text{Video}_{(\text{max})}}{\text{antilog}\,\dfrac{10\text{ dB}}{20}} \tag{14}$$

By limiting the first video amplifier minimum-to-maximum collector voltage to 10 dB, the actual variation in the picture-tube cathode signal will be within these limits because the video amplifiers are linear or may have some slight compression.

Gating is used on the first AGC amplifier. The short duty cycle (0.07 H) horizontal gating pulse reduces the 60 Hz component in the collector filter by its sampling of the vertical block.[a] This permits a reduction in the 60-Hz filtering necessary and makes fast action possible. Fast action helps in reducing the signal fluctuation produced by aircraft reflections and noise pulses. The resulting d-c level of the collector voltage will change by an amount about equal to the voltage peaks, if the capacitor C_4 is large enough to hold a charge for several horizontal lines. The base of the first AGC amplifier is d-c coupled to the collector of the first video amplifier. The impedance of the coupling resistors will be determined by the output characteristics of the video amplifier and the desired isolation. Voltage and current requirements of the video amplifier are 7.35 volts V_C and 6.5 mA I_C and are determined by the operating characteristics of the device and its various loads.

The voltage divider network (Fig. 5) which is used to couple the video amplifier to the AGC amplifier consists of R_{16} and R_{17}. The impedance presented by this network must not load the video amplifier significantly and therefore should have an impedance of about ten times that of the video amplifier or ten times V_C/I_C.

$$\frac{V_C}{I_C} = \frac{7.35\ \text{V}}{6.46\ \text{mA}} = 1.14\ \text{k}\Omega \tag{15}$$

If each resistor (R_{16} and R_{17}) is 10 kΩ the loading effect will be small. The resulting d-c voltages and video levels on the base of the first AGC amplifier will be one-half of those on the collector of the first video amplifier. The d-c level of the base of the first AGC amplifier is therefore $(\frac{1}{2})(7.35\ \text{V}) = 3.67$ volts. The video sync pulses on the base of the first AGC amplifier will have an amplitude of ½ the video amplifier collector limits of +2.8 and +0.887 volt or +1.4 and +0.444 volt, respectively. The AGC amplifier should begin to take effect at the minimum level of +0.444 volts. A small amount of current will begin to flow in the first AGC amplifier when V_{BE} reaches a value of +0.1 volt. This point will be referred to as V_{BE}' and the level of full AGC will be referred to as V_{BE}''.

$$V_E = V_B + V_V' - V_G - V_{BE}' = V_B + V_V'' - V_G - V_{BE}''$$

where

V_E = the fixed bias level of the first AGC amplifier emitter
V_B = the bias level of its base
V_V' = the minimum video signal on the base to begin AGC action
V_V'' = the maximum video signal on the base at the full AGC condition
V_G = a negative horizontal gating pulse applied to the emitter
V_{BE}' = the minimum base-emitter voltage required to begin to turn the AGC amplifier on
V_{BE}'' = the maximum V_{BE} at full AGC

Since

$$V_E + V_G = V_B + V_V' - V_{BE}' = V_B + V_V'' - V_{BE}''$$

then

$$V_{BE}'' = V_V'' - V_V' + V_{BE}'$$

Where

$$V_V'' = +1.4\ \text{volts},\ V_V' = +0.444\ \text{volts, and}\ V_{BE}' = +0.1\ \text{volts,}$$
$$V_{BE}'' = 1.4\ \text{V} - 0.444\ \text{V} + 0.1\ \text{V} = 1.056\ \text{volts}$$

and

$$V_E + V_G = 3.67\ V + 0.444\ V - 0.1\ V = 4.014 \text{ volts}$$

The approximate value of V_G is determined by several things: (Refer to Fig. 6.)

1. V_E must be biased positive with respect to the base but cannot exceed the device BV_{EBO} of +6 volts; therefore, V_E can be no greater than +9.67 volts.
2. From the foregoing calculations, $V_E + V_G = 4.014$ volts. The maximum value of V_G is then $4.014\ V - V_{E(max)}$ or $V_{G(max)} = 4.014\ V - 9.67\ V = -5.66$ volts.
3. V_G must be of sufficient amplitude to allow V_E to be far enough above V_B that the average noise does not forward-bias the emitter-base junction. The noise could be as high as V_V'' or 1.4 volts, therefore, V_G should be greater than −1.4 volts but less than −5.66 volts. The value chosen for V_G, in this example, is −2.0 volts. This is obtained from a winding on the horizontal output transformer. The resulting V_E is therefore 4.014 volts plus 2.0 volts or 6.014 volts. Figures 6(a), (b) and (c) show the relationship of the gating pulse and the video signals resulting from the biasing just described. Overlapping of the sync-tips, present on the base of the first AGC amplifier, with horizontal gating pulses on the emitter will forward-bias the emitter-base junction resulting in conduction.

The emitter should be stiffly biased to hold it at the desired 6-volt level. A bias current of 5 mA, or about ten times the maximum emitter current, may be taken from the low-voltage supply which in this example is +12 volts.

$$R_{15} = \frac{V_E}{I_{bias}} = \frac{6\ V}{5\ mA} = 1.2\ k\Omega \tag{16}$$

Assuming a midrange setting on the AGC bias control R_{14},

$$R_{13} = R_{15} - \frac{1}{2}\left(R_{14}\right)$$

This equation applies because the desired emitter voltage is one-half the supply voltage. The value chosen for the AGC control R_{14} is 500 ohms. This will give good control over the AGC voltage.

$$R_{13} = 1200\ \Omega - \frac{1}{2}(500\ \Omega) = 950\ \Omega$$

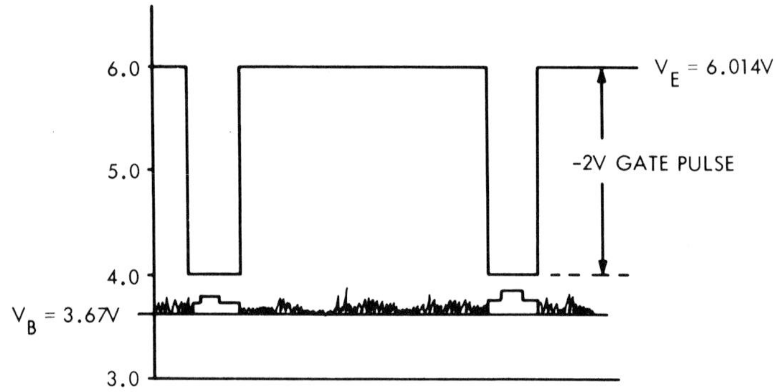

(a) No AGC current. V_{BE} is negative.

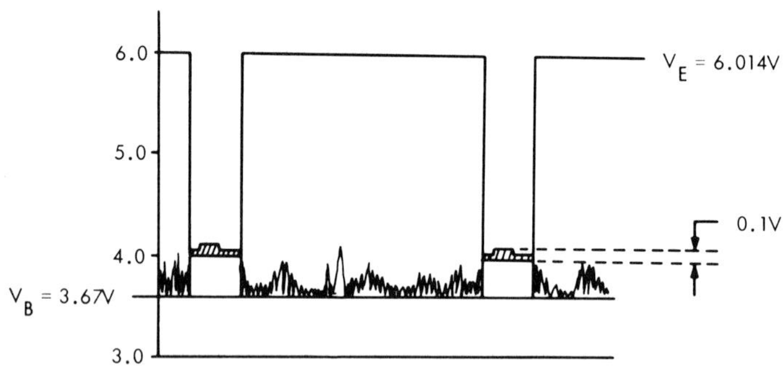

(b) Beginning of AGC action. V_{BE} is +0.1V

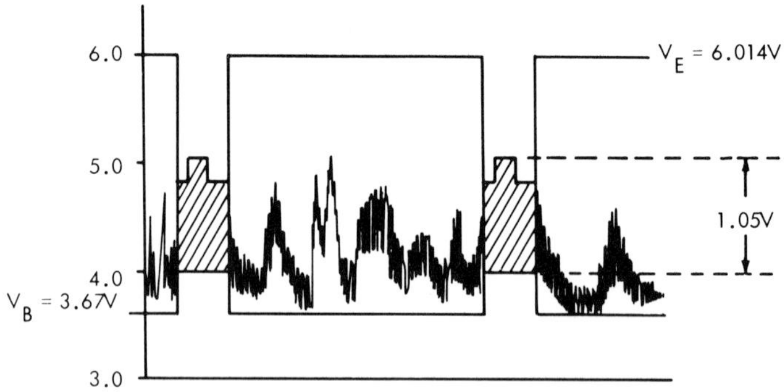

(c) Full AGC action. V_{BE} is +1.05V resulting from a video peak of +1.4 volts.

Fig. 6. Effect of AGC Bias on Gating Pulse and Video Signals

A standard value of 1 kΩ will be satisfactory. The resulting control range of the AGC bias on the emitter is from +5.33 volts to +6.55 volts.

The current through R_{12} is made up of the first AGC amplifier collector current and the AGC limit current from R_5 and R_6. The purpose of the AGC limit current is to prevent a reverse AGC condition from occurring when the AGC amplifier is turned off.

The maximum gain condition of the IF's occurs at +8.7 volts. The value of R_{11}, like R_3, was chosen as 1 kΩ.

The voltage limits $V_{C(min)}$ and $V_{C(max)}$ on the collector of the first AGC amplifier will be the IF AGC limits less the voltage across R_{11}.

$$I_{R11(min)} = 2I_{B(min)} = 2\left[\frac{V_{CC} - (V_{AGC(max)} + V_{BE})}{h_{FE}R_8 + R_{11}}\right] \tag{18}$$

where $I_{B(min)}$ is the minimum base current from an IF stage at minimum AGC.

$$I_{R11(min)} = 2\left[\frac{12\text{ V} - (8.7\text{ V} + 0.3\text{ V})}{50(1\text{k}\Omega) + 1\text{ k}\Omega}\right] = 117.6\,\mu\text{A}$$

$$I_{R11(max)} = 2\left[\frac{V_{CC} - (V_{AGC(min)} + V_{BE})}{h_{FE}R_8 + R_{11}}\right] \tag{19}$$

$$= 2\left[\frac{12\text{ V} - (5\text{V} + 0.3\text{ V})}{50(1\text{ k}\Omega) + 1\text{ k}\Omega}\right] = 263\ \mu\text{A}$$

$$V_{C(min)} = V_{AGC(min)} - V_{R11(max)} \tag{20}$$

where V_C is the voltage on the collector of the first AGC amplifier.

$$V_{C(min)} = 5\text{ V} - 0.263\text{ mA }(1\text{ k}\Omega) = 4.74\text{ V} \tag{21}$$

$$V_{C(max)} = V_{AGC(max)} - V_{R11(min)}$$

$$= 8.7\text{ V} - 0.118\text{ mA }(1\text{ k}\Omega) = 8.58\text{ V}$$

Therefore, during the time the first AGC amplifier is off, the collector voltage should not exceed 8.58 volts. At this time, the base of the second AGC amplifier is at 7.79 volts. Assuming a limiting current of 1 mA, the value of R_6 would be

$$R_6 = \frac{V_{C(max)} - V_{B2(max)}}{I_L + I_{R11(min)}} \tag{22}$$

where V_C is the first AGC amplifier collector voltage, V_{B2} is the second AGC amplifier base voltage, and I_L is the limiting current.

$$R_6 = \frac{8.58\text{ V} - 7.79\text{ V}}{1\text{ mA} + 0.118\text{ mA}} = 707\ \Omega$$

The nearest standard value is 680 ohms. R_5 will have the limit current I_L plus 118 μA from the IF circuit and 38 μA from the second AGC amplifier. Therefore:

$$R_5 = \frac{V_{B2(max)}}{I_L + I_{R11(min)} + I_{B(min)}} \tag{23}$$

$$= \frac{7.79\text{ V}}{1\text{ mA} + 0.118\text{ mA} + 0.038\text{ mA}} = 6.74\text{ k}\Omega$$

The nearest standard value is 6.8 kΩ

Finally, the value of R_{12} for the limit condition when the first AGC amplifier is turned off will be

$$R_{12} = \frac{V_{CC} - V_C}{I_L} \tag{24}$$

$$= \frac{110\text{ V} - 8.58\text{ V}}{1\text{ mA}} = 101.4\text{ k}\Omega$$

The nearest standard value is 100 kΩ.

When a large input signal is applied to the antenna of the receiver the resulting video peak on the collector of the first video amplifier is 2.8 volts. The reactions of the AGC circuit are as follows:

The first AGC amplifier is biased as shown in Fig. 6(c), resulting in a drop in V_C to 4.67 volts. If AGC voltages will drop from 8.7 volts to 5.0 volts, the second AGC amplifier base voltage drops to 3.45 volts, causing the RF AGC voltage to drop to 3.9 volts which results in a maximum forward AGC condition. Figure 7 shows the effect of this type of circuit on an overall TV system.

EQUATION DERIVATIONS

Base Current on RF Device (Equations 7 and 9) (See Fig. 8.)

$$V_{CC} - V_{AGC} = I_B R_{BE} + I_B R_2 + I_E R_1$$

$$I_B R_{BE} = V_{BE}$$

$$I_E = I_C \text{ for large values of } h_{FE}$$

$$I_C = h_{FE} I_B \text{ and } I_E R_1 = h_{FE} I_B R_1$$

$$V_{CC} - V_{AGC} = V_{BE} + h_{FE} I_B R_1 + I_B R_2$$

Solving for I_B

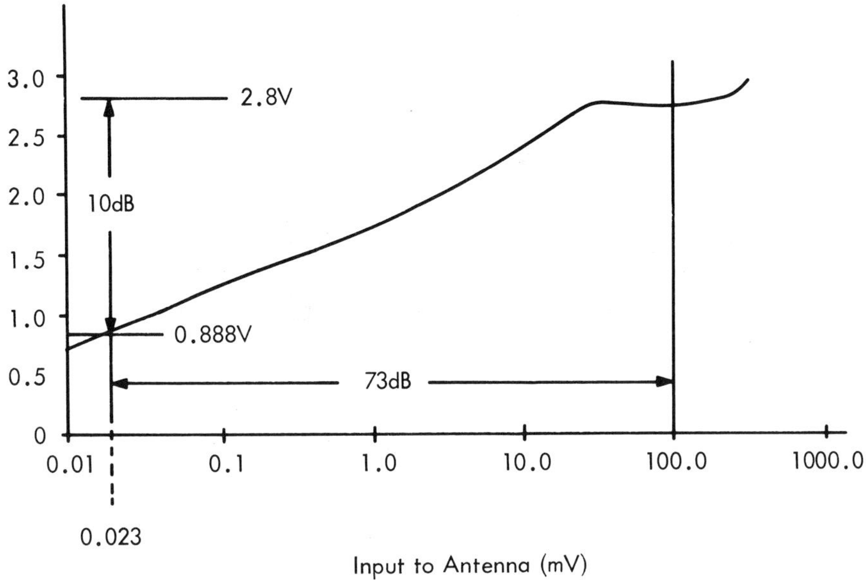

Fig. 7. Effect of AGC Circuit on Overall TV System

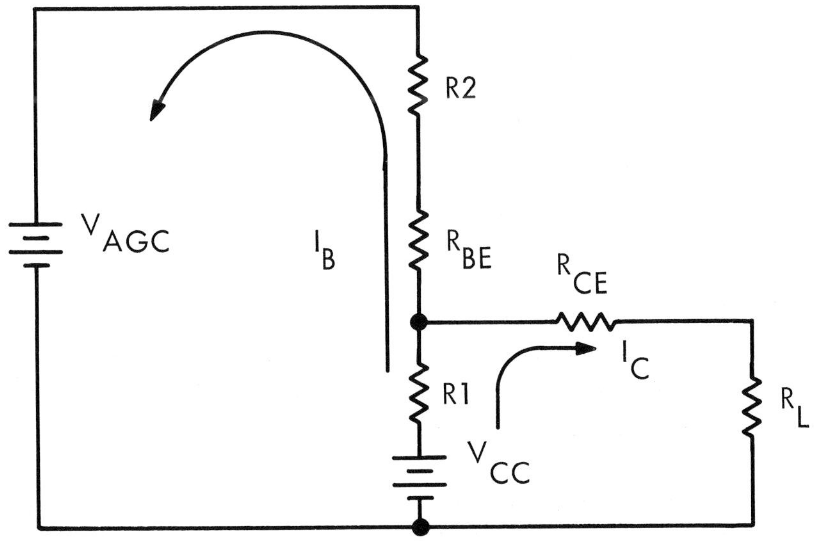

Fig. 8. Base Current on RF Device

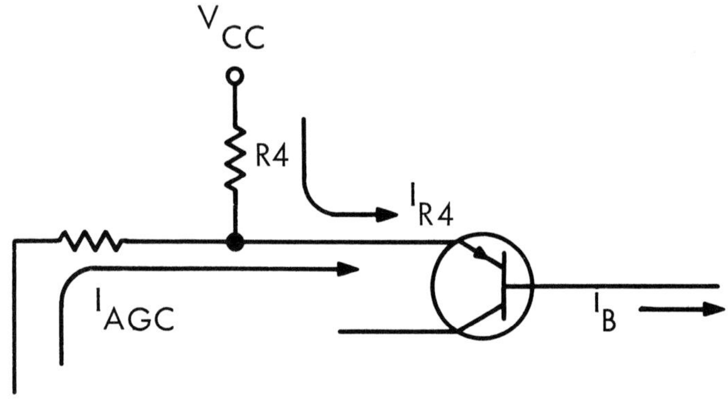

Fig. 9. Base Current on Second AGC Device

$$I_B = \frac{V_{CC} - (V_{AGC} + V_{BE})}{h_{FE}R_1 + R_2}$$

Base Current on Second AGC Device (Equations 13 and 14) (See Fig. 9.)

$$I_E = I_{R4} + I_{AGC}$$

$$I_E = \frac{V_{CC} - V_E}{R_4} + I_{AGC}$$

$$I_B = \frac{I_E}{h_{FE} + 1}$$

$$I_B = \frac{\dfrac{V_{CC} - V_E}{R_4} + I_{AGC}}{h_{FE} + 1}$$

REFERENCES

1. IRE Standards on Television, Section 4.5, 4.6, and 4.7, 1960.
2. "Forward and Reverse AGC Characteristics of VHF Germanium Mesa Transistors," Application Note SC1396, Texas Instruments Incorporated, March, 1961.
3. Fink, D. G.: "Television Engineering Handbook," pp. 146-147, McGraw-Hill Book Company, Inc., New York, 1957.

16

Video Amplifier System

The video consists of that portion of the television receiver circuitry which processes the signal from the video IF amplifier and applies it to the cathode ray tube. In the following discussion, general system requirements are established to indicate the basis of component selection. After the system configuration is established, attention is given to the specific design considerations of each stage in the video amplifier system. From this treatment it may be concluded that the detector and emitter-follower stages may be designed within the performance requirements with relative ease. However, the severity of the video output requirements is such that one or more compensation techniques must be applied.

SYSTEM REQUIREMENTS

Ideally, the video amplifier system should have a flat frequency response from 30 Hz to 4 MHz and the phase response should be linearly proportional to the frequency. These requirements, plus the load of a large-screen picture tube, place rather stringent demands on the video output device. Power dissipation requires a large chip, but low feedback capacitance (for bandwidth) requires a smaller chip. A compromise is usually effected by accepting a reduced bandwidth and using compensation networks to help achieve the desired bandwidth. This practice is more economical than the use of additional stages which would be required if feedback techniques were applied to get the ideal response. An acceptable bandwidth under these circumstances is 60 Hz to 3 MHz.

The signal required to drive the sound IF amplifier must be obtained from the video amplifier system. If it is taken from the video output it may be possible to eliminate a stage in the sound IF amplifier. However, this requires a very good trap to prevent noticeable cross talk between the video and sound signals. It also complicates bandwidth and compensation considerations. Therefore, it is more desirable to trap the sound signal ahead of the video output stage.

In addition to the sound IF signal, drive for the sync separator must be obtained from the video amplifier system. The FCC requires that sync pulses be at least 25% of the total amplitude of the transmitted signal, and allows modulation as low as 15% of the peak level of an all-white picture. It is conceivable that the sync pulse amplitude may be as high as 29.4% of the signal amplitude. In the video amplifier system this signal must be processed linearly. Obviously, the lower the signal amplitude, the lower will be the quiescent power required for linear amplification. If the sync drive is obtained prior to the video output, then sync pulse compression is tolerable in the output stage, and the picture signal amplitude which can be handled is increased by as much as 29.4% for the same device and supply voltage.

The foregoing considerations lead to the video amplifier system shown in Fig. 1. The system is composed of the detector, emitter-follower, and video output stages. The availability of high input impedance, low output impedance, and broad frequency range lead to the use of an emitter-follower stage between the detector and video output stages. The low output impedance provides a good place for extraction of the sync and sound signals without shunting the detector load.

Because the video amplifier system is a relatively broadband system, the low- and high-frequency responses are considered separately. The video output stage is discussed first.

VIDEO OUTPUT STAGE

The system shown in Fig. 1 is designed to obtain maximum use of the video output device. Sound and sync signals are obtained at the emitter of Q_1, thus reducing the bandwidth requirement by about 0.5 MHz and increasing usable video by as much as 29.4%. Additionally, driving the cathode instead of the grid reduces the required drive as much as 15% to 20%.[1] Under these conditions, 130-V peak-to-peak video at the cathode is adequate to drive a 114°, 20-kV picture tube. This includes most large-screen picture tubes.

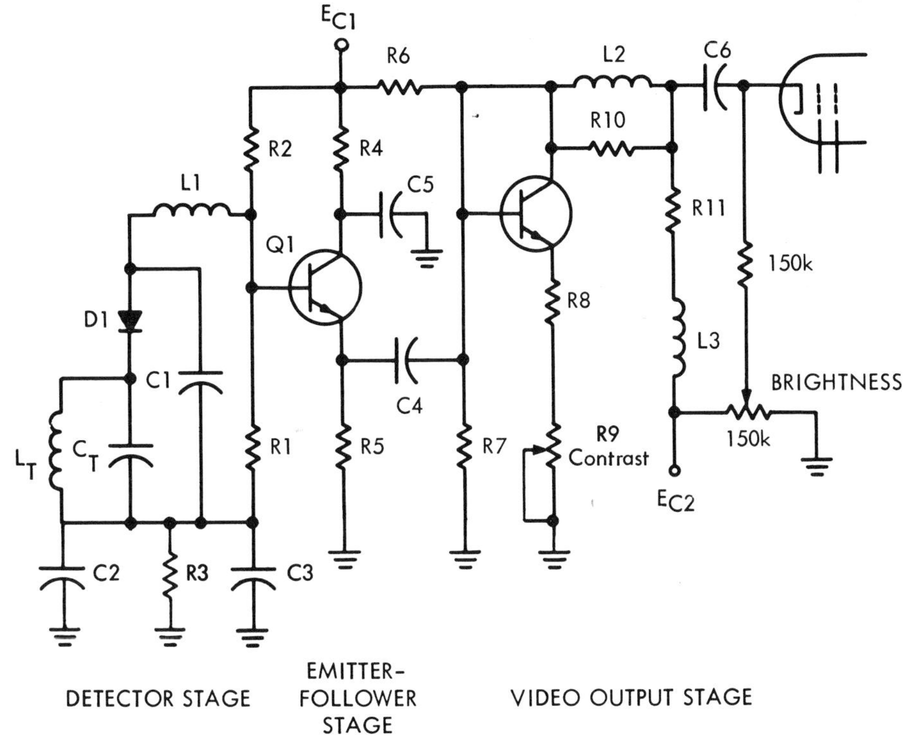

Fig. 1. Video Amplifier System

D-C CONSIDERATIONS

D-C Stability. The choice of bias resistors must be made to ensure bias stability under temperature extremes and when devices are interchanged. It is reasonable to assume an operating free-air temperature 30°C above ambient; that is, about 55°C. Most texts on transistors define a stability factor:

$$S = \frac{dI_C}{dI_{CO}} = \frac{1 + \dfrac{R_b}{R_e}}{1 + \dfrac{R_b}{\beta R_e}} \approx 1 + \frac{R_b}{R_e} \qquad (1)$$

where, in Fig. 1, R_b is the parallel combination of R_6 and R_7, and R_e is the effective resistance of R_8 and R_9. Ideally, $S = 1$; practically, it can only be minimized consistent with limitations such as supply voltages, operating point, and shunting effects due to the bias networks. $S = 20$ is a practical stability factor for low I_{CO} silicon units such as would be used in the video output stage.

Notice in Fig. 1 that the base bias is obtained from E_{C1}. This voltage is usually a fraction of E_{C2} (24 volts compared to 150 volts); therefore, a lower R_b will result at a given operating point. Bias stability is improved by using the lower supply voltage for base biasing.

Supply Voltage. The collector supply voltage will be approximately the same as the peak-to-peak video (about 130 volts for the foregoing CRT's). To preclude non-linear effects due to saturation or cutoff, some additional voltage is necessary. If a 10-volt "guard band" at each end of the signal swing is arbitrarily assumed, then the required supply voltage is 150 volts. The magnitude of the voltage drop across the emitter resistance is related to the stability factor and is assumed negligible. Rigorously, it must be added to the above supply voltage.

Breakdown Voltage. The supply voltage is subject to variations due to transients, variations in line voltage, and component tolerances. It is quite possible that E_{C2} might increase by 10%. Therefore, the minimum BV_{CEO} for a device in this circuit is 165 volts. Secondary breakdown is a common breakdown mode in video output devices. The collector current necessary to define the maximum usable voltage is roughly an order of magnitude larger than that normally used to define breakdown. By specifying the collector current as 15 to 20 mA at $V_{CE} = 165$ volts, one can be assured of a reliable device as far as BV_{CEO} is concerned.

Power Rating. For a given supply voltage, the power rating of the video output device is inversely proportional to the collector load resistor. Unfortunately, the frequency response bears a similar relationship to R_L. Consider the following relations where it is assumed that the voltage across R_8 and R_9 is negligible with respect to E_{CC}:

$$P_{DISS} = \frac{E_{CC}^2}{4R_L} \qquad f_{3dB} = \frac{1}{2\pi R_L C_T} \tag{2,3}$$

$$P_{DISS} = \frac{E_{CC}^2}{2} \pi C_T f_{3dB} \tag{4}$$

R_L is the d-c collector load, f_{3dB} is the upper cutoff frequency, P_{DISS} is the quiescent device dissipation and C_T is the total capacitance in the collector circuit. This illustrates one of the essential compromises to be achieved. E_{CC} and C_T are primarily dictated by the characteristics of the CRT; therefore the engineer may have to sacrifice frequency response in order to use a lower-power, less expensive device. Assuming $E_{CC} = 150$ V, $C_T = 20$ pF, and $f_{3dB} = 4.0$ MHz, Eq. (2), (3), and (4) yield $P_{DISS} = 2.82$ watts and $R_L = 2.0$ kilohms. Note that

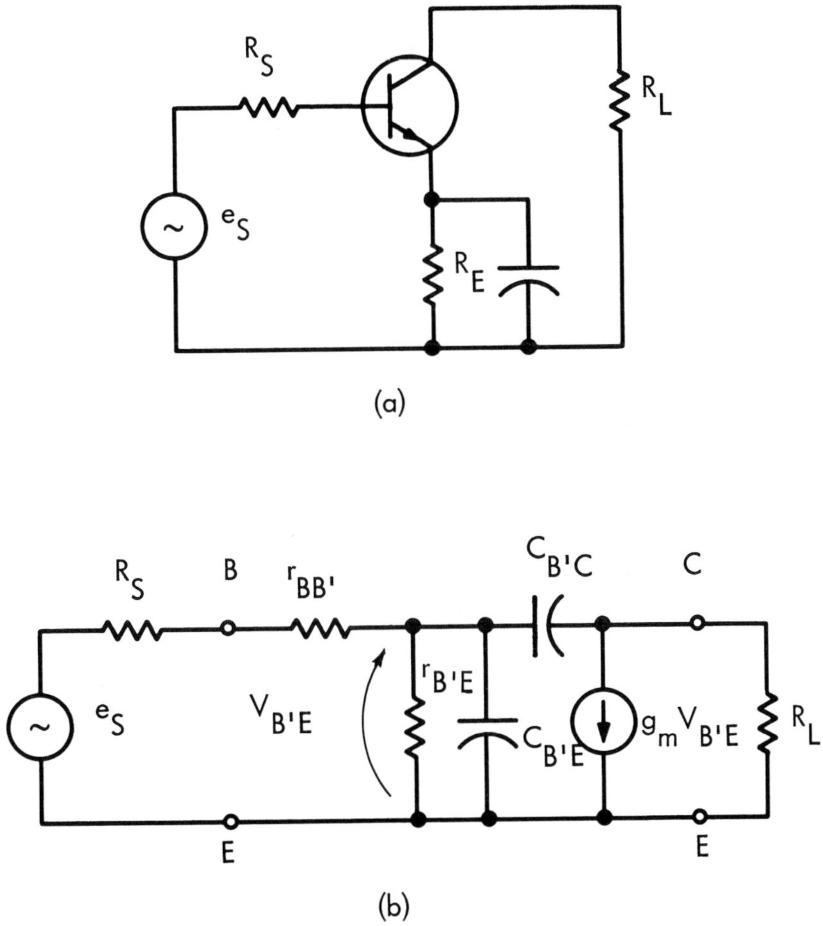

Fig. 2. High-frequency Circuits

this is the power rating required at 55°C. If one of the frequency compensation schemes is used, it may be possible to reduce the power dissipation by using a higher d-c load resistance. It is assumed that this is true here and that the d-c load resistance $R_L = 4.6$ kilohms. The dissipation is reduced to 1.22 watts under the new conditions. Adding 10% to allow for variations due to components and line conditions, a transistor rated at 1.34 watts at 55°C is adequate. The particular compensation scheme used here is discussed in the next section.

HIGH-FREQUENCY CONSIDERATIONS

Consider the circuits of Fig. 2. For the general case, where the product of load impedance and feedback capacitance is not negligible, the 3-dB cutoff frequency is

$$f_{3dB} = \frac{1}{2\pi R_{EQUIV} C_{EQUIV}} \tag{5}$$

where

$$R_{EQUIV} = \frac{r_{B'E}(r_{BB'} + R_S)}{r_{B'E} + r_{BB'} + R_S} \tag{6}$$

and

$$C_{EQUIV} \equiv C_{B'E} + C_{B'C}(1 + g_m R_L) \approx C_{B'E} + C_{B'C}\, g_m R_L \tag{7}$$

Essentially, R_{EQUIV} is the Thevenin's impedance of the terminals B′E. It is desirable to minimize R_{EQUIV}; the designer may do this by minimizing R_S, the generator impedance. For the video output device, the generator impedance is the parallel combination of bias resistors, emitter-follower output impedance, and the impedances of loads such as the sync separator and the sound trap. It is more practical to make empirical adjustments for the latter loads than to try to include them in preliminary calculations.

It is equally desirable to minimize C_{EQUIV}. For a given operating point, this requires higher f_T, lower $C_{B'C}$, or lower R_L. As mentioned before, lowering R_L increases the required power rating of the device. It also reduces the stage voltage gain.

A rough idea of some parameter values ($C_{B'E}$, f_T, etc.) may be found by the following calculations based on an ideal transistor and previously stated assumptions of load, supply voltage, and bandwidth:

$$R_L = 4.6\ \text{k}\Omega;\ I_E = \frac{E_{CC}}{2R_L} = \frac{150}{2(4.6 \times 10^3)} = 16.3\ \text{mA}$$

$$g_m = \frac{I_E}{26} = \frac{16.3}{26} = 0.627\ \text{S}$$

$$r_E = \frac{1}{g_m} = \frac{1}{0.627\ \text{S}} = 1.6\ \Omega$$

when

$$\beta = 100 \qquad r_{B'E} = \beta r_E = 160\ \Omega$$

assume $R_S = 20\ \Omega$; $r_{BB'} = 50\ \Omega$; $R_{EQUIV} = 48.7$ ohms

Further, if $C_{B'E}$ is small enough, the Miller Effect determines most of the effective capacitance of the stage. It is assumed that $C_{B'E}$ is 10% of C_{EQUIV}.

$$C_{EQUIV} = C'_{BE} + C'_{BC}(1 + g_m R_L)$$

$$= \frac{1}{2\pi R_{EQUIV} f_{3dB}} = \frac{1}{6.28(48.7)(4 \times 10^6)} = 820 \text{ pF}$$

$$C'_{BC} = \frac{0.9\ C_{EQUIV}}{1 + g_m R_L} = \frac{0.9(820 \times 10^{-12})}{1 + 0.627(4.6 \times 10^3)} = 0.256 \text{ pF}$$

$$f_{EO} = \frac{1}{2\pi r'_{BE} C'_{BE}} \left(\frac{1}{6.28(160)(82 \text{ x } 10^{-12})} \right) = 12.14 \text{ MHz}$$

Assuming f_{EO} is the common-emitter high-frequency breakpoint, the gain characteristics may be extrapolated at the rate of −20 dB per decade to find f_T of the device. If $\beta = 100$ and $h_{fe} = 20$ dB, then the f_T for this device is 121.4 MHz. The foregoing calculations are for a hypothetical situation, but they shed some light on the problems faced by the device designer. Consideration of the calculated C'_{BC} in conjunction with the size of a chip capable of the necessary dissipation indicates why it is necessary to resort to compensation networks in order to extend the phase and amplitude response of present devices.

Compensation. There are four basic techniques for compensation. They are shunt, series, shunt-series and series-shunt. Design equations and literature concerning the first three techniques are readily available in many electronics publications such as the "Radio Engineering Handbook."[2] The fourth technique is represented by L_2, R_{10}, L_3 and R_{11} in Fig. 1. The following equations may be used in calculating this network:

$$L_2 = 0.29\ R_L{}^2 C_T \tag{8}$$

$$R_{10} = 2.5\ R_L \tag{9}$$

$$L_3 = 0.63\ R_L{}^2 C_T \tag{10}$$

$$R_{11} = \frac{2.3}{2\pi C_T f_{3dB}} = 2.3\ R_L \tag{11}$$

The effect of using this network is to increase the voltage gain and extend the frequency and phase response while using a larger load resistor for the uncompensated case. As previously mentioned, raising the load resistance reduces the required power rating of the device.

For a compensated stage $P_{DISS} = \dfrac{E_{CC}{}^2}{4\ R_{11}} = \dfrac{E_{CC}{}^2}{4(2.3\ R_L)}$. For an uncompensated stage $R_{11} = R_L$ where R_L is that value of resistance required to yield a given 3-dB frequency for a given circuit capacitance C_T. In this example the calculated $R_L = 2.0$ kΩ, but an assumed compensation set $R_{11} = 2.3$ (2 kΩ) = 4.6 kΩ, thereby reducing the power rating of the device.

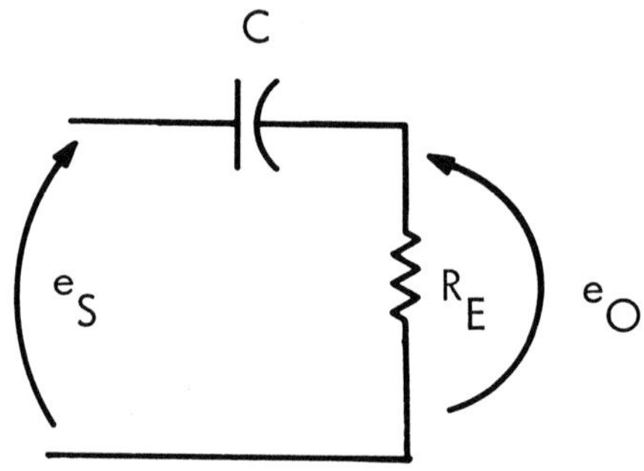

Fig. 3. Coupling Capacitor

Another method of influencing the high-frequency gain is the use of emitter peaking. This method uses an RC combination in the emitter, designed to introduce a reduction in unbypassed emitter resistance and thereby increase the gain at a rate sufficient to compensate for gain reduction with frequency.

Emitter Resistance. The formula for gain reduction due to unbypassed emitter resistance is[3/]

$$F = \frac{1}{1 + g_m R_E} \tag{12}$$

Accompanying the gain reduction is an increase in input impedance as follows:

$$Z_{in} = Z_{in}(1 + g_m R_E) \tag{13}$$

where Z_{in} is the input impedance without the feedback due to the emitter resistor. This increase in input impedance due to the emitter resistance is a highly desirable property. This is apparent when calculating the values of the coupling capacitor. Consider the circuit of Fig. 3.

$$\frac{e_O}{e_S} = \frac{1}{1 - j\dfrac{1}{\omega CR}} = \frac{1}{1 - j\dfrac{1}{2\pi fCR}} \tag{14}$$

Assume $\dfrac{1}{2\pi fC} \leq \dfrac{R}{10}$

Hence,

$$C \geq \frac{10}{2\pi fR} \tag{15}$$

At low frequencies, the input impedance is essentially resistive. Without feedback, it would be r'_{BB} in series with βr_E. For the operating point and related assumption, this might be 200 ohms. Assume $R_E = 50$ ohms, then $g_m R_E = 0.576\ (50) = 28.8$.

$$Z_{in}(1 + g_m R_e) = 200(29.8) = 5.96\ k\Omega$$

therefore

$$C \geq \frac{10}{6.28(60)(5960)} = 4.46\ \mu F$$

Without feedback due to emitter resistance, a 133-μF capacitor would have been required to achieve the same amplitude phase response. The foregoing calculation neglects the shunting effect of the bias network. This effect approximately triples the calculated capacitance. It should be apparent that there are many interrelated compromises to be effected in the design of a video output stage. The calculations here illustrate only one set of solutions. Throughout the foregoing, calculations were made which involved the source impedance which, in turn, was the parallel combination of the emitter-follower output impedance and all bias networks, etc. The validity of the assumption will be better established in the next section.

EMITTER-FOLLOWER STAGE

D-c considerations for this stage are similar to those of the video output and will not be reviewed. Note that the emitter-follower is direct-coupled to the detector. The values of emitter resistance and diode bias resistors must be selected to be compatible. The emitter-follower must be capable of linear large-signal amplification. This suggests that the operating point be selected for maximum symmetrical signal excursion. With terms defined as shown in Fig. 4, Eq. (16) relates the parameters for the noted condition.[4]

$$\frac{R_E}{R_L} = \frac{E_{C1}}{V_{CE}} - 2 \tag{16}$$

If $V_{CE(sat)}$ is appreciable, its effect may be accounted for by subtracting it from E_{C1} in Eq. (16).

In previous calculations the source impedance of the video output stage was assumed to be 20 ohms. The output impedance of the emitter-follower is

$$R_O \approx \frac{R_S}{1 + \beta} + \frac{1}{g_m} \approx \frac{R_S}{\beta} \tag{17}$$

The source impedance of the emitter-follower is approximately the parallel combination of the detector load resistor and the output impedance of the IF

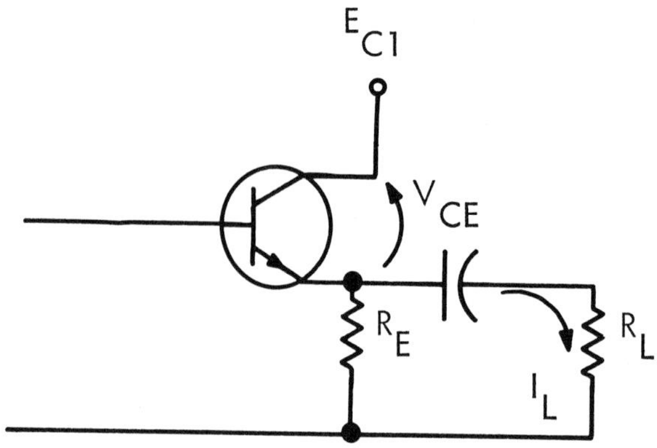

Fig. 4. Emitter-follower Circuit

amplifier strip. The resultant is normally about 2 kilohms. Therefore R_0 is approximately 20 ohms. As in the common emitter stage with an unbypassed resistor, there is a multiplication of the input impedance of the emitter-follower. The input impedance of an emitter-follower is

$$Z_{in} = (1 + \beta)R \tag{18}$$

where R is the net value of unbypassed emitter resistance. Since the net value of R can vary with frequency, it is necessary to use sufficient R so that the emitter-follower does not load the detector at the highest frequency of interest.

A rule-of-thumb for selecting an emitter-follower device is to choose one with an f_T at least a decade above the maximum frequency to be handled. A device with this characteristic will ensure that the foregoing relations are valid.

DETECTOR STAGE

The second video detector stage is essentially an envelope-detector low-pass filter combination. A basic detector circuit is shown in Fig. 5.

C_1 has low reactance at the video IF frequencies and L_1 has high reactance at these frequencies, thus filtering the IF frequencies from the output voltage. Conversely, C_1 should have high reactance at low frequencies while the reactance of L_1 should be negligible compared to the load impedance. The current transfer function for this network is

$$\frac{i_L}{i_S} = \frac{1}{\left[1 - \left(\frac{\omega}{\omega_1}\right)^2\right] + j\,\frac{\omega}{\omega_2}} \tag{19}$$

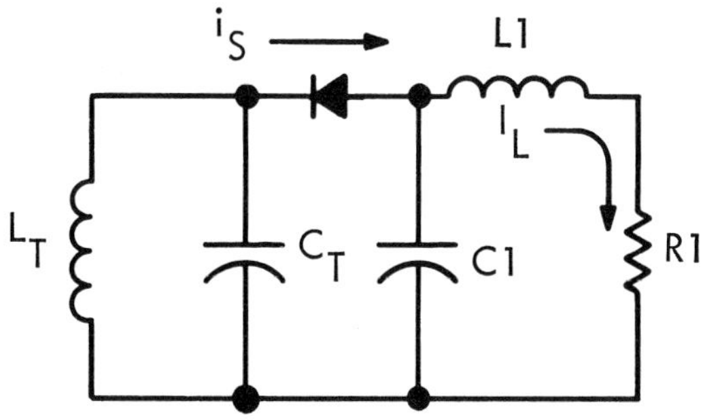

Fig. 5. Detector Circuit

where

$$\omega_1 = \frac{1}{\sqrt{L_1C_1}}$$

$$\omega_2 = \frac{1}{R_1C_1}$$

The transfer function may be examined in three frequency regions:

$$(1) \quad \frac{X_{L1}}{X_{C1}} << 1 \qquad \frac{i_L}{i_S} = \frac{1}{1 + j\omega C_1R_1} \tag{20}$$

In this region, phase shift is directly proportional to frequency, which is satisfactory; however, note that R_1C_1 sets a breakpoint in the frequency response.

$$(2) \quad \frac{X_{L1}}{X_{C1}} = 1 \qquad \frac{i_L}{i_S} = \frac{1}{j\omega C_1R_1} \tag{21}$$

In this region, phase shift is constant (near 90°) but amplitude is falling at 6 dB per octave.

$$(3) \quad \frac{X_{L1}}{X_{C1}} >> 1 \qquad \frac{i_L}{i_S} = 0 \tag{22}$$

This area just shows the ultimate which might be achieved (total rejection at very high frequencies).

It may be concluded from the foregoing that the detector stage satisfies the requirements of linear response and phase shift below the frequency

$$f_2 = \frac{1}{2\pi R_1C_1} \tag{23}$$

To specify the components of the detector stage, first consider the definition of detector efficiency based on a diode in series with the load.

$$\eta = \frac{R_L}{R_L + R_D} \tag{24}$$

where η = diode rectification efficiency, R_L = load resistance and R_D = effective diode resistance. Manipulating Eq. (24) yields

$$R_L = \frac{\eta}{1 - \eta} R_D \tag{25}$$

In the region above $\frac{X_L}{X_C} = 1$, more and more of the source current goes through the capacitor C_1. This, in effect, places the diode in parallel with the tank at IF frequencies. R_D may be calculated as follows:

$$R_D = Q_L X_C \tag{26}$$

The effective diode impedance is determined by the R_1C_1 time constant as well as tank circuit coupling coefficient and may be adjusted by forward biasing the diode. Therefore, it is permissible to assume C_T and a reasonable detection efficiency and calculate the remaining circuit values as follows:

$$C_T = 20 \text{ pF therefore } X_C = 180\ \Omega$$

$$Q_L = \frac{f_o}{BW} = \frac{45 \text{ MHz}}{4.5 \text{ MHz}} = 10$$

$$\eta = 0.65$$

$$R_D = Q_L X_C = 1800\ \Omega$$

$$R_1 = \left(\frac{\eta}{1 - \eta}\right) R_D = \left(\frac{0.65}{0.35}\right) 1800\ \Omega = 3.34 \text{ k}\Omega$$

$$C_1 = \frac{1}{2\pi f_{3dB} R_1} = \frac{1}{6.28(4.5 \text{ MHz})\,(3.34 \text{ k}\Omega)} = 10.6 \text{ pF}$$

$$f_1 = \frac{\omega_1}{2\pi} = 20 \text{ MHz}$$

therefore

$$L_1 = \frac{1}{4\pi^2 f_1^2 C_1} = \frac{1}{4(9.86)(4 \times 10^{14})(1.06 \times 10^{-11})} = 6\ \mu\text{H}$$

The resonant frequency f_1 of L_1 and C_1 is selected rather arbitrarily as 20 MHz. Consideration of the frequency response expressions shows that the phase shift is 90° at the resonant frequency. For phase linearity, it would be desirable to have the resonant frequency as high as possible. For effective IF filtering, it is desirable to have the resonant frequency as low as possible. 20 MHz is roughly two octaves above the video frequencies and is about one octave below the intermediate frequency. As such, it represents a practical compromise.

Detector efficiency is assumed to be 65%. It is possible to achieve higher efficiency by increasing C_1, but this obviously affects the frequency response. Efficiency is also affected by the amount of forward bias caused by the emitter-follower bias current flowing through R_1, R_2 and R_3. About 0.3 volt is a practical value for the diode bias. After the bias is determined, R_2 and R_3 may be selected to establish the desired emitter-follower operating point. Capacitors C_2 and C_3 are selected to establish IF and low-frequency bypasses. The values depend upon the value of R_3.

REFERENCES

1. Salaman, R. G.: Receiver Video Transistor Stages, *IRE Transaction on Broadcast and TV Receivers,* Vol. BTR-4 No. 4, pp. 68-77, September 1958.
2. Henney, Keith (Editor): "Radio Engineering Handbook," Chapter 22, pp. 29-32, McGraw-Hill Book Company, New York, 1959.
3. Kidd, Marshall C.: Transistor Receiver Video Amplifiers, *RCA Review,* Vol. 18, No. 3, pp. 308-321, September 1957.
4. Peark, Bert: Emitter-Follower Nomogram, *EEE*, pp. 58-59, June 1965.

17

Sound IF Amplifier System

In specifying a system to process the TV sound signal and recover the audio program, the television engineer must choose an acceptable compromise between performance requirements and economics. There are several important considerations, but AM rejection is one of the most important requirements for a sound IF system. Cost and system sensitivity are closely related to these requirements. Use of an amplitude-limited amplifier and ratio detector has emerged as a common means of effecting the performance-economic compromise.

A practical appraoch to the design of a sound IF amplifier system is presented on the following pages. Certain approximations are made to simplify calculations. Measured performance compares well with the design and thus supports the validity of the approximations.

SYSTEM REQUIREMENTS

The prime requirement of the sound IF amplifier is to amplitude limit with as small a signal as practical. Limiting sensitivity is defined as that signal level at which the output power has fallen three decibels below maximum. This condition is referred to as "3-dB limiting." A desirable sensitivity for a 3-dB limiting is one mV rms at the input of the system.

Bandwidth of the system should be at least 50 kHz, the frequency deviation for 100% modulation. Increasing the bandwidth reduces the AM resulting from large frequency deviation at low signal levels. The bandwidth of the ratio detector affects the linearity of the audio waveform; increasing the bandwidth improves the linearity. For the system here, the interstage bandwidth is specified as 100 kHz and the ratio detector bandwidth is specified as 200 kHz. A figure of merit for AM rejection is 20 dB measured at 3-dB limiting. This figure represents minimum acceptable performance. All performance measurements should be made under standard conditions. These conditions are: 30% AM and 30% FM (7.5-kHz deviation, modulated at 400 Hz).

AMPLIFIER-LIMITER

Gain. The amplifier is to be designed to limit; the desired input signal level is set; therefore, the required system gain is determined by the quiescent conditions of the output device. The device to be used is the TIXM207. Parameter measurements indicate maximum gain should be expected under quiescent conditions of $V_{CE} = -6.0$ V, $I_C = -2.0$ mA. Required system gain may be estimated by assuming:

1. Output a-c load line is optimized for maximum symmetrical signal excursions.
2. Input voltage is one millivolt developed across 50 ohms.

Required gain is:

$$G = \frac{P_{out}}{P_{in}} = \frac{V_{CE} I_C R_{in}}{2\ V_{in}^2} \tag{1}$$

$$= \frac{6(2 \times 10^{-3})50}{2(1 \times 10^{-6})}$$

$$= 3 \times 10^5 \text{ or } 54.78 \text{ dB}$$

All circuit losses must be added to this gain to determine the gain which one or more transistors must provide.

Stability. Parasitic oscillations may result when a transistor is subjected to signal swings which reverse-bias the collector-base junction. A solution to this problem[1/] is to insert resistance in series with the base or collector. The added impedance dissipates the oscillation energy and minimizes its effect. Resistance values vary with device type and circuit conditions. Practical values range up to 50 ohms in the base and 1000 ohms in the collector.

A second and more fundamental stability problem involves preventing regeneration due to excessive amplifier gain. Practically, this involves the introduction of adequate insertion loss to reduce stage gain to a usable (stable) level. Every device has a theoretical maximum available gain (MAG). This is the stage gain attainable if the device is conjugately matched to generator and load and if the reverse transfer admittance is zero. For a given transistor stage, it is theoretically possible to achieve the foregoing condition by unilateralization, but this is not a realistic approach. Technically, it is also restricted to one device and one set of conditions. In the final analysis, stability is achieved by intentionally designing insertion loss into the transistor stage. The required insertion loss depends upon whether or not the stage is neutralized and upon the magnitude of y_{12} and its net variation from a mean value.

A reasonable figure for the insertion loss of an IF amplifier may be obtained from the following relations.[2]

$$MAG = \frac{|y_{fe}|^2 R_{iep}R_{oep}}{4} \tag{2}$$

$$C_{fb} = \frac{4}{\omega \sqrt{MAG\ R_{iep}R_{oep}}} \tag{3}$$

$$S = \frac{5C_{cb(max)}}{C_{fb}} \tag{4}$$

$$IL = 10 \log_{10} S \tag{5}$$

For a neutralized stage, $C_{cb(max)}$ is the maximum feedback capacitance variation from design center. This includes both variations in C_{cb} and component tolerance. If the stage is not neutralized, $C_{cb(max)}$ is simply the maximum value which may occur in the device.

Transformer. Insertion loss is obtained in the coupling networks, which are usually single-tuned or double-tuned transformers. Only the design of single-tuned transformers is considered in this presentation. Insertion loss of a transformer-coupled stage is expressed as a ratio:

$$IL_{(total)} = \frac{(R_{oep} + R_{iep}')^2}{4(R_{oep}\ R_{iep}')} \left(\frac{Q_U}{Q_U - Q_L}\right)^2 \tag{6}$$

where R_{iep}' is the load impedance reflected to the primary. If this equation is expressed in decibels it is possible to consider interstage insertion loss as being divided between transformer losses and mismatch losses:

$$IL_{(total)} = IL_{(mismatch)} + IL_{(transformer)} \tag{7}$$

Achieving a particular combination of losses while maintaining an optimum a-c load line may be difficult. Utilizing a tapped transformer, as shown in Fig. 1, provides an extra degree of freedom and simplifies realization of the transformer.

At resonance, conductance equations for the circuit in Fig. 1 have the following form:

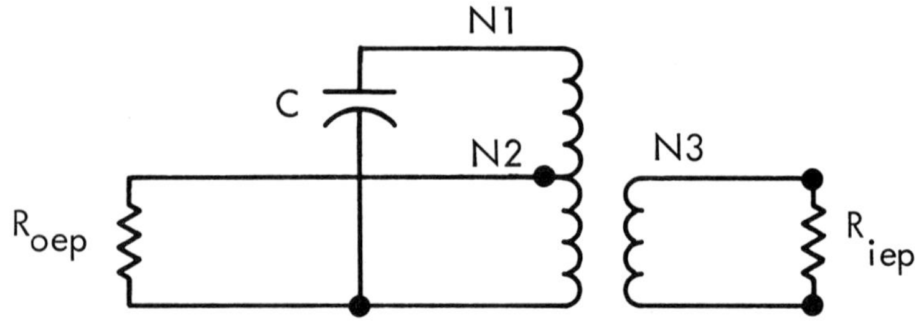

Fig. 1. Tapped Transformer

Referred to N1:

$$A(g_{oep}) + g_t + B(g_{iep}) = \frac{1}{Q_L X_C} \tag{8}$$

Referred to N2:

$$g_{oep} + \frac{g_t}{A} + C(g_{iep}) = \frac{1}{R_C} \tag{9}$$

where

$$A = \left(\frac{N2}{N1}\right)^2, \quad B = \left(\frac{N3}{N1}\right)^2, \quad C = \left(\frac{N3}{N2}\right)^2,$$

$g_t = \dfrac{1}{Q_U X_C}$, and R_C = magnitude of load-line impedance.

If critical coupling is assumed, the following relations specify the transformer:

$$A = \frac{B}{C} \tag{10}$$

Therefore,

$$\frac{N1}{N2} = \sqrt{\frac{Q_L X_C}{R_C}} \tag{11}$$

$$\frac{N2}{N3} = \sqrt{\frac{R_{oep}}{a\ R_{iep}}} \tag{12}$$

$$\frac{N1}{N3} = \left(\frac{N1}{N2}\right)\left(\frac{N2}{N3}\right) \tag{13}$$

where "a" $= \frac{R_{oep}}{R_{iep}'}$. The value of "a" depends upon the desired division of losses in the interstage.

Transformer Design Procedure. A design procedure for an amplifier-limiter interstage transformer is outlined as follows:

1. Determine insertion loss to provide stability.
2. Determine Q_L from the bandwidth:

$$Q_L = f_c/BW \tag{14}$$

3. Divide insertion loss between mismatch and coil loss.
4. Calculate Q_U:

$$Q_U = \frac{Q_L \sqrt{IL_{(coil)}}}{\sqrt{IL_{(coil)}} - 1} \tag{15}$$

5. Calculate "a":

$$IL_{(mismatch)} = 10 \log_{10} \frac{(R_{oep} + R_{iep}')^2}{4(R_{oep}R_{iep}')} = 10 \log_{10} \frac{(1+a)^2}{4a} \tag{16}$$

6. Calculate N2/N3:

$$\frac{N2}{N3} = \sqrt{\frac{R_{oep}}{a\ R_{iep}}} \tag{12}$$

7. Select R_C for optimum load line:

$$R_C = \frac{V_{CE}}{I_C} \tag{17}$$

8. Assume tuning capacitance and calculate N1/N2:

$$\frac{N1}{N2} = \sqrt{\frac{Q_L X_C}{R_C}} \tag{11}$$

9. Calculate N1/N3:

$$\frac{N1}{N3} = \left(\frac{N1}{N2}\right)\left(\frac{N2}{N3}\right) \tag{13}$$

The foregoing procedure considers insertion loss in the interstage transformer and is applicable to a multistage amplifier. The input and output terminations of the first and last stages respectively are not mentioned. Two common approaches to this termination problem are:

1. Consider the input conjugately matched. Then all the insertion loss in the interstage will be associated with the preceding stage. For a multistage amplifier, this requires that the output termination must provide the same insertion loss as the interstage transformers.
2. Associate half the interstage loss with each related device which requires that the input and output terminations provide one-half the necessary losses for their respective devices.

The second approach is taken in the design example:

Design Example using TIXM207. Typical parameters of the TIXM207 are:

$y_{fe} = 72\,000\ \mu S$

$R_{iep}' = 600\ \Omega$

$R_{oep} = 75\,000\ \Omega$

$C_{cb(max)} = 1.7\ pF$

1. Determine insertion loss:

$$MAG = \frac{(0.072)^2(600)(7.5 \times 10^4)}{4} = 5.825 \times 10^4 \text{ or } 47.66 \text{ dB} \quad (2)$$

$$C_{fb} = \frac{4}{6.28(4.5 \times 10^6)\sqrt{(5.825 \times 10^4)(600)(7.5 \times 10^4)}} = 0.0874 \text{ pF} \quad (3)$$

$$S = \frac{5(1.7)}{0.0874} = 97.2 \quad (4)$$

For two stages, multiply S by 2 (3-dB added IL) since the feedback required to produce oscillation diminishes as the number of stages increases.

$$S_{(2\text{-stage})} = 194.4$$

$$IL_{(total)} = 10 \log_{10} 194.4 = 22.88 \text{ dB per stage.} \quad (5)$$

Usable gain (MAG − IL) is 24.78 dB per stage. If the stages were neutralized, usable stage gain would be 30 dB. Stage gain is sacrificed for economy.

2. $$Q_L = \frac{4.5 \times 10^6}{100 \times 10^3} = 45 \quad (14)$$

3. $IL_{(mismatch)} = 10.0\text{ dB} = 10.0$ (ratio)

$IL_{(coil)} = 12.88\text{ dB} = 19.4$ (ratio)

4. $$Q_U = \frac{45\sqrt{19.4}}{\sqrt{19.4} - 1} \approx 60 \tag{15}$$

5. $$10 \log_{10} \frac{(1 + a)^2}{4a} = 10 \tag{16}$$
$$a = 40$$

6. $$\frac{N2}{N3} = \sqrt{\frac{7.5 \times 10^4}{40(6 \times 10^2)}} = 1.77 \tag{12}$$

7. $$R_C = \frac{6}{2 \times 10^{-3}} = 3000\ \Omega \tag{17}$$

8. Assume C = 100 pF:

$$\frac{N1}{N2} = \sqrt{\frac{45}{6.28(4.5 \times 10^6)(100 \times 10^{-12})(3 \times 10^3)}} = 2.31 \tag{10}$$

9. $$\frac{N1}{N3} = 2.31(1.77) = 4.09 \tag{13}$$

Coil data for the interstage is as follows:

Primary: 42 turns #36 Gripeze tapped at 18 turns from the cold end.
Secondary: 5 turns #36 Gripeze closewound over the cold end of primary.
Core material: A-1-13-J
$Q_U = 71.5$
$Q_L = 57$
Primary and secondary wound on 9/32-inch form.

The output termination of the amplifier is the primary of the ratio detector. The design of this termination is a compromise between stability requirements and detector performance. Note that the load presented by the ratio detector is non-linear and varies with signal amplitude; the highest reflected impedance occurs under no-signal conditions. Obviously, this is the worst case from the standpoint of amplifier stability. Due to the non-linearity of the device and to differences in device materials, it is difficult to establish a useful quantitative relationship to handle the low-level stability problem. A worst-case design, that is, assuming that the diodes are open-circuited, is one approach. Practically, this approach is wasteful because the diodes are not truly open circuits and an excessive amount of coil loss would be required. The problem may be solved empirically by designing for stability at the normal operating level and then adjusting coupling or tertiary loading for low-level stability. In some cases it may be necessary to add shunt resistance to the base circuit to finally achieve the desired stability. Design considerations for the primary follow a discussion of the ratio detector.

RATIO DETECTOR

A basic ratio detector circuit is shown in Fig. 2.[3/4/5]

The demodulated audio appears between the junction of C_3, C_4 and R_{L1}, R_{L2}. Terms, referred to Fig. 2, are:

C_1 = total primary tuning capacitance
C_2 = total secondary tuning capacitance
C_3, C_4 = diode load capacitors
C_5 = stabilizing capacitor
L_P = primary inductance
L_T = tertiary inductance
L_S = secondary inductance
D_1, D_2 = detector diodes
R_a, R_b = diode stabilizing resistors
R_T = tertiary resistance

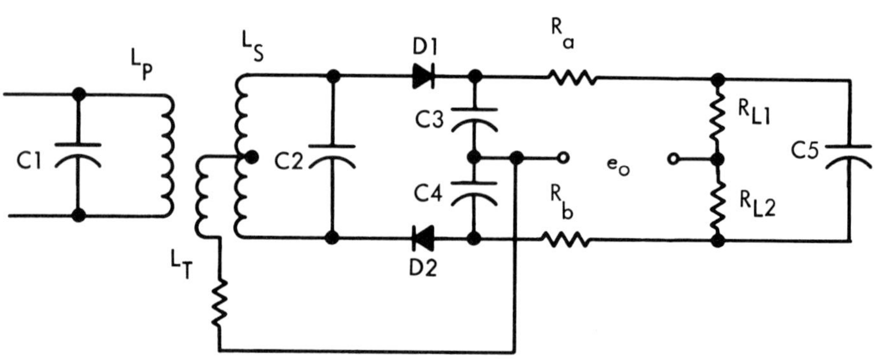

Fig. 2. Ratio Detector Circuit

AM rejection is closely related to the fact that diode impedance changes with signal amplitude. If primary loading is caused substantially by the diodes, the impedance variations tend to maintain a constant voltage across the transformer. A decrease in signal amplitude raises the diode impedance and the tertiary reflects this impedance into the primary. The higher impedance develops more voltage and thus tends to compensate for the lower signal. Operation under increasing signal is analogous. To take full advantage of the

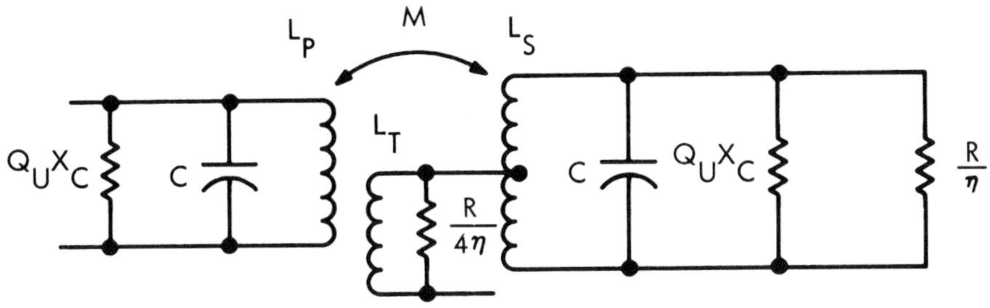

Fig. 3. Equivalent Detector Circuit

"regulating" effect, the primary unloaded Q should be as high as possible. The actual value of Q_U is dictated by the maximum stable gain under no-signal conditions. The secondary unloaded Q should also be as high as practical; good results are obtained with a secondary unloaded-to-loaded Q ratio about 4.0 to 1.0. This implies a secondary unloaded Q between 80 and 100, which is easily obtainable. The important point is that the unloaded-to-loaded Q ratio should be large enough that the diode impedance variations have a significant influence on the circuit impedance. An important parameter of the ratio detector is S/T. Seely and Avins[a/] indicate that best AM rejection is impossible if S/T exceeds unity. S is defined as the voltage reduced across one-half the secondary inductance (half-secondary voltage). T is the induced voltage of the tertiary winding. From the equivalent detector circuit shown in Fig. 3, the following may be derived:

$$\frac{S}{T} = \alpha \sqrt{1 + \frac{n^2 Q_{LS} L_S}{4\, Q_{UP} L_P}} \tag{18}$$

where

α = percent of critical coupling
n = primary-tertiary turns ratio
Q_{LS} = loaded Q of secondary
Q_{UP} = unloaded Q of primary

This equation provides a basis for the specification of n. Manipulation yields:

$$n = \sqrt{\left[\left(\frac{S}{\alpha T}\right)^2 - 1\right] \, 4 \left(\frac{Q_{UP}}{Q_{LS}}\right) \frac{L_P}{L_S}} \tag{19}$$

Operation with critical coupling ($\alpha = 1.0$) gives best sensitivity, but $\alpha = 0.5$ is a more practical value. Although AM rejection may be achieved with S/T = 1.0, circuit adjustments are usually less critical if a lower value such as S/T =

0.8 is used. Once the requirement that $S/T \leq 1.0$ is satisfied, the designer must select the remaining circuit components to achieve best AM rejection.

Amplitude variations in the ratio detector output may be regarded as consisting of two distinct components:

1. A balanced component of AM due to the cyclic impedance variations of the diodes.
2. A positive or negative unbalanced component in the detector output manifested by an apparent frequency shift in the detector.

The balanced component due to changing diode impedance may be reduced by proper choice of resistors R_a and R_b in Fig. 2. It is the series sum, rather than the individual values, of these resistors which affect the rejection. Loughlin[5] considers the case of 100% efficient diodes and shows that AM rejection is improved by artificially reducing the efficiency of the diodes. In essence, R_a and R_b allow adjustment of the effective diode efficiency to optimum value for the desired signal level. The total value of these resistors is generally between 10% and 20% of the sum of the two load resistors.

Positive or negative imbalance in the detector output is caused by either a change in the diode capacitance or insufficient stored energy in the tuned circuits or both. Inadequate bypassing by the diode load capacitance also creates a negative unbalance. The desire is to balance positive and negative effects. Three steps are used in minimizing unbalance in the detector:

1. The bypass capacitors may be optimized. It would seem that a large capacitor would be suitable for this application. Although these capacitors should exhibit a relatively low impedance at the center frequency, using too large a capacitor may eliminate the negative unbalance and leave the positive.
2. The stabilizing resistors, R_a and R_b, may be made unequal and adjusted so that a compensating unbalance is introduced, eliminating or reducing the net unbalance. Again, the sum of R_a and R_b is kept constant.
3. The tertiary series resistor, R_T, may be selected to limit peak currents through the diodes thereby influencing diode impedance variations and limiting the energy delivered from the tank circuit.

In practice, best AM rejection is achieved by operating the detector diodes at the highest feasible level. Loughlin[5] lists six adjustments which increase the operating level of a given ratio detector:

1. Increase primary-secondary coupling
2. Decrease the tertiary turns
3. Increase secondary Q_L

4. Decrease primary Q_L
5. Increase secondary inductance
6. Decrease primary inductance

This information serves as a basis for choosing the relation between primary and secondary inductance. Initial indication is that a voltage step-up is desirable. A practical assumption is that the secondary inductance is two or three times the primary inductance. A higher ratio could be chosen, but there is a practical limit beyond which a negative unbalance will be created and cannot be eliminated.

If one assumes that the secondary in Fig. 3 is uncoupled, the expression for load resistance may be derived:

$$R_L = \eta \left(\frac{Q_{US}Q_{LS}X_{LS}}{Q_{US} - Q_{LS}}\right) \qquad (20)$$

where

η = Diode rectification efficiency
Q_{US} = Unloaded, uncoupled Q of secondary
Q_{LS} = Loaded, uncoupled Q of secondary
$X_{LS} = X_{CS}$ = Reactance of secondary inductance at $f_c = 4.5$ MHz

AMPLIFIER DESIGN PROCEDURE

A design procedure and calculations for the previously described amplifier may now be presented:

Requirements: IF = 4.5 MHz
Bandwidth = 200 kHz
$IL_{(total)} = 11.4$ dB

1. Determine loaded Q of primary:

$$Q_L = \frac{4.5 \times 10^6}{200 \times 10^3} = 22.5 \qquad (14)$$

2. Divide insertion loss and calculate Q_U and "a":

$IL_{(total)} = 11.4$ dB
$IL_{(coil)} = 4.4$ dB = 2.75 (ratio)
$IL_{(mismatch)} = 7.0$ dB = 5.0 (ratio)

$$Q_U = \frac{Q_L \sqrt{IL}}{\sqrt{IL} - 1} = \frac{22.5\sqrt{2.75}}{\sqrt{2.75} - 1} = 56.5 \qquad (15)$$

$$10 \log_{10} \frac{(1+a)^2}{4a} = 7 \tag{16}$$

$$a = 17.95$$

3. Determine a-c load R_L:

$$R_L = \frac{R_{oep}}{a}$$

$$= \frac{75 \times 10^3}{17.95} = 4.17\ k\Omega \tag{21}$$

4. Determine tank impedance, $Q_L X_{LP}$:
   Assuming $R_S = 1\ k\Omega$ (for parasitic suppression)

$$Q_L X_{LP} = R_L - R_S = 3.17\ k\Omega \tag{22}$$

5. Determine tuning reactance, thus the capacitance of C_1:

$$X_{LP} = \frac{R_L - R_S}{Q_L} \tag{23}$$

$$= \frac{3.17(10^3)}{22.5} = 141\ \Omega$$

$$C_1 = \frac{1}{\omega X_{LP}} \tag{24}$$

$$= \frac{1}{6.28(4.5 \times 10^6)141} = 250\ pF$$

Choose $C_1 = 270$ pF

6. Assume L_P to L_S ratio, thus C_2: $L_S = 3L_p$

$$C_2 = 82\ pF$$

7. Determine n, primary-tertiary turns ratio:
   Assume $\alpha = 0.5$
   Assume $S/T = 0.8$

$$n = \sqrt{\left[\left(\frac{S}{\alpha T}\right)^2 - 1\right] 4 \left(\frac{Q_{UP}}{Q_{US}}\right) \frac{L_P}{L_S}} \tag{19}$$

$$= \sqrt{\left[\left(\frac{0.8}{0.5}\right)^2 - 1\right] 4 \left(\frac{56.5}{22.5}\right) \frac{1}{3}} = 2.28$$

8. Calculate R_L, assuming $\eta = 0.8$, $Q_{US} = 80$:

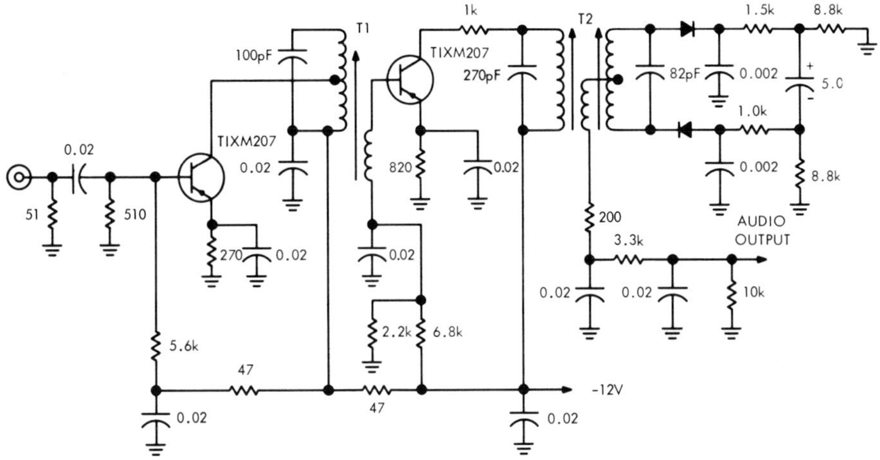

Fig. 4. 4.5-MHz Sound IF Amplifier

$$R_L = \eta \left(\frac{(Q_{US} Q_{LS} X_{LS})}{Q_{US} - Q_{LS}} \right) \tag{20}$$

$$= 0.8 \left(\frac{80(22.5)432}{80 - 22.5} \right) = 10.8 \text{ k}\Omega$$

9. Calculate C_5:

$$C_5 = \frac{0.033 \text{ seconds}}{2 \text{ R}_L}$$

$$= \frac{0.033\text{s}}{21.6 \text{ k}\Omega} = 1.53 \text{ }\mu\text{F}$$

Coil data for the ratio detector is:

Primary: 16 turns #36 Gripeze closewound.

Tertiary: 6 turns #36 Gripeze closewound over cold end of primary.

Secondary: 40 turns (total) #544 Litz bifilar wound.

$Q_{UP} = 54$

$Q_{US} = 72$

MEASURED PERFORMANCE

The circuit shown in Fig. 4 was designed with the considerations discussed in this chapter. The performance characteristics are as follows:

3-dB limiting at 1.1 mV

$V_{p\text{-}p(out)} = 80$ mV at 3-dB limiting

Peak separation of detector = 200 kHz

Linear operating range: greater than 50 kHz deviation at 3-dB limiting

AM rejection: 33 dB

REFERENCES

1. Parasitic Oscillations in IF Stages and Frequency Changers of AM Receivers, *Electronic Applications,* Vol. 20, No. 2, p. 41, Eindhoven, Netherlands, 1959-60.
2. Holmes, D. D. and T. O. Stanley: Stability Considerations in Transistor Intermediate Frequency Amplifiers, *Transistors I,* p. 405, RCA Laboratories, Princeton, N. J., 1956.
3. Seeley, S. W., and J. Avins: The Ratio Detector, *RCA Review 8.2,* p. 201, June 1947.
4. Seeley, S. W.: "Electron Tube Circuits," 2nd Edition, Chapter 19, pp. 624-634, McGraw-Hill Book Co., 1958.
5. Loughlin, B. D.: The Theory of Amplitude-Modulation Rejection in the Ratio Detector, *Proceedings of The IRE,* Vol. 40, No. 3, p. 289, March 1952.

18

Sync Separator

A sync separator removes the video information from the composite signal while retaining both horizontal and vertical sync pulses. Sync pulses must be free of any video or blanking information. Their shape must be maintained with the same rise times and width as is transmitted. A reasonable amount of noise immunity must be supplied either in the form of a noise inverter or noise cancelling circuit ahead of the sync separator stage.

D-C STABILITY

Either germanium or silicon small signal transistors can be used in the sync separator, but the lower leakage characteristics of silicon give it a definite advantage over germanium. Low leakage enhances d-c stability and minimizes design problems.

The sync separator normally conducts during sync pulse periods only. This makes the bias level extremely important because leakage currents can have considerable effect. A change in bias can cause portions of the video signal and blanking pulses to be present in the output; this results in poor sync stability and phase shift. The effect of leakage current on the stability factor S is shown by

$$S = \frac{I_C}{I_{CBO}} \tag{1}$$

where I_{CBO} is the collector-base leakage current. For an ideal case, the stability factor S would be 1. I_{CBO} should always be specified at a voltage level equal to or greater than the supply voltage E_{CC}.

The total collector current is given as

$$I_C = I_{CBO} - h_{FB}I_E \tag{2}$$

where h_{FB} is the grounded-base short-circuit current amplification and I_E is the emitter current.

Since

$$I_{CEO} = \frac{I_{CBO}}{1 + h_{FB}} \tag{3}$$

where I_{CEO} is the collector-to-emitter leakage, the stability factor for a common emitter amplifier is given as

$$S = \frac{\Delta I_{CEO}}{\Delta I_{CBO}} = \frac{1}{1 + h_{FB}} \tag{4}$$

This equation is valid because the leakage between the transistor elements is independent of the actual circuit configuration.

For a good transistor, the h_{FB} is approximately -0.98 giving a stability factor of 50. The I_{CBO} of a transistor can be considered to approximately double with each 10°C increase in ambient temperature above its rated value measured at 25°C. This means that at 55°C ambient, the change in collector current of an unstabilized transistor is

$$\Delta I_C \approx 8\ I_{CBO}\ S \approx 400\ I_{CBO} \tag{5}$$

The simplest and most common form of obtaining d-c stabilization is through the use of current feedback which may be obtained by placing a resistor in the emitter lead. (See Fig. 1.) This gives the following condition[1/]:

$$S = \frac{1}{1 + h_{FB} - h_{FB}\left(\dfrac{R_E}{R_E R_B}\right)} \tag{6}$$

It can be seen that increasing R_E and decreasing R_B greatly increases the stability of the stage. The use of an emitter resistor in a sync separator has its disadvantages:

1. It must be heavily bypassed to ensure preservation of sync output during the vertical interval.
2. Strong noise pulses appearing at the base cause a large amount of emitter current to flow, charging the bypass capacitor and reverse-biasing the emitter-base diode. The stage is then cut off until the charge

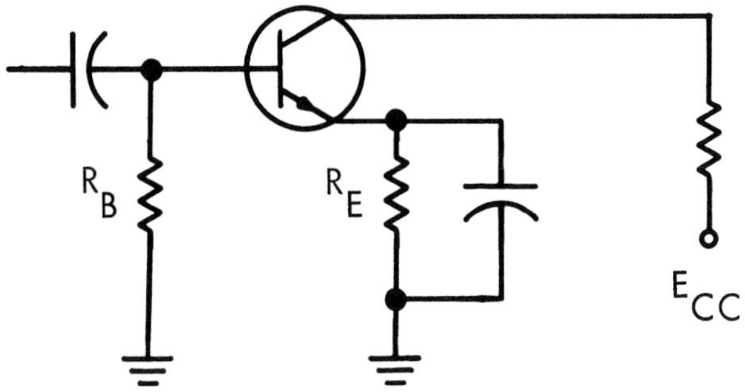

Fig. 1. Current Feedback for D-C Stabilization

drains off and with the large RC time constant required, a momentary loss of sync occurs. If the noise pulses have a repetitive rate as fast as the vertical sync rate, a complete loss of sync can occur.

3. To leave it unbypassed requires an impractical amount of drive signal.

BIAS AND DRIVE METHODS

The use of a double time constant in the emitter is shown in Fig. 2 where C_1R_1 supplies the time constant for horizontal separation and C_2R_2 supplies the vertical time constant. Noise pulses charge C_1 and C_2. The charge on C_2 reverse-biases the diode which blocks this bias from the emitter. C_1 discharges rapidly and allows the transistor to be biased normally by the following sync pulses.

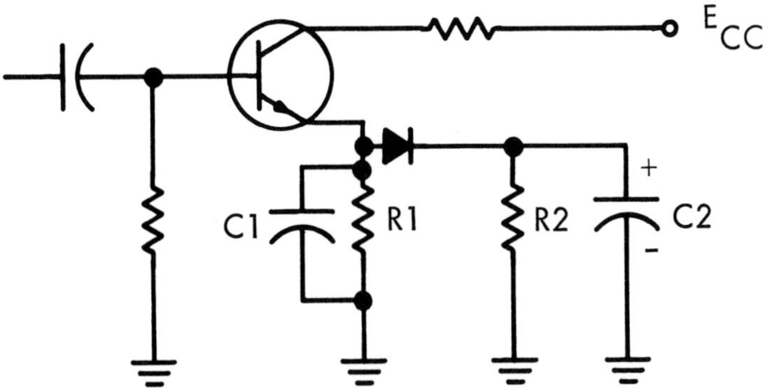

Fig. 2. Double Time Constant in the Emitter

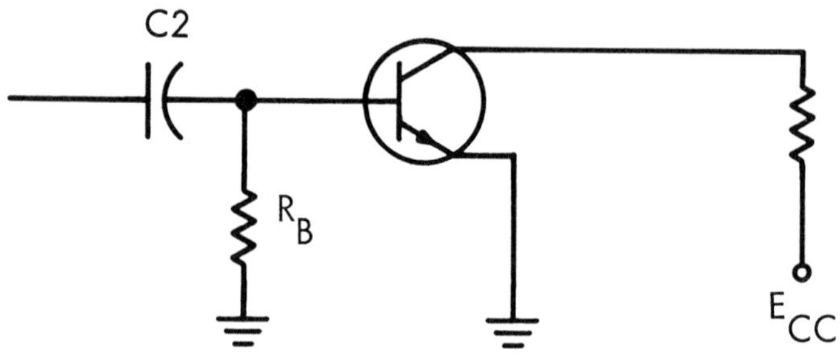

Fig. 3. Grounded Emitter with Base Resistor Returned to Emitter Supply

Figure 3 shows a grounded-emitter sync separator with the base resistor returned to the emitter supply. The transistor used in this circuit must have a very low I_{CBO}. Also, the minimum base current required to drive the transistor into saturation must be much greater than the maximum I_{CBO} encountered at the highest ambient temperature. This is necessary to minimize bias changes due to leakage currents. When the base resistor R_B is returned to ground, the forward bias is dependent upon the drive signal and its effect on the charging current of the coupling capacitor. As has been stated, collector current must flow only during the sync pulse period. At this time, the transistor must be driven into saturation. The common-emitter mode is used for its good current and voltage gain characteristics with its poorer rise time characteristic being offset by selecting a transistor having a reasonably high f_T (5.0 MHz or higher).

The time constant of R_BC_2 must be long enough to ensure base current flow for the duration of the vertical sync interval. Current during this interval must be fairly constant for a minimum amount of tilt and of sufficient amplitude to hold the collector in saturation. Correspondingly, the time constant must not be so long as to allow prolonged biasing by noise pulses. For horizontal noise immunity, this time constant is normally too long if it satisfies the vertical sync separation requirements. In actual practice, the RC time constant varies somewhat depending on individual preference.

A double time constant in the base, similar to that used in tube circuitry, can be used in transistor circuits. (See Fig. 4.) R_1 and C_1 form a time constant which is approximately equal to the horizontal line period. At the vertical frequency, this appears as a series impedance which may represent a large drop in signal voltage to the base during the vertical interval. This effect must be held to a minimum by increasing the value of C_2. A compromise in values used is necessary because increasing C_2 decreases horizontal noise immunity.

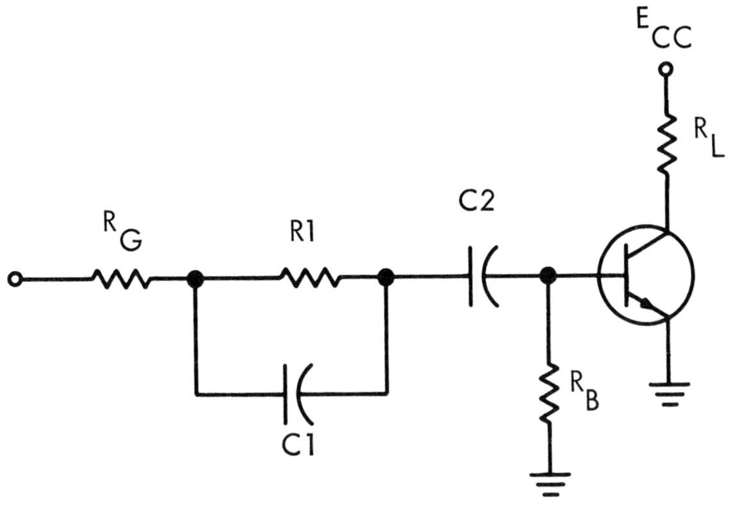

Fig. 4. Double Time Constant in the Base

In addition, to ensure consistant sync separation, R_B and C_2 must satisfy certain requirements. R_B must be compatible with the source impedance R_G for both horizontal and vertical sync pulses. A minimum value for C_2 is required for two reasons. First, it ensures the base returning to a normalized voltage level E_N following both horizontal and vertical sync pulses. Second, C_2 must be large enough to maintain a minimum amount of tilt during the vertical interval of the composite video signal. This will ensure collector saturation during the entire vertical interval.

For minimum tilt during the vertical interval, the following equation has been derived:[2]

$$0.95 = \exp\left(\frac{-t_3}{R_G C_2}\right) \tag{7}$$

where 0.95 is the desired percentage of E_N to which the base voltage is restored from the start of the vertical interval to the end. This allows for a maximum tilt of 5% during the vertical interval. The vertical sync pulse period t_3 is about 190 μs.

The aforementioned normalized voltage E_N is the level to which the sync separator base returns during video portions of the signal. The level for E_N with reference to the composite video signal is shown in Fig. 5. To ensure conduction during sync pulses only, E_N, is set at approximately 80% of the total peak-to-peak video signal E_C. This places it above the blanking level so

Fig. 5. Level of E_N with Reference to Composite Video Signal

that conduction does not occur during the blanking period. Conduction during this time results in a wider output pulse causing a phase shift in the horizontal sync. Basically, the normalized voltage E_N is that level at which the transistor is cut off. Signal voltages above this level turn the transistor on. It is defined as

$$E_N = E_C - E_V \left(\frac{1}{1 + \frac{R_B t_1}{R_G t_2}} \right) \tag{8}$$

where E_V is the average video level from sync tips in percent of total video signal. t_1 is the sync pulse period of 5 μs and t_2 is the remaining period for one horizontal line of approximately 58.5 μs.

Since the source impedance R_G is always given, R_B can be found by rewriting Eq. (8).

$$R_B = \frac{R_G t_2 (E_C - E_V - E_N)}{t_1 (E_N - E_C)} \tag{9}$$

As has been mentioned, E_N is 80% of E_C. E_V is set at 70% of E_C although the actual value is dependent upon the video portion which is complex and varies somewhat. The value of E_C is dependent upon its source and should represent the minimum value of composite video signal with which the sync separator will be required to operate.

From Eq. (7), the following is derived:

$$\frac{t_3}{R_G C_2} = 0.05 \tag{10}$$

and solving for C_2 gives

$$C_2 = \frac{20\ t_3}{R_G} \tag{11}$$

This then is the minimum value of C_2 which can be used with a given source impedance. The charging current I_C of C_2 is derived from the voltage of the applied composite video signal. Its magnitude is dependent upon the source impedance R_G, base resistance R_B, and emitter-base impedance of the transistor R_D. A portion of this charging current also represents the base current I_B which will turn the transistor on. This current is derived by the voltage difference between E_C and E_N. This is shown by

$$I_C = \frac{E_C - E_N}{R_G + \dfrac{R_B R_D}{R_B + R_D}} \tag{12}$$

As previously stated, the level of E_N is the value where the transistor is just at cutoff. In a grounded-emitter stage this is about 0.2 volt for germanium and 0.5 volt for silicon. During the voltage level $(E_C - E_N)$, the emitter-base diode is forward-biased and R_D is negligible compared to R_B. At this time the base current will be

$$I_B = \frac{E_C - E_N}{R_G} \tag{13}$$

Using this equation, the minimum base current can be determined. It can be seen that a low source impedance is desirable. A high source impedance can be achieved with a high value of drive signal E_C but this raises the emitter-base reverse breakdown voltage limit since the video portion of the signal reverse biases the emitter-base junction. Another reason a high source impedance is not desirable can be seen in Eq. (8). The base resistance R_B and the source impedance R_G are directly related and any increase in R_G necessitates an increase in R_B which, in conjunction with the transistor input capacitance, increases the rise time of the output sync pulses.

OUTPUT CHARACTERISTICS

Since the transistor is driven from cutoff to saturation during the sync pulse period, the peak voltage pulse across the collector load is

$$E_P = E_{CC} - V_{CE(sat)} \tag{14}$$

with $V_{CE(sat)}$ equal to the voltage drop across the transistor during conduction.

The peak collector current will be

$$I_P = \frac{E_P}{R_L} \tag{15}$$

The load resistance R_L is selected to match the input impedances of the vertical and horizontal oscillator circuits and may be divided into sections accordingly as shown in Fig. 6.

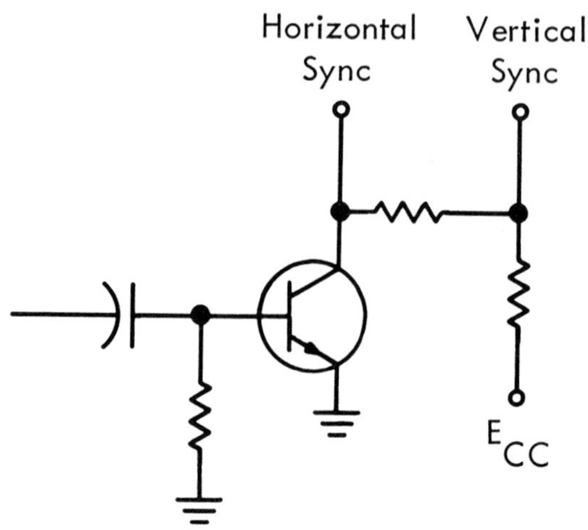

Fig. 6. Division of R_L for Horizontal and Vertical Oscillator Circuits

Minimum beta requirements of the sync separator transistor can be found by

$$h_{FE(min)} = \frac{I_P}{I_{B(min)}} \tag{16}$$

The output of the sync separator is applied to the horizontal AFC circuit and a vertical integrating network. The integrating network removes the horizontal sync pulses from the vertical pulses. This is required to obtain good vertical sync and interlaced scanning. Figure 7 shows an integrating circuit consisting of two sections of an RC low-pass filter. The time constant of the network is given as

$$t = R_1C_1 + R_2C_2 \tag{17}$$

An integrating network does not eliminate the d-c component of the sync signal but it reduces the a-c component of the horizontal sync pulse to a value dependent upon the time constant of the circuit. The time constant must be long enough to maintain maximum output during the vertical sync period. Sidney Deutsch[3/] has shown that a two-section integrating network having a time constant of 40 μs reduces the horizontal pulse to less than 4% of its input while maintaining the vertical pulse output to about 90% of its level. Increasing the time constant further reduces the horizontal pulse output if necessary, but this also reduces the vertical output. The only major problem

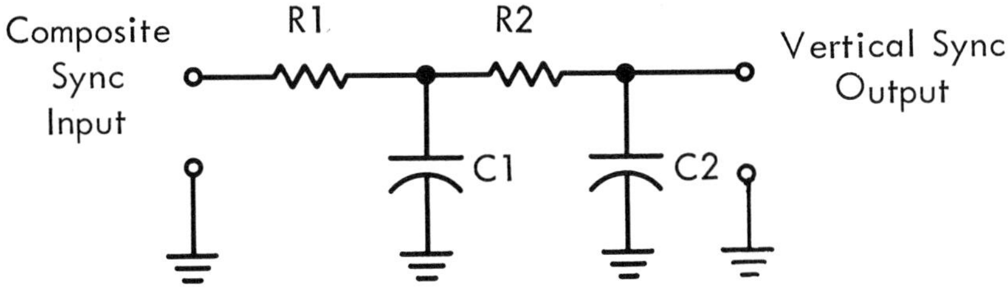

Fig. 7. Integrating Circuit

in designing the integrating circuit is the low impedances encountered in transistor circuitry. The output impedance of the sync separator can be met with little difficulty but the transistor vertical oscillator circuits may have very low input impedances. Under these circumstances, it is practical to use an emitter-follower between the integrating circuit and the oscillator as shown in Fig. 8.

CIRCUIT EXAMPLE

Figure 9 shows a complete sync separator and integrating network for use in a set with a supply voltage of 65 volts. The source impedance is one kilohm with a video signal of two volts peak. From Eq. (9), R_B is found to be

$$R_B = \frac{10^3(58.5 \times 10^{-6})(2 - 1.4 - 1.6)}{5 \times 10^{-6}(1.6 - 2)} = 29.25 \text{ k}\Omega$$

C_2 is computed using Eq. (11).

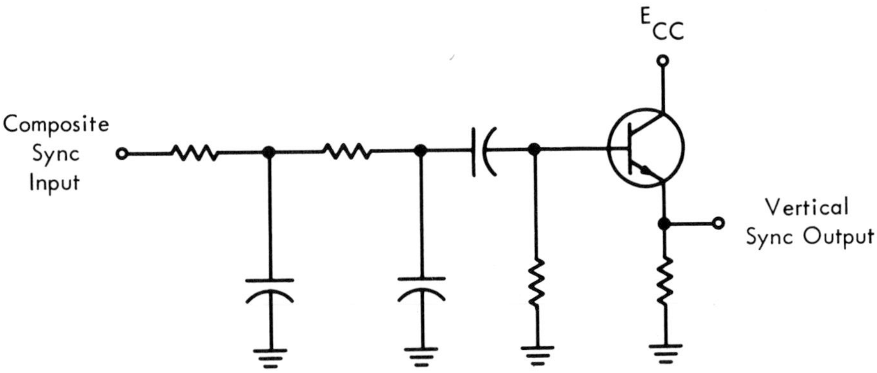

Fig. 8. Emitter Follower between Integrating and Oscillator Circuits

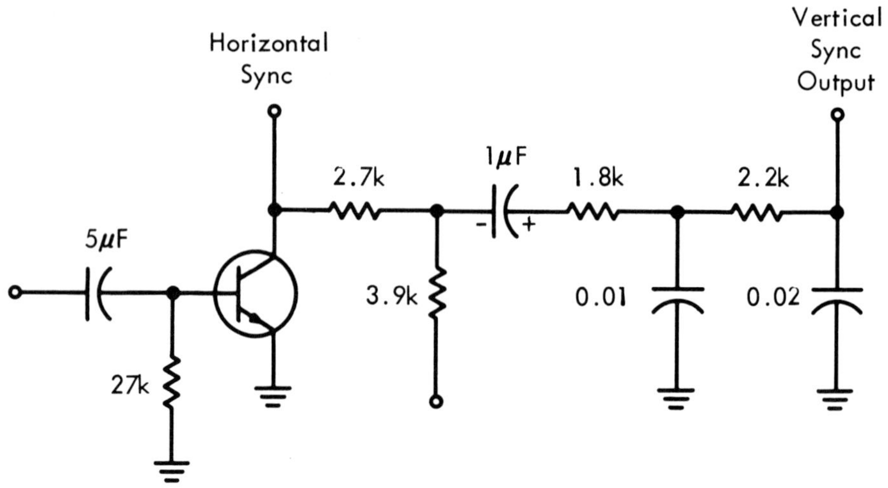

Fig. 9. Sync Separator and Integrating Network

$$C_{2(min)} = \frac{20(190 \times 10^{-6})}{10^3} = 3.8\ \mu F$$

Actual values used in the circuit are 27 kΩ and 5 μF.

Minimum base current is found using Eq. (13).

$$I_{B(min)} = \frac{2 - 1.6}{10^3} = 0.4 \text{ mA}$$

A 2N3708 type transistor with a $V_{CE(sat)}$ of 1 volt maximum, an I_{CBO} of 0.1 μA maximum and an $h_{FE(min)}$ of 30 is chosen. For a 65-volt supply, the peak collector voltage output will be 64 volts. A minimum h_{FE} of 30 will yield a peak collector current of

$$I_P = 30(0.4 \times 10^{-3}) = 12 \text{ mA}$$

The collector load R_L then becomes

$$R_L = \frac{64}{12 \times 10^{-3}} = 5.3 \text{ k}\Omega$$

In Fig. 9, R_1 consists of two series resistors in parallel with the horizontal AFC and vertical integrating network.

For the vertical integrator network, a 1.8-kΩ resistor and a 0.01-μF capacitor are used in the first section followed by a 2.2-kΩ resistor and a 0.02-μF capacitor in the second section.

REFERENCES

1. Riddle and Ristenblatt: "Transistor Physics and Circuits."
2. Heiser, W.: Sync Separator Analysis, *Electronics*, July, 1950.
3. Deutsch, Sidney: "Theory and Design of Television Receivers."

19

Vertical Oscillator and Sweep Output

The vertical oscillator can be either a multivibrator or a blocking oscillator. The multivibrator eliminates the use of a transformer and thereby eliminates the possibility of poor interlace due to radiating horizontal pulses being inductively coupled into the vertical oscillator. There is some interaction between vertical hold, size, and linearity controls which must be kept to a minimum.

Figure 1 shows a multivibrator type circuit. Q_1 and Q_3 are the two stages of the multivibrator with Q_2 supplying an impedance match between the stages. Q_2 could be eliminated although its use greatly improves linearity and overall operation of Q_3 which operates as a linear amplifier supplying the required current for the deflection yoke. The basic multivibrator design is described under the horizontal oscillator in Chapter 9. While there are some differences, due to the use in the vertical oscillator of the output stage as a linear amplifier, which requires a ramp voltage drive, the multivibrator design for the horizontal oscillator is basically the same.

BLOCKING OSCILLATOR

The blocking oscillator can be designed as a common-base, common-collector, or common-emitter circuit. The common-emitter configuration is shown in Fig. 2. The greatest advantage of a blocking oscillator is its power capabilities. The design centers around the transformer and the base-bias time constant. The transformer is not too critical, the main consideration being the conduction time t_{ON} of the transistor. If a transistor with a reasonably high current gain is used, then t_{ON} is given as[1/]

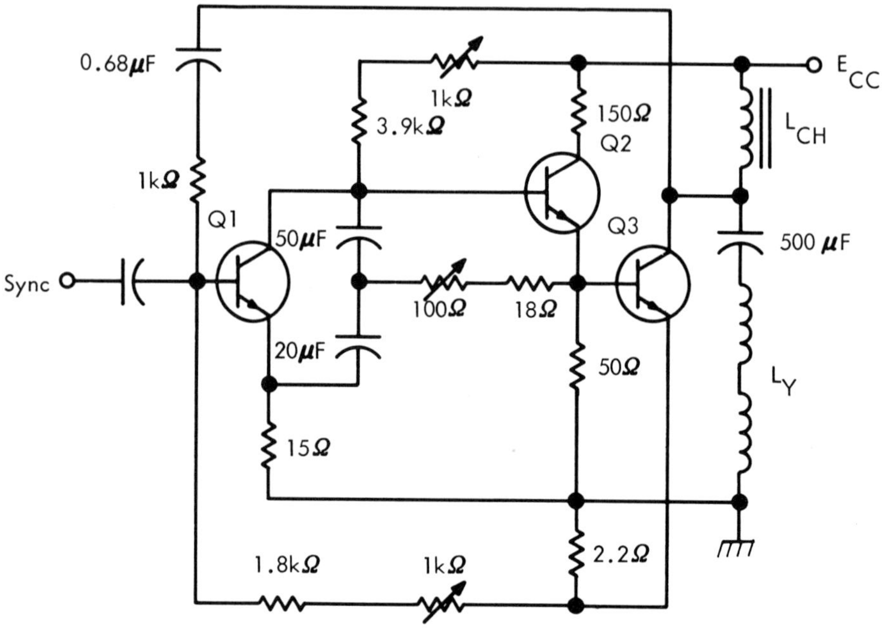

Fig. 1. Multivibrator Vertical Sweep Circuit

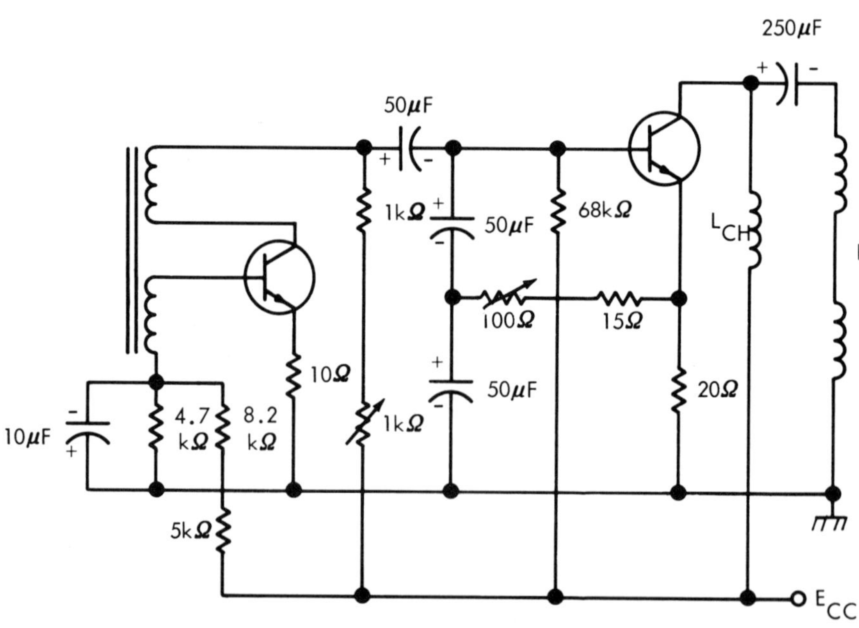

Fig. 2. Blocking Oscillator Vertical Sweep Circuit

$$t_{ON} \approx \frac{L}{r_1 + r_2} \ln\left(1 + \frac{r_1 + r_2}{N\, r_2}\right) \tag{1}$$

where r_1 is the d-c resistance of the transformer plus the R_{CS} of the transistor; r_2 is the emitter saturation resistance r_e plus the external emitter resistance R_E. N is the turns ratio between primary and secondary and is generally about 5:1 and L is the magnetizing inductance. The inductance L is

$$L = \frac{t_{ON}(r_1 + r_2)}{\ln\left[1 + \left(\frac{r_1 + r_2}{N\, r_2}\right)\right]} \tag{2}$$

To ensure good interlace, the conduction time of the transistor should not be longer than three horizontal lines or 190 μs. The resistance of the transformer is inversely proportional to the conduction time t_{ON}. As the transformer resistance is increased E_{CC} must increase accordingly to maintain the required power output. Hence, in low-voltage supply circuits, it is desirable to keep both the inductance and the d-c resistance of the transformer as low as possible.

The tranformer selected for the oscillator must be capable of delivering a peak current during t_{ON}, which will develop a large voltage pulse across the load resistor. This pulse is integrated to supply the drive signal for the vertical output stage.

When the transistor has been turned off by the normal blocking oscillator action, the off time is determined by the RC time constant in the base. For stability and transistor interchangeability, the base circuit impedance should be low. At low supply voltages, the capacitor must be an electrolytic type and, considering their wide tolerances, this poses a production problem. As the supply voltage is increased and associated impedances increase, this problem can be eliminated with the use of paper or ceramic capacitors.

TRANSISTOR REQUIREMENTS

The transistor collector-base and emitter-base breakdown voltages must not be exceeded during the off time. If the voltages developed in the circuit exceed their rated values, clamping diodes must be used to limit the voltages across the transformer windings to a safe value.

VERTICAL OUTPUT DRIVER

The collector of the oscillator is a-c coupled to either the base of the vertical output stage or a driver stage. An RC network with a time constant greater than the vertical sweep period generates a sawtooth voltage for driving the output stage. Circuit impedances dictate the relative values of R and C. Again, in low-voltage circuits, electrolytic capacitors must be used and for production quantities their tolerances must be narrowed down. The coupling capacitor must present a low impedance to the 60-Hz pulse and here again an electrolytic capacitor is used.

The use of a driver stage is required mainly for impedance matching between the oscillator and output. This improves linearity and reduces the required amount of feedback from the output stage. An emitter-follower is used for this purpose since it allows direct coupling between the driver and output stage as shown in Fig. 1. By making the emitter resistor R_E (which is also the base resistor for the output stage) low, the breakdown voltage requirements of the output stage can be reduced. However, the lower the resistance of R_E, the higher will be the required dissipation of the driver stage since the required sawtooth voltage drive for the output does not decrease with the decrease in base resistance.

VERTICAL OUTPUT CONSIDERATIONS

The output stage must supply a reasonably linear sawtooth of current to the deflection yoke, the amplitude of which is dependent upon the CRT and the deflection yoke inductance. The vertical deflection angle is smaller than that of the horizontal and the frequency is much lower, thus power requirements are much less for vertical deflection.

During the scanning period, the deflection yoke represents an almost purely resistive load thus requiring the use of a Class A amplifier for the vertical output stage. The actual load line is not linear over the complete cycle due to the load appearing inductive during retrace time. Although a very low frequency response is required to ensure a linear scan, an amplifier designed to have a low-frequency response flat to less than five Hz is not practical. Transformer coupling to the deflection yoke would require a large expensive transformer due to the low supply voltages used in transistor circuits. A more practical method is to design the amplifier with excessive gain and utilize feedback to obtain the desired linearity. Choke coupling as shown in Fig. 1 and 2 is more commonly used. The choke should be designed to have maximum inductance and minimum d-c resistance. A high a-c impedance reduces the power requirements of the transistor while the low d-c resistance allows the maximum voltage swing available across the yoke. The design of the choke is usually limited by physical size and cost; still, its inductance must be much higher than that of the yoke.

The slope of the sawtooth voltage applied to the base of the amplifier must be in the forward biasing direction of the transistor; i.e., for a PNP transistor, a negative going slope and for an NPN transistor, the slope must be positive. A sawtooth current flows in the yoke, the peak value being dependent upon the d-c resistance of the yoke and the voltage across it. At the end of scan, the drive voltage reverses direction and the collector current falls rapidly. During retrace time, the load appears inductive due to the shorter fall time. As the collector moves toward a cutoff condition, the yoke and its stray capacitance rings for one-half cycle of its self-resonant frequency (reflected and stray capacities of the circuit also form part of the total capacity). Half way through this cycle, the voltage reaches a peak value (vertical retrace pulse). Since the stage is operating Class A, it never completely cuts off. This supplies some damping and limits the value of the peak voltage during retrace time. Since it also increases retrace time, conduction during retrace should be minimized but the stage must never cut off as this will cause poor linearity at the start of scan.

The values of yoke inductance and resistance are dependent upon the current and voltage limitations of presently available transistors. Reasonably high current gain (beta), which must be constant from a low value of collector current to the highest peak value, is required by the circuit. This characteristic refers to beta linearity and generally should be 10% or better. Because the amplifier is operating Class A, its voltage breakdown is rated as BV_{CER} which is collector-to-emitter voltage with a resistance from base to emitter as specified by the designer.

For best power gain, it would be desirable to use a high-impedance yoke and a high supply voltage but, during retrace time, the peak voltage pulse would require a very high BV_{CER}. On the other hand, a low-impedance yoke would result in lower power gain which would have to be offset by running higher peak collector currents. This requires beta linearity over a much wider range of currents which, to date, is not easily obtained. Poor beta linearity contributes to poor vertical linearity and requires more feedback to correct for it. Presently, yokes having from 30 to 70 ohms d-c resistance and 35 to 75 mH inductance have been used with good results for CRT's rated at up to 20 kV with a deflection angle of 114°. The required peak-to-peak value of current for deflection can be found by the same formula as is used in Chapter 10 for horizontal deflection.

$$I_Y \approx 0.053 \sqrt{\frac{E_{kV} D_A F_\beta}{L_Y}} \tag{3}$$

where E_{KV} is high voltage, D_A is yoke shell diameter, L_Y is yoke inductance and F_β is the function of the deflection angle as given by

$$F_\beta = \left(\cos\frac{\beta}{4}\right)\left(\sin\frac{\beta}{4}\right)^3 \tag{4}$$

Peak collector voltage during retrace time is

$$E_P \approx E_{CC} + \frac{LI_Y}{t_R} \tag{5}$$

where E_{CC} is the supply voltage, and t_R is the retrace time. The required amount of supply voltage E_{CC} is determined by

$$E_{CC} = I_{AV}R_{CH} + I_{AV}R_E + I_YR_Y \tag{6}$$

where I_{AV} is the average collector current, R_{CH} is the d-c resistance of the choke, R_E is the emitter resistance and R_Y is the d-c resistance of the deflection yoke.

The d-c stability of the amplifier is important; a shift in operating point with temperature will cause a noticeable change in picture size or linearity. The addition of an emitter resistor will supply current feedback and serve to stabilize the amplifier. The amount of resistance permissible in the emitter is dependent upon the available power and the required power for the yoke and transistor. Maximum available power is given by[2/]

$$P_{(max)} = \frac{E_{CC}^2}{4(R_E + R_{CH})} = \frac{E_{CC}^2}{4R_{d\text{-}c}} \tag{7}$$

The yoke power requirements are

$$P_Y = \frac{I_Y^2R_Y}{12} = \frac{E_{CC}^2}{4R_{d\text{-}c}} \tag{8}$$

A-c coupling is used to the deflection yoke to minimize centering problems and in this case, the transistor requires twice the deflection yoke power, thus transistor power requirement is

$$P_T = \frac{I_Y^2R_Y}{6} \tag{9}$$

By equating the available power to the sum of the required power of the transistor and yoke, the maximum amount of d-c resistance which can be tolerated with a given supply voltage E_{CC} can be found.

$$\frac{E_{CC}^2}{4R_{d\text{-}c(max)}} = \frac{I_Y^2R_Y}{4} \tag{10}$$

$R_{d\text{-}c(max)}$ is defined as

$$R_{d\text{-}c(max)} = \frac{E_{CC}^2}{I_Y^2 R_Y} \tag{11}$$

The value of this resistance is the maximum allowable d-c resistance which can be employed in the circuit and is the sum of the emitter resistor and the d-c resistance of the choke. For best stability it is desirable to make the emitter resistor large but this increases the drive requirements. Increasing the d-c resistance of the choke reduces the voltage across the transistor and yoke. With a given yoke resistance, this voltage must be maintained at that level which supplies the required sweep output. As higher supply voltages are used, circuit impedance and $R_{d\text{-}c(max)}$ increases, then circuit stability problems are minimized.

To ensure thermal stability, the maximum specified junction temperature of the device must not be exceeded. The vertical output device must be mounted on a heat sink to meet this requirement. To ensure sufficient heat sink area, the following formula is used:

$$P_T = \frac{T_J - T_A}{\theta_{J\text{-}C} + \theta_{C\text{-}HS} + \theta_{HS\text{-}A}} \tag{12}$$

where P_T is the average power of the transistor, T_J is the average temperature at which the junction will stabilize, T_A is the maximum ambient temperature to be encountered, $\theta_{J\text{-}C}$ is the junction-to-case thermal resistance, $\theta_{C\text{-}HS}$ is the case-to-heat-sink thermal resistance and may be specified with or without the use of a silicon grease to improve thermal contact, and $\theta_{HS\text{-}A}$ is the heat-sink-to-ambient thermal resistance. Inclusion of a safety factor and allowance for adverse conditions such as oscillator off-frequency operation and high line voltage (for unregulated supplies), should be 20% to 30% below the maximum junction temperature given for the transistor used.

DESIGN EXAMPLE

In the circuit of Fig. 2, the blocking oscillator transformer has a turns ratio of 4:1. Maximum d-c resistance of the transformer is 25 ohms and t_{ON} is 200 μs. An emitter resistance of 10 ohms is used making r_e negligible. R_{CS} of the transistor used is also small compared to the transformer resistance. Using Eq. (2), the inductance is given as

$$L = \frac{200 \times 10^{-6}(25 + 10)}{\ln\left(1 + \frac{35}{50}\right)} = 13 \text{ mH}$$

The charging capacitor for the sawtooth drive voltage is split using two 50-μF capacitors. Feedback for linearity adjustment is connected to the junction of these capacitors from the emitter of the amplifier.

The required yoke current for a CRT having a deflection angle of 114° with a high voltage of 18 kV is found by using Eq. (3). The deflection yoke has an inductance of 75 mH and a d-c resistance of 70 ohms; thus

$$I_Y \approx 0.053 \sqrt{\frac{18(5.3)(0.08)}{75 \times 10^{-3}}} \approx 0.5 \text{ A}$$

This is the peak current that the transistor must supply with a beta linearity of 10% or less. For a Class A amplifier, the average current is approximately one half the peak current. When operating as a vertical output stage, added power requirements for feedback (for linearity correction) and circuit losses increase the average current to 55% or 60%. Thus, for computing the supply voltage E_{CC} an average current of 300 mA is used. The choke d-c resistance is 25 ohms and the emitter resistance is 20 ohms. From Eq. (6) the supply voltage is

$$E_{CC} = 0.3(25) + 0.3(20) + 0.5(70) = 48.5 \text{ V}$$

From Eq. (5) the peak voltage during retrace is

$$E_P \approx 48.5 + \frac{75 \times 10^{-3}(0.5)}{0.4 \times 10^{-3}} \approx 143 \text{ V}$$

The total d-c resistance of the emitter resistor and choke resistance is well under the maximum value allowable for the circuit as derived by Eq. (11).

$$R_{d\text{-}c(max)} = \frac{48.5^2}{0.5^2(70)} = 135 \ \Omega$$

The vertical output transistor is a selected silicon transistor with a BV_{CER} ($R_{BE} = 2.2$ kΩ) of 180 V and a minimum h_{FE} of 40 with good beta linearity up to 750 mA. The TO-5 package is pressed in a stud heat sink to simplify mounting to the metal plate serving as a heat sink.

From Eq. (9) the transistor power dissipation (neglecting base drive which is small compared to the collector power) is

$$P_T = \frac{0.5^2(70)}{6} \approx 3 \text{ W}$$

Maximum junction temperature of the device is 100° C. Junction-to-case thermal resistance $\theta_{J\text{-}C}$ is 35° C/W. Case-to-heat-sink thermal impedance

$\theta_{C\text{-}HS}$ is 0.7, and maximum ambient temperature is 50°C. Using a junction temperature of 85°C and rewriting Eq. (12), the heat-sink-to-ambient thermal impedance can be found:

$$\theta_{HS\text{-}A} = \left(\frac{85 - 50}{3}\right) - 3.5 - 0.7 = 7.5°C/W$$

Tables of heat sink types and their thermal resistance to air can be found in many transistor data sheets. These tables show than an aluminum heat sink three inches square and ⅛ inch thick will meet the 7.5°C/W requirement.

With transistors becoming available with higher voltage breakdown ratings, the use of toroid vertical yokes is desirable. These are being used almost exclusively in present-day tube sets. The toroid coil has less power lost in winding resistance for a given magnetic field due to shorter length with little wasted wire at the ends. The time constant of a saddle yoke is fairly constant at about one ms while the toroid coil offers an increase to about two ms. This offers greater sensitivity in the toroid. The only major problem in using a toroid vertical yoke is the higher retrace pulse which occurs due to higher inductance of the toroid at the same d-c resistance. Toroid yokes are used in many small-screen transistor sets now and will soon be seen in larger screen sets. The basic operation of the circuit will not change due to this change in yoke design.

REFERENCES

1. "Transistor Circuit Design," Texas Instruments Incorporated, McGraw-Hill Book Company, New York, 1963.
2. Hellstrom, M. J.: Design of Transistor Vertical Deflection Output and Driver Stages, *IRE Transactions on Broadcast and TV Receivers.*

20

Horizontal AFC and Oscillator

The use of a differentiating circuit to separate the horizontal pulses from the vertical pulses is not desirable because of poor noise immunity and lack of phase control. For this reason automatic frequency control circuits are used exclusively to control the horizontal oscillator. Composite sync is fed directly to the AFC circuit.

AUTOMATIC FREQUENCY CONTROL

AFC circuits compare the frequency of the oscillator with the frequency of the horizontal sync pulse. When these frequencies are the same and in phase, the d-c output voltage will be that corresponding to oscillator operation at 15.75 kHz. This voltage may be either positive or negative depending upon the oscillator circuit. If the frequencies differ, a change in voltage output occurs which corrects for the change in frequency. There are two basic types of AFC circuits with slight variations in each. One is a push-pull input type as shown in Fig. 1 where two sync pulse signals are supplied from a phase splitter. The sync pulses are 180° out of phase with each other and generally are equal in amplitude. The other type is a single-ended input type as shown in Fig. 2. This type requires one sync pulse input which may be either positive or negative with the polarity of the diodes being connected accordingly.

The performance of an AFC system is largely dependent upon the following parameters: noise bandwidth, loop gain, damping factor, and cutoff frequency. The merit of its performance is specified by horizontal oscillator pull-in and hold-in characteristics and its stability during weak signal and strong noise conditions.

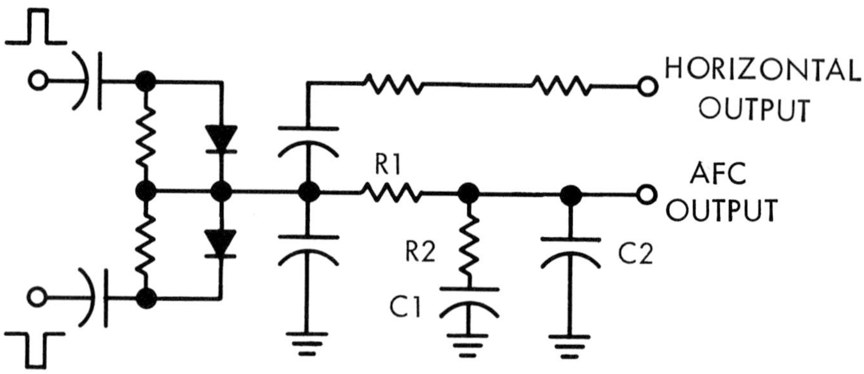

Fig. 1. AFC Circuit with Push-pull Input

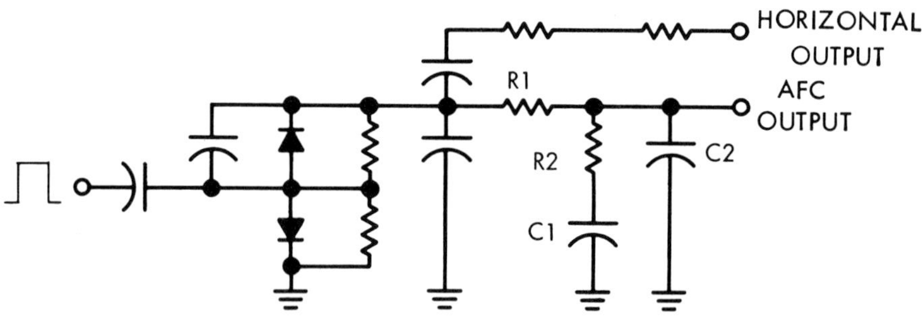

Fig. 2. AFC Circuit with Single-ended Input

The reference voltage is a sawtooth voltage obtained by integrating the horizontal retrace pulse. The integrating network supplies a low-impedance point for the reference voltage. The sync pulse is fed into a relatively high-impedance point. Sync pulse amplitude should be greater than the sawtooth voltage.

AFC diodes should be matched and the parallel resistors equal to maintain a balanced system. Following the diode section is an RC filter system. It is here that the performance of the AFC system is largely controlled.

If the noise bandwidth B, the pull-in frequency range f_P, and the damping factor d are given, the filter system can be designed. The damping factor prohibits a sudden change in oscillator frequency with fast changes of voltage output from the AFC system. Insufficient damping causes the oscillator to "hunt." This appears as a "weaving" in the picture. Over-damping reduces the effectiveness of the AFC.

The noise bandwidth B, loop gain K, and pull in frequency f_P are all interrelated. For a wide pull-in range, K should be large. However, for noise immunity, the noise bandwidth B must be narrow. This calls for a low value of loop gain. The final performance requirement is largely an engineering choice. The damping factor is given as

$$d = \frac{T_1 \omega_N}{2} + \frac{\omega_N}{2K} \tag{1}$$

where T_1 is the time constant of R_2C_1 (Fig. 1 and 2) and K is the AFC loop gain. The resonant frequency of the filter ω_N is given as

$$\omega_N = 2\pi f_C \tag{2}$$

with f_C being the cutoff frequency of the RC filter network.

The noise bandwidth B has two values depending upon the value of $\frac{\omega_N}{K}$.

These are defined by W. J. Gruen[1] and given as

$$B \left| \frac{\omega_N}{K} \rightarrow 1 = \pi\omega_N = \pi K \text{ (rad/s)} \right. \tag{3}$$

and

$$B \left| \frac{\omega_N}{K} \rightarrow 0 = 2\pi\omega_N \text{ (rad/s)} \right. \tag{4}$$

These hold true for a damping ractor of 0.5 and under this condition the pull-in frequency f_P becomes

$$f_P \approx \sqrt{f_C \frac{K}{\pi}} \tag{5}$$

When the reference voltage for the AFC is a sawtooth wave form derived by integrating a portion of the horizontal retrace pulse, the loop gain K becomes

$$K \approx 2E_{(saw)}S_{(osc)} \tag{6}$$

where $E_{(saw)}$ is the peak voltage of the sawtooth and $S_{(osc)}$ is the sensitivity of the oscillator in hertz per volt. The time constants T_1 and T_2 are given as

$$T_1 = R_2C_1 \tag{7}$$

and

$$T_2 = (R_1 + R_2)\,C_1 \tag{8}$$

T_2 is found by the equation

$$T_2 = \frac{K}{\omega_N{}^2} \tag{9}$$

The circuit shown in Fig. 2 has the following given parameters: noise bandwidth 600 Hz, pull-in frequency ±200 Hz, damping factor 0.5, and oscillator sensitivity of 600 Hz/V. From Eq. (4) the desired resonant frequency is found.

$$f_R = \frac{2\pi(600)}{2\pi} = 600 \text{ Hz}$$

The cutoff frequency is found from Eq. (2).

$$f_C = \frac{600}{2\pi} = 95 \text{ Hz}$$

The required loop gain K which will meet the pull-in requirements is then computed by solving Eq. (5) for K.

$$K = \frac{400^2\ \pi}{95} = 5300 \text{ rad/s}$$

Rewriting Eq. (1), solve for the time constant T_1.

$$T_1 = \frac{2(5300)\,(0.5) - 600}{5300(600)} = 1.48 \text{ ms}$$

From Eq. (9) the time constant T_2 is

$$T_2 = \frac{5300}{600^2} = 14.7 \text{ ms}$$

Values for R_1 and R_2 are found by the equations

$$T_1 = R_2C_1 \qquad R_2 = 1.5 \text{ k}\,\Omega$$

and

$$T_2 = (R_1 + R_2)\,C_1 \qquad R_1 = 12 \text{ k}\,\Omega$$

where C_1 is chosen as one μF.

Their actual values are selected on a basis of impedance matching. C_2 is added to improve the noise bandwidth which in the actual circuit is too wide. Pull-in and hold-in are both good being in excess of ±200 Hz and ±300 Hz, respectively.

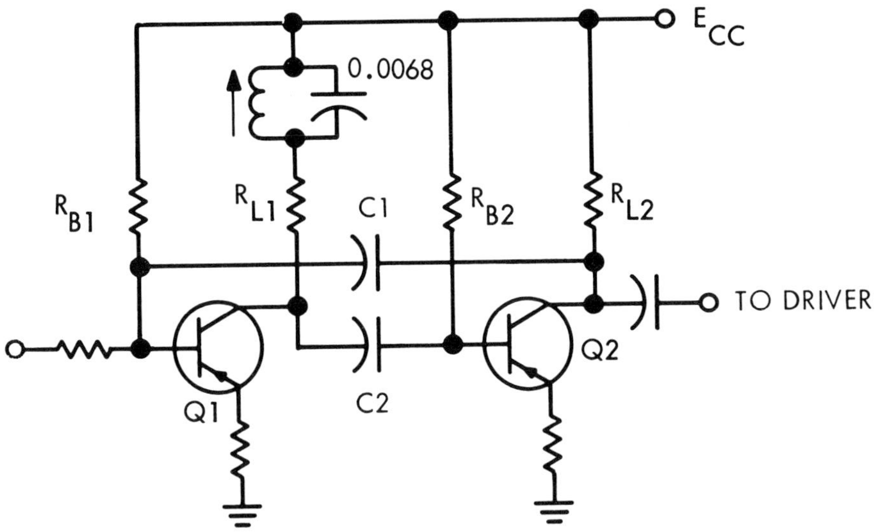

Fig. 3. Horizontal Oscillator Using Multivibrator

HORIZONTAL OSCILLATOR

The horizontal oscillator can be either a blocking oscillator, sine wave (modified Hartley), or a multivibrator. The circuit shown in Fig. 3 is a multivibrator. It is held in sync by the d-c output of the AFC circuit. A ringing coil is added to improve stability. The multivibrator design is dependent upon RC time constants and transistor parameters. The RC time constants control the duty cycle of the transistors. It can be designed as a symmetrical or unsymmetrical oscillator.

If the ON time of Q_2, $T_{2(ON)}$, is shorter than the ON time $T_{1(ON)}$ of Q_1, then

$$T_{2(ON)} > 3R_{L1}C_2 \tag{10}$$

where R_{L1} is the collector load resistor and C_2 is the collector-base coupling capacity.

The OFF time of Q_2, $T_{2(OFF)}$ is given by

$$T_{2(OFF)} \approx R_{B2}C_2\ln 2 \tag{11}$$

R_{B2} is the base resistance of Q_2. The OFF time of Q_1 is

$$T_{1(OFF)} \approx R_{B1}C_1\ln 2 \tag{12}$$

R_{B1} is the base resistor of Q_1 and C_1 is the coupling capacitor to the base. The collector load resistor R_L can be found by the following equation:[2]

$$R_{L(min)} = \frac{R_B\,[E_{CC} - V_{CE(sat)(min)}]}{h_{FE(min)}\,[E_{CC} - V_{BE(ON)}]} \qquad (13)$$

$h_{FE(min)}$ is the minimum gain required to drive the transistor into saturation, E_{CC} is the supply voltage, $V_{BE(ON)}$ is the voltage required to turn the transistor on, and $V_{CE(sat)(min)}$ is the voltage drop across the transistor when it is in saturation.

The transistors must have a collector-base breakdown BV_{CBO} and emitter-base breakdown BV_{EBO} in excess of the supply voltage. When the BV_{EBO} is exceeded, the device may withstand the added dissipation from leakage; however, the frequency and stability of the oscillator will be affected.

In the circuit of Fig. 3 the following time constants are set:

$$T_{2(OFF)} = 53\ \mu s$$

$$T_{1(OFF)} = 10.5\ \mu s$$

The coupling capacitors are both 0.001 μF. Q_1 and Q_2 are 2N1303 types and the supply voltage is 12 volts.

Using Eq. (11) the base resistance of Q_2 is found.

$$R_{B2} \approx \frac{53 \times 10^{-6}}{0.001 \times 10^{-6}(0.693)} \approx 75\ k\Omega$$

This is divided into a fixed and a variable resistance, the variable R being the hold control.

The base resistance for Q_1 is

$$R_{B1} \approx \frac{10.5 \times 10^{-6}}{0.001 \times 10^{-6}(0.693)} \approx 14\ k\Omega$$

In the actual circuit R_{B1} consists of the parallel combination of R_{B1} and the AFC circuit. Minimum h_{FE} is set at 30 and the minimum value of collector load resistance for each transistor is found from

$$R_{L1(min)} = \frac{14 \times 10^3(12-0.2)}{30(12-0.4)} \approx 470\ \Omega$$

and

$$R_{L2(min)} = \frac{75 \times 10^3(12-0.2)}{30(12-0.4)} \approx 2.5\ k\Omega$$

In the actual circuit R_{L1} and R_{L2} are both 5.6-kΩ resistors.

To improve stability a 47-Ω emitter resistor is added in each section. This does not change the design appreciably and stability is good up to 55°C ambient.

REFERENCES

1. Gruen, Wolf J.: Theory of AFC Synchronization, *Proc. of IRE,* August, 1953.
2. "Transistor Circuit Design," Texas Instruments Incorporated, p. 379, McGraw-Hill Book Company, New York, 1963.

21

Horizontal Driver and Sweep Output

The high driver power requirements of available semiconductor devices usually dictate a driver stage between the horizontal oscillator and output stage. The driver stage can be eliminated in small-screen sets (four- to eight-inch picture tube sizes) where drive requirements can be met by the oscillator without impairing the circuit stability.

The driver improves performance by supplying isolation between the oscillator and output stage, thereby allowing improved oscillator stability as well as better control of the required driver power. Either transformer or RC coupling can be used between the oscillator and driver stages, but the driver collector is usually transformer-coupled to the horizontal output stage.

Although the transformer design is not critical, its primary inductance and turns ratio depend on the supply voltage, the type of devices in the driver and output stages, and the proper impedance match for meeting the drive power requirements of the output stage. Primary and secondary may be connected in phase or out of phase depending on the source of drive required.

The type of output device will determine the desirability of a constant-voltage or constant-current source of drive. With reference to the OFF time of the output device, a constant-voltage drive is used when the driver stage is ON during this time.

Figure 1 shows a driver stage using RC input coupling and transformer output coupling. When the driver turns ON, capacitor C_1 supplies the energy for a fast-rising high-negative current pulse for three or four microseconds after which the resistor R supplies a current-limiting factor for the remainder of the driver ON time. The high peak-current pulse provides the turnoff power after

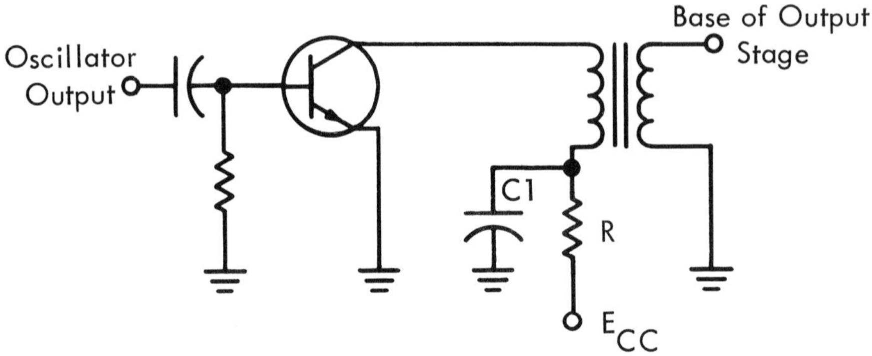

Fig. 1. Driver Stage with RC Input and Transformer Output

which it is only necessary to reverse bias the output device for the remainder of the OFF cycle. High current here contributes only to increased device dissipation and does nothing to improve operation.

The driver transistor should have a fairly high f_T (5 MHz or more) to ensure fast rise and fall times fo the drive currents. Its collector current rating must be capable of meeting the drive requirements and a low $V_{CE(sat)}$ is needed to minimize power loss across the device. Its breakdown voltage BV_{CES} can be approximated by

$$E_P \approx \left(\frac{\pi}{2}\right) \frac{t_{ON}}{t_{OFF}} \left(E_{CC}\right) \tag{1}$$

where t_{ON} and t_{OFF} are the respective ON and OFF times of the transistor and E_{CC} is the supply. Either germanium or silicon may be used. Where E_{CC} is low (12 to 24 volts) germanium can be used with a definite advantage in $V_{CE(sat)}$. For higher values of E_{CC}, silicon has better high-voltage characteristics.

Average power dissipation of the driver will vary from less than one watt to more than two watts depending upon the type of output device. The driver stage is driven into saturation and then cut off. Its ON time will vary from 30% to 50% of the duty cycle. Depending on the transformer design and the output used, the collector waveforms may be complex with some ringing occurring. This makes it difficult to define the exact power dissipation but a usable approximation can be obtained.

The horizontal output stage operates as a switch. The basic sweep circuit is shown in Fig. 2(a) with the equivalent circuit shown in Fig. 2(b). The transistor

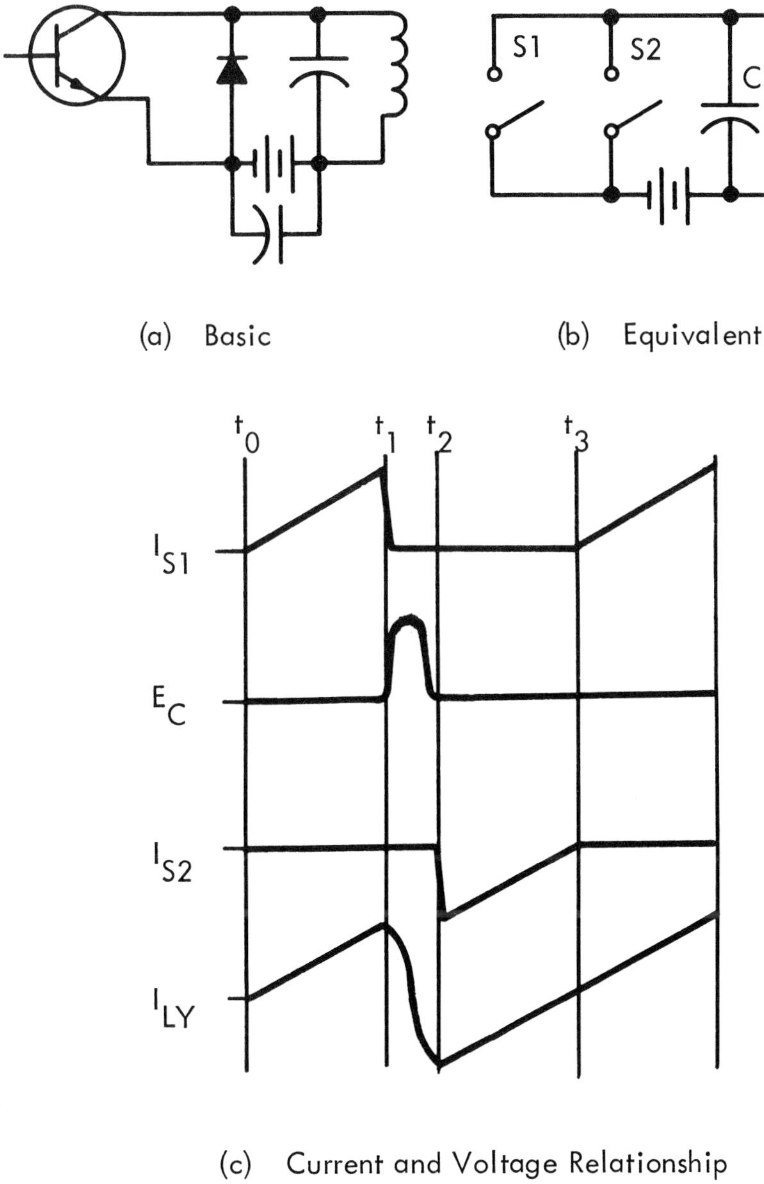

Fig. 2. Sweep Circuit

and damper diodes are analogous to S_1 and S_2, respectively. Figure 2(c) shows the current and voltage relationship with respect to time.

When S_1 is closed (i.e., the transistor is turned ON) at time t_0, the supply voltage, E_{CC}, is impressed across the inductance, L_Y. A current through the inductance rises linearly, the slope depending upon the d-c resistance of the yoke and the saturation resistance of the transistor switch S_1.

At the time t_1, current through the switch has reached a peak value which is defined as

$$I_C = \frac{E_{CC}}{L} t_{ON} \tag{2}$$

where t_{ON} is the time duration from t_0 to t_1 of Fig. 2(c). In a typical sweep circuit, t_{ON} is approximately 30 μs. When S_1 is opened, the circuit consists of the inductance and capacitor. Energy is maximum in the inductance and minimum in the capacitor. Current in the inductance reverses direction and starts to flow into the capacitor. A high-voltage pulse builds up across the capacitor and its peak value is derived from

$$E_C \approx \left(1.7 + \frac{\pi}{2}\frac{t_{ON}}{t_R}\right) E_{CC} \tag{3}$$

where t_{ON} is the combined conduction times of both the transistor and the damper (t_0 to t_1 and t_2 to t_3). The retrace time t_R is the time from t_1 to t_2. During retrace time, the voltage across the capacitor rises to the peak value and falls back to zero as the LC current starts to ring at its resonant frequency. Current in the inductance during retrace time is sinusoidal for one-half cycle of the resonant frequency. The frequency is dependent upon the retrace time which is usually 9 to 11 μs but may be longer to reduce the value of E_P. As the voltage across the capacitor falls and reverses polarity, S_2 is turned ON. This is the damper diode in the actual circuit and it is forward biased at this time. Current flows through the damper diode and inductance until the energy in the capacitor is dissipated. This current represents the first portion of the horizontal scan. It has been shown* that the peak damper current I_D can be found with very little error by the equation

$$I_D \approx 0.4\, I_Y \left(1 + \frac{T}{t_D}\right) \tag{4}$$

where I_Y is the peak-to-peak value of yoke current. T is the rise time of the damper current and t_D is the damper conduction time. Since the damper rise time is very short compared to t_D, the value of $\frac{T}{t_D}$ can be neglected and the peak current will be 0.4 I_Y.

---

*Deutsch, S.: "Theory and Design of Television Receivers."

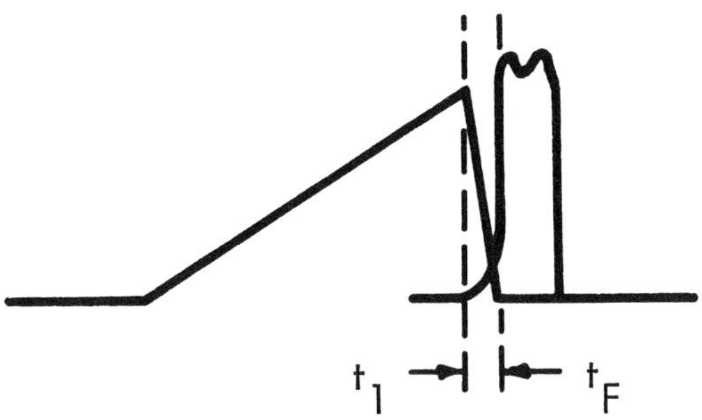

Fig. 3. Current and Voltage Relationship

While the damper diode is conducting, the voltage across the transistor and damper has a negative value which prohibits collector current from flowing. This allows the base to be forward-biased prior to the end of damper conduction. When the damper ceases to conduct, the voltage across the damper and transistor rises to the value of the supply and, as this voltage swings positive, the transistor starts conducting again. There is no apparent interruption in yoke current during the transition from current supplied by the damper to that supplied by the transistor. Horizontal linearity distortion primarily rises due to the fact that the $V_{CE(sat)}$ of the transistor and the V_F of the damper may not be equal. This results in a change in the slope of the sawtooth when the transistor takes over as the current supply for the yoke.

The speed with which the transistor turns off (i.e., its collector current drops from maximum to zero) is of extreme importance. As the current falls toward zero, the voltage across the capacitor (and across the transistor) is rising. Figure 3 shows the relationship between the voltage and current and it can be seen that with a slow fall time for the collector current, high peak power dissipation can occur. To minimize this peak power, the turnoff time of the transistor must be as fast as possible (preferably less than one μs).

At time t_1, simply removing the base drive would allow the transistor to turn off but the time required would be far greater than the one μs desired. To speed up the turnoff time, the base of the output stage is driven into a reverse conduction mode. This reverse current depletes the base of its forward biasing carriers and cuts off collector current flow. The amount of reverse current flow is dependent upon the type of output device used and the value of the peak collector current to be turned off. Available devices require as high as one ampere reverse current to successfully turn off the collector current in one μs.

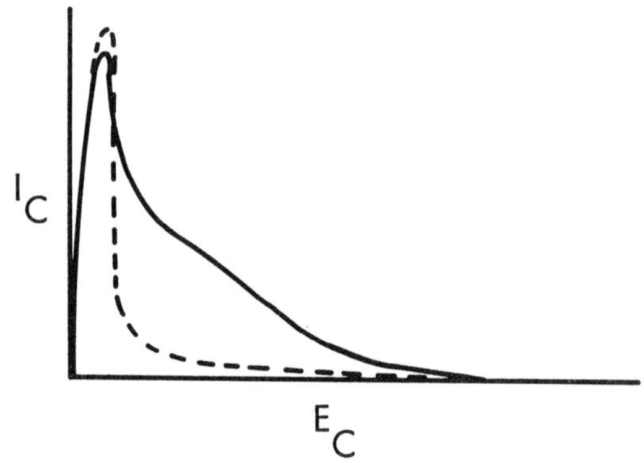

Fig. 4. Load Line Path

The rise time of the reverse current cannot be too slow nor does it need to be less than one μs. This is due to a device characteristic called storage time. This is the time lapse which occurs between the time when the base is driven into reverse bias and the collector current starts to fall. It is due to the energy developed in the transistor caused by high current flow which tends to continue the flow after the drive has been removed. Storage time is dependent upon transistor design and the amount of peak current required from the device. Some devices have storage times as long as three μs. It is desirable to limit this time to less than two μs. If storage time is two μs, then it is desirable to have a rise time of two μs for the reverse base current. This ensures a fast turnoff time.

Slow switching speeds also affect the breakdown voltage requirement of the transistor. Figure 4 shows the path of the load line of the transistor as it switches from its peak current mode to the high voltage peak which occurs during the retrace time. The slower the switching speed, the more the line approaches a linear (or Class A type) load line. This contributes to very high power dissipation, the ideal load line being the dotted line which represents a switching time of less than 0.5 μs. As this load line extends farther out, the $BV_{CEO(sustain)}$ rating of the transistor must be raised. Figure 5 shows the BV_{CEO} and $BV_{CEO(sustain)}$ rating in relation to the load line. If the load line passes outside the $BV_{CEO(sustain)}$ region, the device will destroy itself. In actual operation, the peak voltage during retrace can approach the BV_{CEV} rating (shown also in Fig. 5). This represents a much higher voltage than BV_{CEO} in the device type used here.

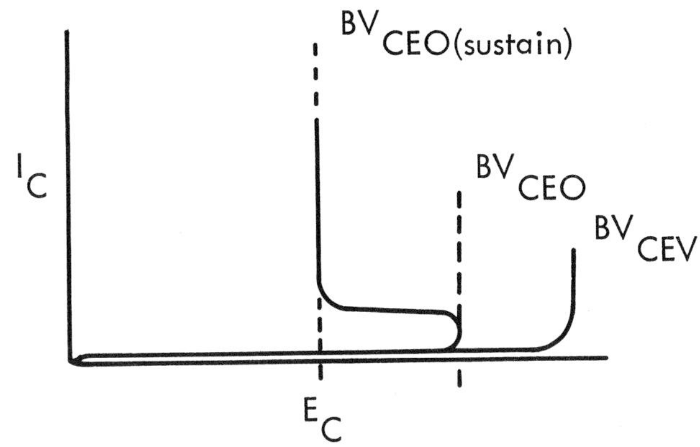

Fig. 5. Breakdown Voltage in Relation to Load Line

Device dissipation is the sum of the base drive power and the collector power. Collector dissipation is divided into two segments with respect to time, one being the ON time for the device from t_0 to t_1 and the other being the turnoff time of the device (t_1 to t_F in Fig. 3). This neglects most of the retrace time and all of the damper conduction period. However, since dissipation during these periods is dependent upon leakage currents, it can be neglected because low leakage is a must in the design of the device and this has been true here.

Average collector dissipation from t_0 to t_1 is given by

$$P_{ON} = \frac{1}{63.5} \int_{t_0}^{t_1} i_C{}^2 R_{CS} dt \qquad (5)$$

where i_C (assuming a linear rise in current) is given by:

$$i_C = \frac{I_C t}{t_1 - t_0} \qquad (6)$$

I_C is the peak value of the collector current at time t_1 and R_{CS} is the saturation resistance of the device. $i_C e_C$ could be used in place $i_C{}^2 R_{CS}$ where e_C is the $V_{CE(sat)}$ rating of the transistor. However, $V_{CE(sat)}$ is not constant from t_0 to t_1 whereas R_{CS} is more constant over the current range. Average dissipation during the turn off time t_{OFF} (refer to Fig. 3) is found by

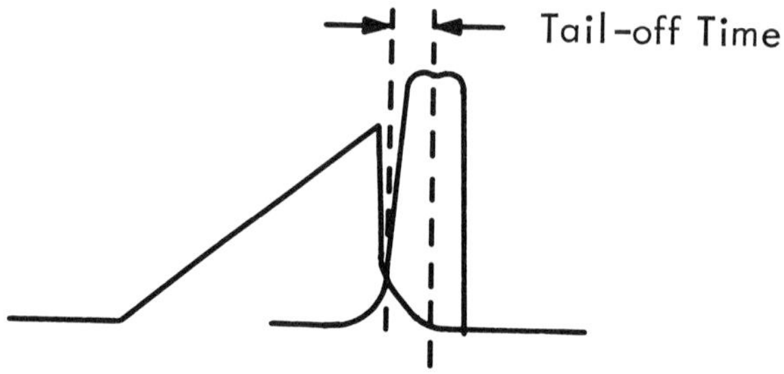

Fig. 6. Tail-off Time

$$P_{OFF} = \frac{1}{63.5} \int_{t_1}^{t_F} i_C e_C dt \qquad (7)$$

Again, linear functions of i_C and e_C are assumed and i_C is a decaying function while e_C is rising. i_C is given as

$$i_C = I_C\left(1 - \frac{t}{t_F}\right) \qquad (8)$$

and with a turnoff time of one μs, e_C is found to be

$$e_C = \frac{t}{t_F} E \qquad (9)$$

where E is the value which E_C reaches at the end of the falltime t_F. In some devices, the collector current exhibits a sudden slowing up in fall time when it reaches 10 to 20% of its peak value. The period from this point to the approximate zero current level has been referred to as "tail-off" time (shown in Fig. 6). Here, it becomes necessary to divide the current into a fall time and a tail-off time and to compute the dissipation for each time separately.

The basic Eq. (7) is used to compute the dissipation during the tail-off time and Eq. (8) holds true for the current; however, since tail-off time may extend for more than three μs, the voltage may reach its peak and e_C must then be shown as

$$e_C = E_C \sin \frac{\pi t}{t_A} \qquad (10)$$

where t_A is one alternation of the second harmonic of the flyback frequency (or 0.5 t_R).

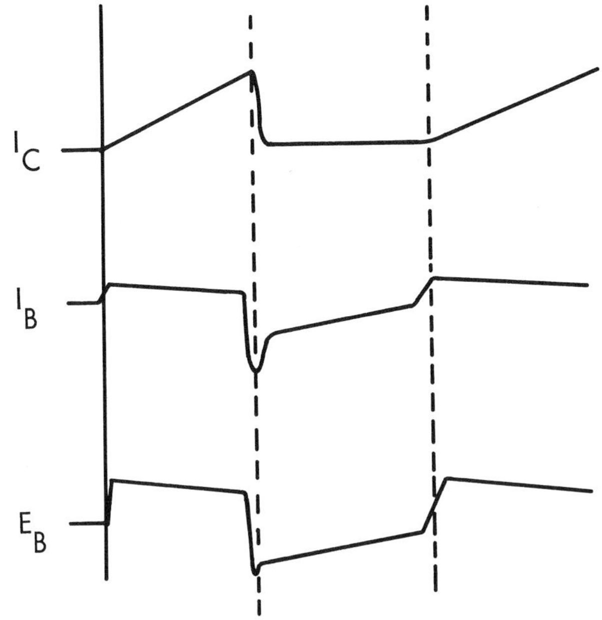

Fig. 7. Current and Voltage Waveforms of the Base

Base drive power can be calculated in the same manner. Figure 7 shows the basic current and voltage waveforms of the base. Power is calculated for the ON time and OFF time separately using the same basic Eq. (5) and (7). Due to device characteristics and reflected impedances, the waveforms may be more complex and a rough approximation is adequate.

The design of a horizontal output stage is primarily dependent upon three factors relating to the type of cathode ray tube to be used. These are yoke shell diameter, high voltage, and angle of deflection. The required yoke current for supplying full scan for a given CRT is

$$I_Y = 0.053 \sqrt{\frac{E_{kV} D_A F_\beta}{L_Y}} \tag{11}$$

where I_Y is the peak-to-peak yoke current, L_Y is the yoke inductance, D_A is the yoke shell diameter in centimeters and E_{kV} is the high voltage in kV. F_β is the function of the deflection angle β and is given by

$$F_\beta = \left(\cos \frac{\beta}{4}\right) \left(\sin \frac{\beta}{4}\right)^3 \tag{12}$$

The peak current capabilities of the transistor and damper define the maximum yoke current I_Y and the deflection yoke. Thus the deflection yoke must be selected accordingly.

Other important factors which must be considered are the voltage breakdown capabilities of the transistor and damper. The high-voltage pulse which is developed across the capacitor during retrace time must not exceed the voltage breakdown limits of the transistor and damper. This voltage is defined as

$$E_P \approx \left[1.7 + \frac{\pi}{2}\left(\frac{t_{ON}}{t_{OFF}}\right)\right] E_{CC} \tag{13}$$

where t_{ON} is the combined ON time of both the transistor and damper, t_{OFF} is the retrace time, and E_{CC} is the supply voltage available across the transistor. The supply voltage is found by the equation

$$E_{CC} \approx \frac{I_Y L_Y}{t_{ON}} \tag{14}$$

Normal retrace time as is transmitted by the station is usually from 9 to 10.5 μs. This can be increased in the receiver to reduce the voltage pulse during retrace time thus lowering the voltage rating required of the transistor and damper. However, retrace blanking must be used to eliminate a horizontal foldover effect in the picture. A method of rating the output transistor is to multiply the peak current and maximum blocking voltage to obtain a volt-ampere (VA) rating. This rating must not be exceeded even under the most adverse conditions such as high line-voltage, transients, and/or overload due to a high-voltage arcing. It is not to be used as a power rating.

Figure 8 shows a horizontal sweep and high-voltage circuit. The high voltage is generated by transforming the retrace voltage pulse up to the desired value for rectification. The secondary is tuned to the third harmonic of the resonant frequency of the LC circuit. Its distributed capacity must be kept at a minimum to get maximum voltage step-up by the turns ratio. Leakage inductance must be kept at a minimum to reduce ringing currents. The primary of the high-voltage transformer serves as the d-c path for the collector supply voltage. The deflection yoke is then returned to ground through a coupling capacitor which supplies some "S" correction to improve horizontal linearity.

The use of third-harmonic tuning reduces the peak value of the retrace voltage pulse by about 15% thus lowering the voltage breakdown requirements for the output transistor. On the other hand, it increases the rise time of the voltage pulse at the beginning of retrace and this increases the peak power

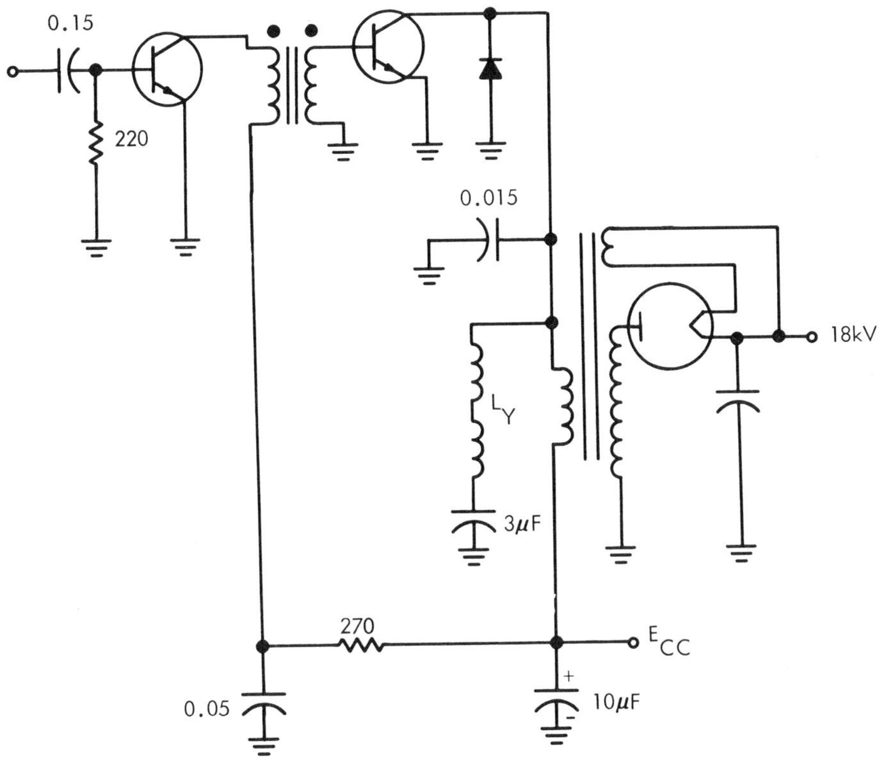

Fig. 8. Horizontal Sweep and High-voltage Circuit

dissipation during turnoff. In Fig. 8, the driver transistor used is a 2N1718. It supplies a peak turnoff current of one ampere to the base of the output which is an experimental silicon device. The driver transformer is wound on a pot core with a turns ratio of 4:1. The primary inductance is 10 mH. The picture tube used is a 19-inch 114° type with a maximum high voltage of 18 kV. The horizontal deflection angle of 110° gives an F_β of 0.087 (Eq. (12)). Peak-to-peak yoke current is found from Eq. (11).

$$I_Y = 0.053 \sqrt{\frac{18(5.3)(0.087)}{650 \times 10^{-6}}} \approx 6 \text{ A}$$

The supply voltage is derived from Eq. (14).

$$E_{CC} \approx \frac{6(650 \times 10^{-6})}{53.5 \times 10^{-6}} \approx 70 \text{ V}$$

A retrace time of 10 μs is used, thus from Eq. (13) the peak voltage during retrace time is

$$E_P \approx \left[1.7 + \frac{\pi}{2}\left(\frac{53.5}{10}\right)\right] 70 \approx 700 \text{ V}$$

With the use of third-harmonic tuning, this is reduced to approximately 600 volts. The high-voltage transformer uses type T-1 core material. The primary winding has 90 turns and the secondary 2700 turns. High-voltage output is 18kV at 500 μA of beam current.

Index